Flow of Industrial Fluids—
Theory and Equations

Flow of Industrial Fluids—
Theory and Equations

By Raymond Mulley

Library of Congress Cataloging-in-Publication Data

Catalog record is available from the Library of Congress

This book contains information obtained from authentic and highly regarded sources. Reprinted material is quoted with permission, and sources are indicated. A wide variety of references are listed. Reasonable efforts have been made to publish reliable data and information, but the author and the publisher cannot assume responsibility for the validity of all materials or for the consequences of their use.

Neither this book nor any part may be reproduced or transmitted in any form or by any means, electronic or mechanical, including photocopying, microfilming, and recording, or by any information storage or retrieval system, without prior permission in writing from the publisher.

The consent of CRC Press LLC does not extend to copying for general distribution, for promotion, for creating new works, or for resale. Specific permission must be obtained in writing from CRC Press LLC for such copying.

Direct all inquiries to CRC Press LLC, 2000 N.W. Corporate Blvd., Boca Raton, Florida 33431.

Trademark Notice: Product or corporate names may be trademarks or registered trademarks, and are used only for identification and explanation, without intent to infringe.

Visit the CRC Press Web site at www.crcpress.com

© 2004 by CRC Press LLC

No claim to original U.S. Government works
International Standard Book Number 0-84932-767-9
Printed in the United States of America 1 2 3 4 5 6 7 8 9 0
Printed on acid-free paper

DEDICATION

This book is dedicated to my patient wife, Ginette, to my grandchildren, Walker, Olivia and Fiona and to Jamil who adopted me as his Tonton. To me these children represent the future. Who knows? Perhaps there is a Sadi Carnot hidden in one of them.

Raymond Mulley

Contents

List of Illustrations X

List of Tables XII

About the Author XIII

Preface XV

Chapter I: Flow of Incompressible Fluids 1
    I-1:   Scope of Chapter – Basic Concepts 1
    I-2:   Flow of Incompressible Fluids in Conduits 2
    I-3:   Flow Regimes – Reynolds' Contributions 6
    I-4:   Flow Profiles – Velocity Distributions 11
    I-5:   Fluid Flow – An "Irreversible" Process 16
    I-6:   Fundamental Relationships of Fluid Flow 17
    I-7:   The Role of Viscosity 22
    I-8:   "Friction Losses" 33
    I-9:   Bernoulli Equation and the Darcy Equation Combined 36
    I-10:   Conservation of Energy in Hydraulics Practice 40
    I-11:   Worked Examples 44
    I-12:   Chapter Summary 49

Chapter II: Incompressible Fluid Flow – Losses of Mechanical Energy 51
    II-1:   Scope of Chapter – Applying Basic Concepts 51
    II-2:   Reasoned Approach to Design – A Little Personal Philosophy 52
    II-3:   The Bernoulli Equation Revisited 53
    II-4:   Irreversibilities Due to Pipe and Fittings 57
    II-5:   Examples of Estimations of Irreversibilities 80
    II-6:   Chapter Summary 98

Chapter III: Pumps Theory and Equations 99
    III-1:   Scope of Chapter – Pumps and Their Performance Capabilities 99
    III-2:   Functions of Pumps 100
    III-3:   A Brief History of Pumps 104
    III-4:   Classification of Pumps 105
    III-5:   Characteristics of Pumps 109
    III-6:   Inherent and Installed Characteristics of Pumps 130
    III-7:   Controlling Flow Through Pumps 135
    III-8:   Hydraulic Turbines 141
    III-9:   Worked Examples 143
    III-10: Chapter Summary 148

Chapter IV: Compressible Fluid Flow **151**
  IV-1: Scope of Chapter – Comprehending Compressible Flow. 151
  IV-2: Differences between Compressible and Incompressible Flow 152
  IV-3: Using Models 154
  IV-4: Treating Mixtures 156
  IV-5: Equations of Compressible Flow of an Ideal Gas 157
  IV-6: Ideal and Non-Ideal Gases – Comparison of Some Equations-of-State 166
  IV-7: Model Processes for Compressible Flow 174
  IV-8: Choked Flow and the Mach Number. 182
  IV-9: Equations for Adiabatic Flow with Irreversibilities not Involving the Mach Number –
        the Peter Paige Equation. 184
  IV-10: Equations for Isothermal Flow with Irreversibilities 188
  IV-11: Chapter Summary. 192

Chapter V: Compressible Fluid Flow – Complex Systems **195**
  V-1: Scope of Chapter – Computations for Complicated Compressible Flow Systems 195
  V-2: Describing the Piping Network 196
  V-3: Describing the Flow Regime 198
  V-4: Plan of Attack 199
  V-5: Manifold Flow. 200
  V-6: Data Collection and Verification 206
  V-7: Chapter Summary. 207

Appendix AI: Equations of Incompressible Fluid Flow and Their Derivations **209**
  AI-1: Purpose – Providing Chapter I Details 209
  AI-2: SI and Customary U.S. Units 210
  AI-3: Pressure at a Point within a Fluid 215
  AI-4: Hydrostatic Equilibrium 219
  AI-5: Friction Losses Explained 221
  AI-6: Force-Momentum Considerations for Variable Mass Systems 224
  AI-7: Derivation of the Darcy Equation. 229
  AI-8: Derivation of the Bernoulli Equation Including Irreversibilities. 236
  AI-9: Laminar Flow and the Hagen-Poiseuille Equation 245
  AI-10: Summary of Appendix AI 252

Appendix AII: Losses in Incompressible Fluid Flow **253**
  AII-1: Purpose – Providing Chapter II Details 253
  AII-2: Relation of Valve Coefficient, C_V to Loss Coefficient, K 254
  AII-3: Relationship between Energy per Unit, Mass Units, Head Units and Pressure Units 256
  AII-4: Churchill-Usagi Friction Factor Equations. 257
  AII-5: Pressure Drop versus "Friction Losses". 259
  AII-6: K Factors – Loss Coefficients. 260
  AII-7: Summary of Appendix AII 270

Appendix AIII: Computations Involving Pumps for Liquids **273**
    AIII-1: Purpose – Providing Chapter III Details... **273**
    AIII-2: Theory of Centrifugal Pumps... **273**
    AIII-3: Performance of Real Centrifugal Pumps **284**
    AIII-4: Real Centrifugal Pumps – Suction Lift, Cavitation and NPSH **294**
    AIII-5: Positive Displacement Pumps.. **298**
    AIII-6: Theory and Analysis of Jet Pumps.. **299**
    AIII-7: Worked Problems... **312**
    AIII-8: Summary of Appendix AIII **318**

Appendix AIV: Equations of Compressible Flow, Derivations and Applications **319**
    AIV-1: Purpose – Providing Chapter IV Equation Details **319**
    AIV-2: Using Thermodynamic Variables – in Particular, Enthalpy.......................... **320**
    AIV-3: Adiabatic and Irreversible Flow in Uniform Conduits – Basic Equations **321**
    AIV-4: The Peter Paige Equation, Choked Flow **323**
    AIV-5: Choked Flow Using the Ideal Gas Equation **329**
    AIV-6: Adiabatic Choked Flow; P, v, T Relationships Using the Redlich-Kwong Equation **350**
    AIV-7: Summary of Appendix AIV **362**

Appendix AV: Compressible Fluid Flow – Complex Systems **363**
    AV-1: Scope – Estimating Complicated Pressure Drops and Flows **363**
    AV-2: Describing the Piping Network... **363**
    AV-3: Describing the Flow Regime **368**
    AV-4: Component Input Data, Eleven Sources **369**
    AV-5: Plan of Attack **370**
    AV-6: Irreversibilities Due to Form (and Mixing) Effects **371**
    AV-7: Manifold Flow **374**
    AV-8: Viscosity Considerations... **376**
    AV-9: Simulation Results (Analytic Method)... **388**
    AV-10: Summary of Appendix AV **390**

Appendix B: Endnotes **391**

Appendix C: Table of Principle Symbols and Glossary of Principal Terms and Units **395**

Appendix D: Table of Caveats **399**

Appendix E: Selected Bibliography... **401**

Index **403**

Illustrations

Figure	Title	Page
I-1.	Reynolds apparatus	6
I-2.	Laminar flow	6
I-3.	Turbulent flow	7
I-4.	Moody friction factor versus Reynolds' number	10
I-5.	Distribution of velocity across pipe, fully developed flow of Newtonian fluid.	11
I-6.	Velocity components downstream of close-coupled elbows in different planes	13
I-7.	Multiple sensor averaging techniques	14
I-8.	Average, maximum and point velocities	15
I-9.	Developing profile	15
I-10	Shear diagrams	23
I-11.	Definition of viscosity	24
I-12.	Thixotropic behaviour	29
I-13.	Generalized viscosity chart	30
I-14.	Interpretation of hydraulic "head"	43
I-15.	Manometers in hydraulic practice	44
II-1.	Moody friction factor versus Reynolds' number and pipe schedule	60
II-2.	Component loss coefficient correlation	67
II-3.	Component loss coefficients	68
II-4.	Establishment of loss coefficients for a 90 degree mitered elbow	71
II-5.	Sudden contraction	73
II-6.	Sudden expansion	73
II-7.	Recovery and permanent losses across an orifice	76
II-8.	Example of system irreversibilities	87
II-9.	Extension to system irreversibility example	94
III-1.	Simple centrifugal pump	106
III-2.	Axial-flow elbow-type propeller pump	106
III-3.	Seven-stage diffuser-type pump	107
III-4.	Simplified sketch of an air lift, showing submergence and total head.	107
III-5.	Ejectors	108
III-6.	Typical head-capacity curves for centrifugal pumps	111
III-7.	Typical characteristic curves for reciprocating pumps	124
III-8.	Self-controlled pump	136
IV-1.	Three models of compressible flow	174
IV-2.	Mass flow rate through nozzle	178
IV-3.	Variation of pressure ratio with distance from nozzle inlet	178

V-1.	First example of isometric sketch	197
V-2.	Second example of isometric sketch	197
V-3.	Manifold flows	201
V-4.	Miller's K Factors	202
AI-1.	Pressure tetrahedron	216
AI-2.	Projected areas	217
AI-3.	Hydrostatic equilibrium	219
AI-4.	Variable mass systems	225
AI-5.	Restrained system	227
AI-6.	Derivation of the Darcy equation	233
AI-7.	Development of Bernoulli equation	237
AI-8.	Horizontal laminar flow	246
AI-9.	Hollow cylinder model for Poiseuille's law	249
AIII-1.	Typical single-stage centrifugal pump	273
AIII-2.	Centrifugal pump showing Bernoulli stations	276
AIII-3.	Velocity vectors at entrance and discharge of centrifugal pump vane	277
AIII-4.	Vector diagram at tip of centrifugal pump vane	277
AIII-5.	Vector diagrams describing perfect and imperfect guidance	285
AIII-6.	Reduction in theoretical head due to various causes	286
AIII-7.	Impeller and stationary vanes in diffuser pump	287
AIII-8.	Power versus volumetric flow in a centrifugal pump	288
AIII-9.	Ejector analysis	300
AIII-10.	Jet pump configurations	304
AIII-11.	Suction lift	312
AIV-1.	Multiple choke points	338
AIV-2.	Simultaneous equations – functions of two variables	346
AV-1.	Isometric sketch of complex vent header	365
AV-2.	Gas mixture viscosities	386

Tables

Figure	Title	Page
I-1.	Fluid categorization by viscous behavior	24
I-2.	Mechanically useful forms of specific energy	40
I-3.	Irreversibilities in turbulent flow	42
II-1.	Absolute roughness of various materials (Data from Crane)	59
II-2.	Absolute roughness of various materials (Data from Miller)	61
II-3.	Results of computations	94
IV-1.	Comparison of equations-of-state for CO_2 with experimental data	169
AII-1.	K factors	262
AII-2.	Classification of components	266

About the Author

Raymond Mulley was the original author of *Control System Documentation Applying Symbols and Identification* (ISA, 1994). He joined ISA in 1973 and was chairman of ISA SP5.1 the committee that developed ANSI/ISA-S5.1-1984, *Instrumentation Symbols and Identification*.

During his more than 30 years of professional experience, he worked for Fluor Daniel (Mississauga, Ontario, Canada, Rochester, New York, and Irvine, California); Dravo Chemplants (Pittsburgh, Pennsylvania) and International Nickel (Sudbury, Ontario, Canada). Part of this experience was in private practice as a professional engineer in the field of process hazards analysis and control. His experience ranges from instrument and electrical supervisor to start-up engineer, instrument design engineer, chief instrument engineer, director of design engineering and hazards analyst. His principal professional interests have been the flow of industrial fluids and thermodynamics.

Preface

P-1: PURPOSE OF BOOK – INSIGHT

The purpose of this book is to supply adequate context for those who really want an understanding of fluid flow. Every subject must be studied within a certain context. If the context is too narrow, our knowledge of the subject is limited, wrong assumptions are made, and errors result. Specialization tends to narrow context. Generalization can make it so large that we lose sight of important details. In this text, we hope to supply enough context to give insight into the phenomena of flow. The insight gained will hopefully aid in the solution of real problems.

This book was written with a specific audience in mind – all those who wish to understand fluid flow. The book's purpose is to link fluid flow theory to practice in sufficient detail to give its chosen audience an understanding of both theory and practice. The theoretical detail is limited to that necessary to understand practical problems – to the application of equipment and devices, not to their design. Insight is the primary goal.

There are many texts that deal with instruments to measure and control flow of fluids. They describe how devices work and only discuss fluid mechanics and dynamics incidentally. This text is intended to be complementary to the above mentioned ones. It takes a different approach. It discusses fluid flow, so as to give the reader a clear grasp of the fundamentals that impact his or her work. The fundamentals are then linked to entire fluid systems.

Sometimes, the language of subject matter serves as an impediment to insight into the subject. This is contrary to the intent of a textbook, but it cannot be helped because of the necessity of using accepted terminology. This book tries to make sure all terminology used is adequately explained.

The pressure losses due to fluid irreversibilities are covered. The general characteristics of pumps, blowers, compressors, vacuum devices and other prime movers are given. These characteristics are discussed in sufficient detail necessary for the reader to understand some of the dynamic aspects of fluid flow. In general, we try to show that a prime mover that adds energy to a stream, or a device that extracts energy from a stream, should be studied in the context of the total system through which the fluid is flowing. In particular, we analyze the limitations imposed on a system and on its associated instrumentation by the inherent characteristics of prime movers as these characteristics interact with those of the system.

P-2: Why Study Fluid Flow in Detail?

It seems to be taken for granted that process engineers should have a good grasp of fluid flow. Mechanical engineers also need to understand fluid flow. But, why should control systems personnel study the flow of industrial fluids? They could concentrate on the study of measuring instruments and control valves. This introduction will attempt to answer that question. The rest of the book will try to supply understanding, methodology, theory and equations useful to measure and control industrial fluid flow.

The practical importance of studying any subject is directly linked to the frequency with which problems in the field of the study are encountered. The most frequently encountered operation in most industrial processing is the transfer of fluids from one place to another. Sometimes the transfer is from point to point. Sometimes the flow diverges from one point to many. Sometimes, the flow converges from many points to one. Each case presents different sets of phenomena to be understood. Each case has different sets of constraints imposed upon it.

Fluid flow is always associated with a driving potential. This potential can be constant, as when a head tank is used. It can vary with flow, as when a pump, compressor, blower or vacuum-producing device supplies the driving potential. Each prime mover provides its own set of characteristics that must be understood if we are to correctly apply measurement and control technology. Sometimes the purpose of the flowing stream is to allow energy to be absorbed, as by a turbine. The turbine also imposes its own characteristics on the flowing stream.

Basic characteristics of the prime movers must be understood in order to establish the feasibility of any proposed application. The turndown capabilities of the system and those of the control devices are independent, but they must be studied in tandem. Why pay for a flow meter with a 100:1 turndown if the system will operate only over a 3:1 range? (There may be other reasons than turndown for using a particular meter).

P-3: Flow in Closed Channels

We tend to think of fluid as flowing only in pipes with circular sections. This is the case for the majority of industrial applications. However, fluids do flow in open channels and in ducts of non-circular sections. The pipes, channels and ducts frequently vary in cross section. They have fittings and transitions that cause a pressure drop different from that of the uniform section. Each combination presents different problems to solve.

Flow also occurs remotely from a channel wall – in the atmosphere, for instance. Because of the necessity to put limits to all endeavors, the scope of the book will be limited to that of fluid flowing through closed channels. This choice of scope excludes the consideration of related subjects such as open channel flow, weirs and flumes and environmental dispersion modeling.

Problems associated with the flow of liquids, slurries, vapors and gases through ducts and channels are common in industry. The power requirements of pumps, blowers and compressors to cause a given flow are frequently needed. The pressure drops between two points at different elevations in a conduit is often required. The flow for a given pressure drop is also required.

One of the most difficult flow computations associated with pressure relief devices is that of computing the optimum size of a relief header. Here, the driving pressure at each relief device discharge is unknown. In addition, the pressure at the discharge pipe into the sink may be unknown because of the possibility of choked flow. So, there is a minimum flow requirement in a network of pipes with unknown upstream and downstream pressures. This was the very problem that started the author on his search for a better grasp of fluid flow phenomena.

Each fluid or mixture of fluids comes with its own peculiarities. Viscosities and densities have to be estimated for pure fluids and for mixtures. Vapor pressures of pure fluids and mixtures have to be considered if flashing is to be avoided or accommodated and if cavitation is to be avoided. The choking phenomenon occurs with gases and flashing liquids. Slurries behave differently from clean fluids. Some fluids are extremely aggressive to industrial piping, to human beings, and to the environment. The characteristics of each fluid must be understood if we are to measure, control and contain the fluid.

The permissible velocity of a fluid depends on the service. Allowable erosion effects, cavitation, choking, flashing, slugging or mist flow in two-phase situations must all be considered. Vortexing, pulsating flow, noise generated by flow, flow regimes (single-, two- or three-phase flow) all have to be predicted, if not measured and controlled.

P-4: A Word About Units

A misunderstanding of units is a common source of errors, both arithmetic and conceptual. It would be easy to decide to use only one set of units in a book and to allow the reader to sort out the real world differences. This approach, however, would defeat the purpose of the book.

There are two sets of units prevalent in the industrial world. The set that is given most ink is the SI system (Système international d'unités). The one that is most common in the industrial context of North America is the Customary U.S. System.

In the field of engineering and design work, mixed units are often used due, frequently, to sectorial preferences. Conversion factors are applied within formulae. Since most of us are faced with the problem of mixed units on a daily basis, we have to deal with them.

In this book, we will use both SI and the customary U.S. units as the basic units. In addition, we will give some examples of mixed units. Rather than have a section on units, we will define them when initially presented. At the risk of boring some, we will repeat the definitions from time to time when the context seems to demand it.

P-5: Motivation

In pursuing a deeper knowledge of fluid flow, the reader should be motivated by this thought: If fluid flow is not understood, how can it be measured or controlled?

The practices of some engineering companies, in terms of division of work, limit control systems personnel to operating on data produced by process engineers. If such is the case, it is to be hoped that the process engineer has access to the knowledge that this book will attempt to impart. In that case, at a minimum, this book will allow the control systems personnel to ask appropriate questions regarding the data that was supplied.

The best motivation for the reader to pursue his or her quest for knowledge of fluid flow should be that fluid flow and its measurement and control represent a fascinating study. Our present knowledge of fluid flow is based on the pioneering work of some intellectual giants. Bernoulli[i], Carnot[ii], Coriolis[iii], Kelvin[iv], Reynolds[v], Venturi[vi] are some of the names of people whose contributions will be discussed and, hopefully, explained in this book.

P-6: Plan of Attack

The book is divided into two parts. The first part, consisting of the chapter material, gives a common sense, logical description and explanation of fluid flow phenomena and of fluid machinery. The basic theory and equations associated with flow and fluid machinery are given. As each concept is introduced, an attempt will be made to make it relevant to practical engineering problems.

The second part of the book, consisting of the material in the appendices, takes a more formal approach and treats more complex problems. It gives the detailed derivations of the equations found in the first part. It parallels the first part in that each chapter of part one has a corresponding appendix in part two. This approach will allow the reader to follow more easily the logic in the first part. He or she may then use the second half for reference or for more intensive, in-depth study.

In addition to the separation of the book into two parts based on a common sense, logical approach, versus a more formal one with increased complexity of detail, it is convenient to deal with so-called incompressible fluids first and then with compressible ones. This allows a firm grasp of more basic concepts before the more complex ones are presented. It makes the more complex concepts easier to tackle. However, it is necessary to deal with some aspects of compressible flow during the discussions of incompressible flow. This helps avoid the pitfalls of allowing the reader to make false assumptions about when to apply equations that were developed for incompressible flow

The historical contributions of some of the actors in the field of fluid mechanics, dynamics, measurement and control are of interest. They are sprinkled throughout the book.

CHAPTER ONE

Flow of Incompressible Fluids

I-1: SCOPE OF CHAPTER — BASIC CONCEPTS

Chapter I introduces basic concepts necessary to understand incompressible fluid flow in closed conduits. These concepts will be introduced and described in logical, not necessarily historical, sequence, as they are needed. The culmination of Chapter I will be introducing the generalized Bernoulli[i] equation – probably the single most useful equation for solving fluid flow problems. Detailed computations will be left to subsequent chapters and appendices. In particular, the concepts described in this chapter will be developed in greater detail in Appendix AI.

Chapter I will show:
- dividing fluids into incompressible ones and compressible ones is an arbitrary construct that helps organize knowledge, but sometimes clouds other insights;
- there is a more rational way to describe mechanical energy losses than by the term "fluid friction", which is a logical inconsistency;
- in pipes and channels, there are two basic flow regimes — laminar and turbulent — with an unstable regime between;
- these regimes are well categorized by an easily computed number, the Reynolds[v] number;
- there exists an equation, the Bernoulli[i] equation, which can be used to compute flow rates and pressure drop relationships;
- there exists another equation, the Darcy[vii] equation, which relates irreversibilities ("friction losses") to fluid viscosity, density and velocity. The Darcy equation can be easily incorporated into the Bernoulli equation to give a system for solving fluid flow, measurement and control problems.

We will start with a brief discussion of flow and what causes or influences it. Then we will discuss the pioneering work of Sir Osborne Reynolds[v], introduce the Darcy equation and follow with the work of Daniel Bernoulli. This chapter will consider only round ducts flowing liquid full. Greater detail will be given in Appendix AI. Compressible fluids and other cross sections will be discussed in subsequent chapters and detailed in subsequent appendices.

CHAPTER ONE | Flow of Incompressible Fluids

I-2: FLOW OF INCOMPRESSIBLE FLUIDS IN CONDUITS

Some material in this section may seem self-evident to recent engineering graduates. However, it is worth establishing some basic concepts as a springboard to what follows. It does not hurt to review basic material from time to time. It is surprising what one forgets – or never learned.

Compressibility of all fluids

Many terms used in the field of hydraulics are simplified descriptions of reality. Sometimes these didactic simplifications are useful; sometimes they are not. They help younger people over early hurdles and hinder them in their later comprehension. The name of the term, itself, can lead to conceptual errors. For instance, the distinction between an incompressible fluid and a compressible one is just a matter of degree. There may be orders of magnitude between the compressibility of a gas and that of a liquid; nevertheless, it is wise to bear in mind that all fluids are compressible to some degree — even the ones we call "incompressible". Later, we will show how compressibility helps explain another logical inconsistency – fluid friction.

Mechanism of flow

It is obvious to anyone who has ever used a garden hose that water flows when there is a higher pressure upstream than downstream. When we open a tap, or faucet, the higher pressure in the pipe causes water to flow. This common observation leads to the assumption that a higher pressure upstream than downstream is always associated with flow.

The assumption is not correct. It is true a higher pressure is required upstream to start a fluid flowing across a restriction. However, once a fluid is flowing, it has momentum. As will be developed in greater detail when we discuss the Bernoulli[i] equation, pressure is associated with one form of energy, static energy, and energy can take on different forms. One of these forms is kinetic energy. When a high velocity fluid flows from a small diameter pipe through a swage to a larger diameter pipe, the fluid slows down because of the continuity principle. In slowing down, some of its kinetic energy is changed to static energy. This change can cause the pressure downstream to be greater than the pressure upstream of the swage. This phenomenon is called pressure recovery.

It is not necessary to have a higher pressure upstream in order to initiate flow. If a pipe is connected to an open reservoir – for instance, a water

Flow of Incompressible Fluids | CHAPTER ONE

tower in a city water distribution network — flow will begin once a valve is opened. The pressure on the surface of the water in the reservoir is approximately equal to the pressure in a kitchen sink. In this case, there is no difference in pressure between the ends of the piping system, but flow occurs. There is more than just pressure difference involved in the flow of water. An unbalanced force in the direction of the flow is necessary.

Re-establishing equilibrium

The phenomenon of flow in conduits is best thought of as an attempt to re-establish an equilibrium state in a fluid constrained by conduit walls. In a closed system under pressure, when all valves are shut, all forces are accompanied by equal and opposite forces. The fluid is stationary. In an open system, when valves are open, there is an imbalance of forces and the fluid moves in a direction that will re-establish equilibrium. If mass and energy are added continuously to an open system, flow will be continuous.

The mass added could be water flowing to a reservoir behind a dam, or water pumped to a head tank. Once there, the water has energy by virtue of its position. This energy is potential energy or the potential to do work. Upstream of a closed tap, faucet or valve, the water will be subject to the weight (force) of the water above it. It will exert a pressure (force per unit area) to equilibrate this force. This pressure is the result of repulsion arising when molecules approach one another. The pressure is associated with a static fluid and with a form of energy called static energy.

If the valve is opened, the water will flow through it by virtue of the pressure difference across it. The pressure upstream of the valve will decrease for two reasons. First, some of the pressure energy has been converted to kinetic energy, the energy a body has by virtue of its motion. Second, some of the energy of the water has been converted to internal energy because of what is commonly called "fluid friction" in the upstream piping. The increase in internal energy is accompanied by an increase in temperature — hence, the loose term, "thermal energy". This loose term shortly will be explained more fully.

The kinetic energy is recoverable. It can be transformed back to static (pressure) energy by closing the valve or to potential energy by connecting an open, vertical pipe to an elevation equal to that of the reservoir. The incremental internal energy is not completely recoverable to one of the mechanical forms of energy. This is the reason the water in the

CHAPTER ONE | Flow of Incompressible Fluids

previously mentioned vertical pipe cannot quite reach the original elevation of the surface of the reservoir. The incremental internal energy flows as heat through the pipe walls and is lost to practical use. It warms up the atmosphere. In a well-insulated pipe, it remains with the fluid for a longer period of time and is seen as a higher fluid temperature.

Logical inconsistency of fluid friction

The term "fluid friction" is used commonly to explain losses in mechanical forms of energy associated with fluid flow. A little knowledge of physics shows it is a poorly chosen term. Friction is a macroscopic phenomenon – such as when two blocks of wood slide over one another, or when the rubber hits the road. There is a definite resisting force to be overcome – the force of friction.

Fluid molecules do not touch

Fluids are made up of molecules, and molecules do not touch in the fluid state. Collisions of gas molecules are actually repulsions that prevent the molecules from touching. Liquid molecules are less energetic than gas molecules. They slide past one another or remain at an equilibrium distance. The equilibrium distance is established by a balance between an attractive force and a repulsive force. If molecules do not touch, how can there be "friction"?

The answer is found in the compressive forces that occur when local flow is slowed down by obstructions of any kind. These obstructions can be large objects such as valves or fittings; they can be eddies within the flowing stream; they can be other molecules that have diffused between two layers of fluid moving at different velocities.

Compression involves work

Whenever a portion of a fluid is compressed, work is done on it by an external force. This work increases the internal energy, and the temperature increases. There has been a conversion of mechanical energy to internal energy. Temperature increase is not uniform throughout a fluid, especially in turbulent flow. The direct contact of different temperature fluids is a classical case of an inefficiency: a loss in the capability to convert heat flow to useful work. In thermodynamics, the term used to describe this natural phenomenon is "irreversibility".

The concept of irreversibility describes a law of thermodynamics which states that work can be converted completely to internal energy or to heat flow. However, heat flow cannot be completely converted to work energy in any cyclic process – no matter how perfect. Some of the heat energy must flow to a lower temperature (the environment) or the energy associated with it must remain in the system as internal energy. The bulk temperature will be higher than that of the fluid before it was compressed.

Heat and work are energy in transit across a boundary. Temperature is a measure of the intensity of energy, not of its quantity. Once heat and work no longer cross a boundary, we cannot recognize them as heat or work. They simply take some other form – internal energy, for instance. Many engineers still talk of heat as if it were a substance, like caloric, as if it were a property of matter. It is not.

In this book, we will try to avoid talking loosely about heat as if it were a property of matter. We will also try to avoid the logical inconsistency by using the term "irreversibility" in place of "friction loss" and the plural "irreversibilities" in place of "losses". This concept will be more fully developed in Appendix AI.

Equilibrium and energy transformations

To recapitulate, flow can be thought of as a means of re-establishing an equilibrium state in energy levels. Energy exists in various forms, some of which can be converted relatively easily to other forms. Potential energy can be converted to static energy and to kinetic energy. Any of the three forms of mechanical energy can be interconverted by applying suitable constraints. Kinetic energy can also be converted to internal energy by turbulence or any other deceleration. Once in this form, there is no means of converting it completely back to one of the more useful forms of mechanical energy. This is a law of nature.

The study of fluid flow is really the study of energy transformations. Energy is indestructible; its total amount remains unchanged. Mechanical energy converted to energy flowing as heat between two temperatures represents a loss of useful mechanical energy, not of total energy. It is this unidirectional change of mechanical to internal energy and the subsequent flow of heat to a lower temperature sink that explains the irreversibilities associated with flowing fluids.

CHAPTER ONE | **Flow of Incompressible Fluids**

I-3: FLOW REGIMES – REYNOLDS' CONTRIBUTIONS

Reynolds[v] performed a series of correlations in 1883 based on the equipment shown in Figure I-1. His apparatus consisted of a transparent tank in which a constant level of water was maintained. The source of the make-up water is not shown in the figure. A transparent tube, open within the tank, was used to bleed off water at controlled rates. A fine tube was used to introduce a colored stream to the mouth of the larger transparent tube. The apparatus allowed Reynolds to see clearly the difference between two flow regimes and to see the transitional regime between them.

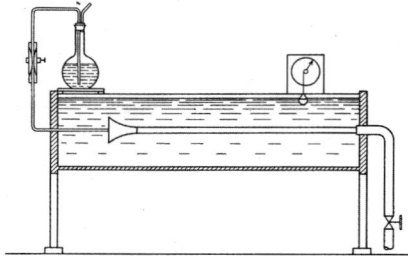

Figure I-1. Reynolds' apparatus

Source: L. McCabe, Unit Operations in Chemical Engineering, 1967, reprinted by permission McGraw-Hill Companies.

Laminar flow

In the first flow regime, laminar flow, the colored stream occupies the very central core of the transparent tube. There is no lateral mixing over the full length of the tube. This regime occurs at very low flow rates (Figure I-2).

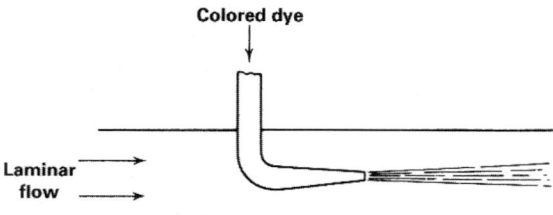

Figure I-2. Laminar flow

Source: Weber and Meissner

Turbulent flow

In the second flow regime, turbulent flow, the colored stream quickly becomes fully mixed with the clear stream. There is obviously much lateral mixing due to eddies in the turbulent stream. This regime occurs at higher flow rates (Figure I-3).

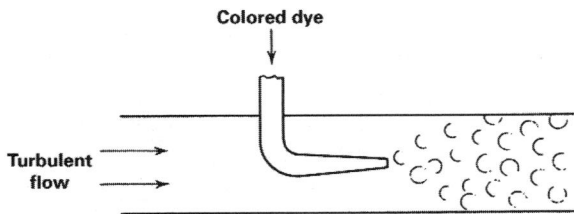

Figure 1-3. Turbulent flow

Transitional flow

The change from laminar flow to turbulent flow occurs over a very small increment in flow. The regime within this increment is termed transitional flow. The transitional flow regime is unstable; it is a regime to be avoided. The characteristics are not uniform nor reproducible.

Recognizable and quantifiable flow regimes

The contribution of Reynolds[v], his co-workers and their successors, was in the definition of recognizable and quantifiable flow regimes. They also gave us an index that can be computed from the conditions of flow and from the fluid characteristics, the Reynolds number. The Reynolds number can be used to identify the flow regime. Further, this index can be used to calculate a friction (mechanical energy loss) factor that, in turn, allows calculating pressure drop and irreversibilities. The Reynolds number is also important in establishing the turndown capabilities and operating accuracies of metering devices.

Transitional flow, an unstable flow regime, occurs between the laminar and turbulent flow regimes. In the transitional region, turbulent flow and laminar flow alternate in an irregular manner.

The flow regime impacts our ability to accurately measure flow. It also changes the sizes required of control valves. The pressure drop in turbulent flow is proportional to flow squared. In laminar flow it is directly proportional to the flow rate. The equations used are different.

Reynolds number

Fortunately, the index of the flow regimes mentioned above, the Reynolds number, N_{Re}, allows identification of the flow regime. It enables us to tell whether we will cross from one regime into another. It allows us to choose the correct equations to use for both measuring devices and control valves. It also allows us to compute with a high degree of accuracy the mechanical energy losses in conduits and the associated pressure drop.

CHAPTER ONE | **Flow of Incompressible Fluids**

The Reynolds[v] number is a dimensionless ratio used extensively in flow correlations and "loss" computations. It is a very important number.

The Reynolds number is often regarded as the ratio of inertial forces to viscous forces – the tendency to keep moving versus the tendency to slow down. It is a useful correlating parameter for all industrial liquid flows. It also applies to industrial vapor and gaseous flows. The Reynolds number is usually given as:

$$N_{Re} = \frac{DU\rho}{\mu} = \frac{DG}{\mu} = 6.31\frac{W}{d\mu_{cP}} \tag{I-1}$$

The Reynolds number is a pure number – all dimensions cancel. D is usually a diameter or some other length dimension, U is the average velocity across the section, rho is density, mu is absolute viscosity (to be defined in I-7), and G is mass velocity (velocity times density or mass flow rate divided by area). The mean velocity across a section, U, used in most computations, must be correlated with factors influencing the velocity profile. These factors have been found mainly to be μ, ρ, D, the viscosity, the density and the internal diameter.

The numerator of the first equation could have dimensions of pipe diameter, D, in feet, average fluid velocity, U, in feet per second, and fluid density, ρ, in pounds-mass per foot cubed. Multiplying the dimensions and cancelling gives units of pounds-mass over seconds times feet in the numerator.

The denominator has units of "absolute" viscosity, which in the customary U.S. system are also given in pounds-mass over seconds times feet. The units of the numerator and the denominator cancel leaving a dimensionless, "pure" number.

The last equation of the set, I-1, is frequently seen. It is a mixed bag of older metric (cgs or centimeter-gram-second) units and English-American units. The viscosity is in centipoise or one hundredth of a poise. The poise is defined as one gram divided by the product of one centimeter and one second. We give the conversion in detail in Appendix AI.

The coefficient of the last term of Equation I-1 has dimensions. These dimensions cancel the mixed units of W, lb_m/h, d, pipe internal diameter, inches, and μ, centipoise. The expression is commonly used in North American flow computations.

Flow of Incompressible Fluids | CHAPTER ONE

All three equations produce the same Reynolds[v] number for the same conditions of flow.

Viscosity is defined in section I-7 of this chapter and it is covered in more detail in Appendix AI.

Different types of Reynolds numbers

It should be noted a Reynolds number can be defined for different circumstances and the appropriate dimensions should be used. The reason there are several Reynolds numbers is the number is used for purposes of correlation of data. For convenience, the length dimension, D, may change in a given correlation. It is defined for flow in pipes (where D is usually the internal diameter of the pipe). It is defined for flow past bluff bodies (where D is usually the projected width of the body normal to the flow). It is defined for flow past aeroplane wings (where D is usually the dimension of the wing in the direction of the flow). The velocity used is sometimes the average velocity across a pipe, sometimes the average velocity across an orifice plate bore, sometimes a point velocity.

For instance, a Reynolds number defined for a pipe requires that D be the pipe's internal diameter. Velocity, U, is usually the average velocity across that section, but may be a point velocity. Density, rho, is the average density at the section. Viscosity, mu, is the average viscosity at the section. All units must be compatible so they cancel.

Another Reynolds number is defined in terms of hole diameter of an orifice plate. In this case, U is the average velocity across the section of the orifice. The appropriate numbers must be used if the correct conclusions are to be drawn from correlations. Subscripts are usually employed to differentiate among the various Reynolds numbers.

Caveat — Reynolds numbers

Always check very carefully the definition of the Reynolds number for the correlation being used.

In this book we will use primarily the pipe (based on the internal diameter of the pipe) Reynolds number symbolized by N_{ReD}. The velocity, U, will be the average velocity across the section of pipe, not the point velocity or the centerline velocity. We will specifically define other Reynolds numbers when they are introduced.

CHAPTER ONE | Flow of Incompressible Fluids

Any units can be used for the variables of the Reynolds[v] number provided they are consistent, they cancel and produce a dimensionless number. Mixed units may be used, provided the correct conversion factors are included to allow the group of numbers to be dimensionless. We will deal thoroughly with real numbers later in the book.

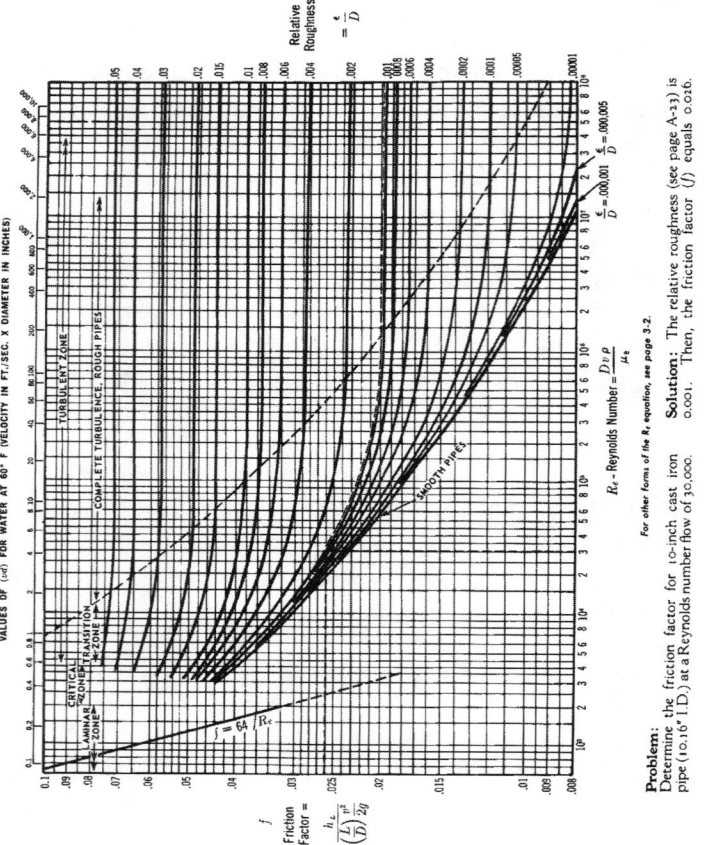

Figure I-4. Moody[viii] friction factor versus Reynolds number

In circular pipe at a pipe Reynolds number below 2,000, flow is laminar. At Reynolds numbers greater than 4,000, flow is generally considered turbulent. However, to be sure, since laminar conditions sometimes persist at greater flow rates, it is best to consider 10,000 the lower limit of the turbulent zone. The range of Reynolds numbers between 2,000 and 4,000 flow is considered to be in the transitional zone. Remember that transitional flow is not stable, it fluctuates. So, when you have control over the design of a system, for most metering purposes, make sure you are in one regime or the other over the full range of flows. If you do not have control over the design, then compute the limitations imposed on the system by the actual range of flows.

Figure I-4 is a plot of the Moody[viii] friction factor versus the Reynolds[v] number. Relative roughness of the pipe wall is a parameter. It shows the distinctions among the three regimes.

I-4: FLOW PROFILES – VELOCITY DISTRIBUTIONS

A flow profile is the plot of velocities across a section of pipe. Flow profiles are influenced by the behavior of the Reynolds number and particularly by its viscosity component. Not all profiles are as smooth as those described in Figure I-5. As we shall see a little later, fluids are generally characterized by their viscosities. One type of fluid, a Bingham plastic (typical of sewage) can have a central core that flows as a plug. The surrounding annulus has a more normal velocity distribution.

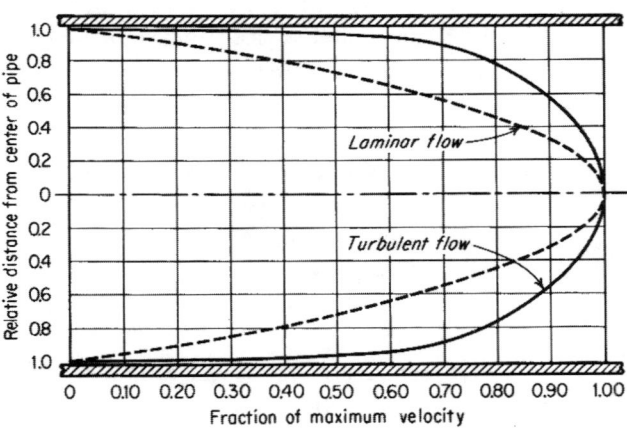

Figure I-5. Distribution of velocity across pipe, fully developed flow of Newtonian fluid

Source: L. McCabe, Unit Operations in Chemical Engineering, 1967, reprinted by permission McGraw-Hill Companies.

Influences on flow profiles

Proximity to the pipe inlet from a vessel is another factor influencing the velocity distribution. At least 30 pipe diameters of length must be allowed to develop a steady profile in incompressible flow.

Bends and partial obstructions can distort the profile to such a degree that the average velocity, U, cannot be inferred with any degree of accuracy from a measuring device immediately downstream of the disturbance.

CHAPTER ONE | **Flow of Incompressible Fluids**

This is the reason upstream and downstream straight lengths are specified for almost all measuring devices. It is also the reason for the existence of straightening vanes.

Straightening vanes are axial inserts in pipes. They are designed to cause flow to be parallel to the axis so, shortly downstream of the insert, the flow resumes its fully developed profile – the one for which the measuring device was designed.

Irreversibility

In turbulent flow, large eddies break up into smaller ones, and the smaller ones into even smaller eddies, until they finally disappear. Bulk kinetic energy is transformed into rotational energy. This is a transformation of one form of mechanical energy to another. What is generally termed "fluid friction" is mechanical energy transformed finally to thermal (internal) energy by compression as the fluid is decelerated. It is this conversion from mechanically useful kinetic energy to mechanically useless internal energy that results in the concept of "lost energy". This term would be better stated as "lost mechanical energy". We will use the thermodynamic term "irreversibility" to represent this conversion.

In laminar flow in pipes, concentric annuli flow past one another with no net normal flow. Molecules, however, do bounce back and forth across the imaginary boundaries of the annuli. Compression takes place as faster moving molecules in the bulk flow are decelerated by the proximity of the slower moving ones. Mechanical energy is converted to internal energy as work is done in decelerating the molecules.

The loss in mechanical energy due to irreversibilities must be made up by the prime mover or by other means if steady-state flows are to be maintained.

Flow profiles

Some instruments are more sensitive than others to the flow profile in a conduit. The flow profile is the plot of the axial components of the point velocities as the conduit is traversed. Figure I-5 is an example of fully developed flow in a circular pipe, remote from any disturbance, including that due to the conduit inlet. The figure depicts both laminar flow and turbulent flow, but note they cannot both be present at the same instant. Note the flattening of the profile in turbulent flow and note there is no attempt made to show a transitional flow profile because transitional flow fluctuates.

Flow of Incompressible Fluids | CHAPTER ONE

The point velocity at the center of the pipe is twice the average velocity in laminar flow. It is a factor, alpha (greater than, but close to one), times the average velocity in turbulent flow. We will develop the relationships more exactly later in the book. The velocity profile must be taken into consideration when measuring flow. This is done by computing the range of Reynolds[v] numbers over which flow will be measured. Controlling the pipe geometry with straightening vanes or using minimum lengths of straight pipe are the usual means of fixing the limits of the flow profiles. The point velocities can be obtained by traversing a Pitot tube across the conduit and taking point readings.

It is to be noted that both curves of Figure I-5 have zero velocity at the conduit wall. Both curves have a maximum velocity on the center line. The curves are drawn so the maximum velocities coincide. Conclusions should not be reached regarding the same maximum velocity for the same pressure drop in both types of flow. Figure I-6 shows that, after different types of obstructions, the velocity profile can become very distorted. The profile can not only be distorted over a longitudinal section, it can even rotate axially. The axis of rotation is not necessarily the pipe axis.

Influence of flow profiles on measuring instruments.

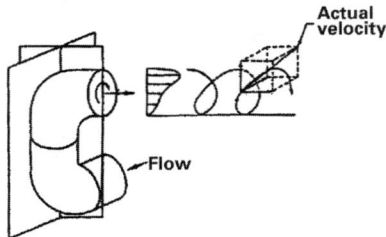

Source: ISA T240

Figure I-6. Velocity components downstream of close-coupled elbows in different planes

Most measuring instruments are used to *infer* the average flow in a duct. They do not measure flow directly. These devices are sometimes called inferential meters.

A device such as a Pitot tube measures the impact pressure and the static pressure at a point. Impact pressure is the static pressure at the point plus the pressure difference created by decelerating the fluid to zero velocity. The point velocity is computed from the difference between this impact pressure and the static pressure at the section. This difference is often called the velocity head. The word "head" arises from the use of open manometers to measure differential pressure in early hydraulic work.

CHAPTER ONE | **Flow of Incompressible Fluids**

Sensor location

Most of the time, we are interested in the average flow through the section; therefore, it should be obvious the location of the sensor is critical to the accuracy of the inference. From Figure I-6, it should be evident that a point sensor located within a distorted profile will give erroneous readings. In addition, if the profile rotates, as it will after two close pipe bends, a pulsating signal will be generated which may not be integrated correctly. A predictable, reproducible velocity profile is necessary for accurate measurement with most flow meters.

Kegel[ix] (Flow Measurement, ISA) gives a good description of the problem of measuring flow when profiles are distorted. Figure I-7 compares the results of seven different measurement points across the same section of pipe with the same flow rates but with three different profiles. He makes the argument for using averaging techniques when dealing with distorted flows. He also references ISO standards for specific techniques.

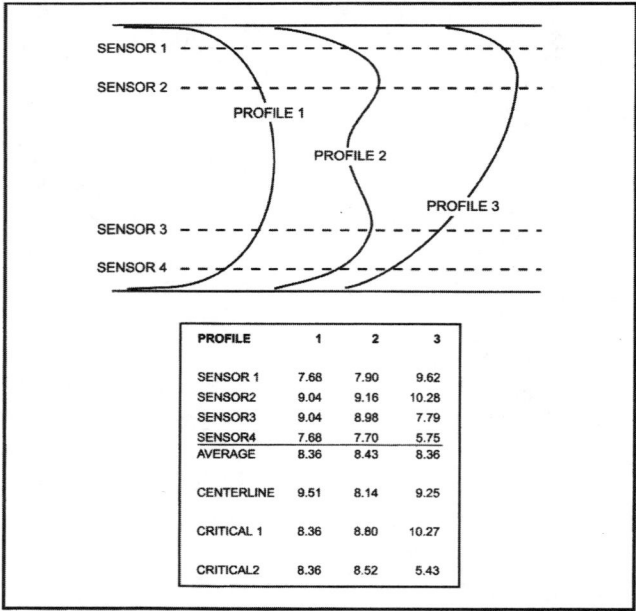

Figure I-7. Multiple sensor averaging techniques

Figure I-8 explains the differences among mean, maximum and point velocities. It should be noticed that the profile is uniformly smooth. Distortion of the profile will give numbers that cannot easily be used in a correlation.

Flow of Incompressible Fluids | CHAPTER ONE

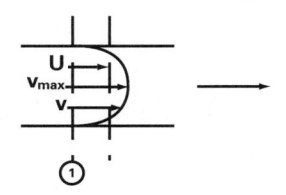

Figure I-8. Average, maximum and point velocities

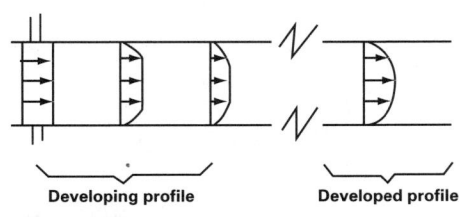

Figure I-9. Developing profile

Figure I-9 shows a profile that may not be fully developed for about thirty pipe diameters.

Devices such as orifice plates, which infer flow rates from differential pressure, are subject to profile limitations. In other words, if the profile is not known or controlled, the accuracy is not trustworthy. Some devices, such as magnetic flow meters, are less susceptible to flow profile influences. However, even a magnetic flow meter will lose some of its accuracy, due to flow profile disturbance, if it is connected directly to a pipe elbow.

The extreme, non-continuous flow profile of Bingham plastic flow must obviously be considered if we wish to obtain metering accuracy on Bingham plastics.

A Coriolis mass flow meter, which is a true mass flow meter, is not at all influenced by the flow profile, but is limited in available diameters.

The influence of the flow profile on metering accuracy is the reason so much time is spent establishing the required upstream and downstream straight runs for piping around a metering device. When these minimum requirements cannot be met, flow straightening vanes are used to correct the flow profile, or one simply gives up on the accuracy requirement and uses the metered flow rate as an approximation.

CHAPTER ONE | Flow of Incompressible Fluids

I-5: FLUID FLOW – AN 'IRREVERSIBLE' PROCESS

Thermodynamics teaches that fluid flow is an irreversible process. The sense of the word "irreversible", in this case, is useful available mechanical energy has been transformed into internal energy, or heat flow, without useful work being extracted. The system and the environment cannot be restored to their original states without additional work being done. The irreversibility is generally the unavoidable flow of heat energy generated by turbulence to a lower temperature sink, either in the environment or in the same flowing fluid. There is no practical way of recovering this incremental internal energy as mechanical energy.

Flow of a fluid through a horizontal duct of constant cross section is always associated with a pressure drop. If the fluid flow through this duct were "ideal", there would be no losses, and there would be no pressure drop due to irreversibilities. "Ideal" flow would not require adding energy to maintain itself. Once established, flow would continue without external help. Real flow requires adding mechanical energy to make up for the constant conversion to internal energy and, ultimately, to heat flow.

If we divide the pressure drop, using customary U.S. units, by the fluid density, we see the units are foot-pounds-force per pound-mass or energy per unit mass. In the SI system, the units are newton-meters per kg or joules per kg. In irreversible (normal) processes, the energy associated with pressure drop due to irreversibilities is commonly described as having "disappeared" or as having been "lost". It does not disappear and it is only lost to mechanical use. It is transformed to internal energy, which results in an increase in temperature. The increase in temperature usually causes the internal energy to be dissipated to the environment as a heat flow.

Irreversible factors in fluid flow

What follows is a list of some factors that cause losses of mechanical energy (irreversibilities) in fluid flow.

Viscous drag forces

Force has to be applied to keep fluid layers moving past one another, even at steady state. The reluctance of fluid layers to slide past one another is termed viscosity. Later, we shall develop the equations that describe how some static energy is converted to kinetic energy and how the rest is converted to incremental internal energy. Viscous drag forces are the principal forces that we deal with in solving fluid flow problems.

Fluid mixing

When two different streams are mixed, mechanical energy is dissipated (transformed to internal energy) due to turbulence. Turbulence results from energy gradients within the flowing fluid. It results from the fluid attempting to redistribute energy, to achieve equilibrium. Ideally, this energy could have been recovered in an imaginary turbine or a hypothetical Carnot[ii] engine, but was not. It is often stated that nature abhors a vacuum. It would be more accurate to say than nature abhors an energy gradient. Effectively, the natural state is one of equilibrium, as was discussed in Section I-2.

An example of fluid mixing is when two different pressure safety valves relieve simultaneously into the same header. Fluid is also mixed when it first separates around obstructions and then the separated streams rejoin. Vortex shedding sensors (bluff bodies) and control valves are examples of such mixing processes. The mechanical energy conversion to internal energy due to this cause is often called "form friction" as opposed to the "skin friction" of normal pipe flows.

Pressure and temperature inequalities

Pressure gradients in a flowing stream usually cause a dissipation of useful mechanical energy to non-recoverable internal energy. Temperature gradients usually result in loss of potential to do work. (A Carnot[ii] engine could have recovered some of the internal energy).

Chemical reactions

Industrial chemical reactions usually are associated with irreversibilities. Normally, chemical reactions are confined to specific reactors and we are only concerned with the flow of the reactants and products. Sometimes, however, the reaction takes place in a pipe or in a distillation column and the physical properties of the flowing fluids change along the flow path. You may be asked to calculate or measure or control the flow.

I-6: FUNDAMENTAL RELATIONSHIPS OF FLUID FLOW

There are three fundamental relationships to which we will return time and time again to solve fluid flow problems. These three fundamental relationships involve the conservation of mass, the conservation of energy and the Bernoulli[i] relationship. Actually, the Bernoulli relationship is a specific application of the more fundamental relationship involving Newton's second law.

CHAPTER ONE | **Flow of Incompressible Fluids**

Conservation of mass

Mass is not destroyed. It can flow into, accumulate within, and flow out of a system. In the steady state – and fortunately many flow problems can be reduced to steady state problems – what flows in equals what flows out. There is no accumulation. If there is only one incoming flow line and one outgoing one, the steady state situation can be described as follows,

$$\dot{m}_1 = \dot{m}_2$$
$$A_1 U_1 \rho_1 = A_2 U_2 \rho_2$$
$$G \equiv U\rho = U/v$$
$$G_1 A_1 = G_2 A_2$$

(I-2)

The first equation, above, simply states that the mass flow rates into and out of the system at the steady state are equal when there is no accumulation.

The second equation is equivalent to the first, but introduces three different variables: area of the conduit, average velocity, and density. If the density is constant, the second equation reduces to a statement that the volumetric flow in and out is constant. This is the case with most liquid flows, but not with gases or vapors.

The third equation defines mass velocity or mass flux as being the product of the average velocity across a section and density (or the quotient of average velocity and mass volume). It has units of mass flow per unit area. The fourth equation makes use of mass velocity. It is a useful form under certain circumstances (compressible fluids) which will be explained later, when we discuss compressible fluids.

Caveat — Average or point velocities

Always be aware of the velocity that is being considered, discussed or defined. Whether we are talking about average or point velocities can make a huge difference in computations of kinetic energy, friction factors, Reynolds[v] numbers and losses of mechanical energy. We will try to be specific each time this problem arises.

Conservation of energy

The first law of thermodynamics deals with the conservation of energy. It cannot be bypassed. It applies to all systems, reversible and irreversible. However, it is not sufficient to answer all problems (This will become clear in subsequent developments). For open (flowing), steady state systems,

Flow of Incompressible Fluids | CHAPTER ONE

the first law is expressed in two forms. One form is the differential form useful for the developing relationships. The second form is the integrated form that is more useful when relationships already developed have to be applied in practice. Both forms will be put on a per-unit-mass basis. All practical equations will ultimately be in the integrated form.
Differential equations are necessary to completely understand the development of solutions.

$$\delta q + \delta w_n = dh + \frac{UdU}{g_c} + \frac{gdX}{g_c} \quad \text{(I-3)}$$

$$q + w_n = \Delta h + \frac{\Delta(U^2)}{2g_c} + \frac{g\Delta X}{g_c}$$

In the first equation of the set I-3, the Greek deltas on the differential heat and work terms signify that these differentials are not "exact", they are dependent on the "path" between equilibrium states. The dimensional constant, g_c, has been retained for those who deal with American customary units. The units of g and g_c are different – f/s^2 and lb_m-ft/lb_f-s^2. If you are in the habit of only working with SI units, just regard the constant as equal to one and as being dimensionless. The deltas on the terms on the right of the integrated form indicate we are dealing with differences in quantities measured between two points. The heat and work quantities will be the corresponding heat and work flows per unit mass between the same two points.

The above forms of the equations also assume that the user knows enough to convert mechanical units associated with the work, velocity and elevation terms to energy units associated with the heat and enthalpy terms. This will be explained more fully later. In this book we will use the convention that energy added to a body is positive, energy leaving is negative (a different convention requires that work added be negative).

The first law for open systems, as described in the set I-3, simply states that energy in transit across the boundaries of a system as heat flow and work flow per unit mass results in changes in three identifiable properties of the flowing fluid. The first is the enthalpy, which is the internal energy of the fluid plus the pressure-volume work it took to get the fluid into and out of the system. The second is the kinetic energy term, which is the specific energy due to bulk motion and which will be used to derive the equations of flow through head meters. The last term is simply the change in energy due to change in elevation – potential energy. All terms are on a per unit mass basis.

CHAPTER ONE | Flow of Incompressible Fluids

The first law requires calculating heat energy transfer, q, per-unit-mass or, Q, total heat energy transfer between two states. This calculation is often not feasible in the case of an industrial piping system. Therefore, the first law has limited direct application in fluid flow computations.

Force balance equation (Bernoulli's equation)

The Bernoulli[i] equation is one of the most useful equations ever developed as far as fluid flow is concerned. Its derivation will be given in Appendix AI. Bernoulli's original equation was for the ideal situation. It did not include a mechanical energy loss term.

A more modern version of the Bernoulli equation is given below.

$$-\int_{1}^{2} v dP = h_f + \frac{\Delta(U^2)}{2g_c} + \frac{g\Delta X}{g_c} \quad \text{(I-4)}$$

The term on the left is the difference in specific energy change between the two points due to changes in static pressure and mass volume. The negative sign is due to the fact that the pressure downstream is less than that upstream. If the fluid can be considered incompressible, the term may be integrated immediately to $v(P_1 - P_2)$, mass volume times pressure difference, or the equivalent expression $(P_1 - P_2)/\rho$, pressure difference divided by density. Note that in this case the pressure difference is the upstream less the downstream pressure because of the negative sign.

The term, h_f, is the mechanical energy converted to thermal energy (irreversibility). The next term represents kinetic energy. The last term is the change in energy per unit mass associated with the change in elevation between sections 1 and 2.

The deltas are simply the differences between the variables, U^2 and X, at the two points represented by the limits of integration 1 and 2. The deltas mean the differences between downstream and upstream conditions. This is the normal convention.

If a pump, blower or a turbine is included between points 1 and 2, an appropriate work term is added on the left. The sign would be positive for a pump or blower, negative for a turbine.

$$w_n - \int_{1}^{2} v dP = h_f + \frac{\Delta(U^2)}{2g_c} + \frac{g\Delta X}{g_c} \quad \text{(I-5)}$$

Flow of Incompressible Fluids | CHAPTER ONE

Units of each term

Each term in the above equations has units of energy per unit mass flowing. The term, w_n, is the net work added to the fluid by a pump or blower per unit mass of flowing fluid. It is the net work extracted in the case of a turbine. The term "net" may be misleading. It is actually the work added to or subtracted from the fluid. In the incompressible case the term under the integral simply reduces to the negative of the pressure difference at the two sections divided by the density. The downstream section carries the number 2.

The first term on the right is not enthalpy. It is a loss term. It is that part of the available energy converted to internal energy or heat flow because of irreversibilities (shock, compression, "fluid friction"). The next term is the change in kinetic energy and the last term is the change in elevation head or potential energy.

The loss term, h_f, is always positive. Energy may be transferred in either direction among the other terms. Energy can only be transferred to the loss term, not from it. Equations for the loss term, h_f, have been developed experimentally for both laminar flow and turbulent flow. The Darcy[vii] equation is the result. The Darcy equation will be described later in this chapter and in detail in Appendix AI.

The Bernoulli[i] equation in the form of equation I-5 makes up for some of the shortcomings of the first law when calculating fluid flow. From it are derived the flow equations for head meters and control valves. It allows the computation of conditions at each end of a conduit when the dimensions, the change in elevation, the configuration and quality of the wall are known. Alternatively, it allows the computation of dimensions when the flowing quantities and routing are known. It warrants memorizing.

Caveat — Bernoulli equation versus the first law

The resemblance of the first law equation to the Bernoulli equation and the use of the symbol, h, to represent two different quantities, enthalpy and lost mechanical energy, can lead to confusion.

CHAPTER ONE | Flow of Incompressible Fluids

I-7: THE ROLE OF VISCOSITY

The ideal model for the concept of viscosity is the newtonian model that will be discussed shortly. For the moment, the reader is invited to regard viscosity as simply a coefficient that is the ratio of two terms. The numerator is the force per unit area applied to the moving element of two coaxial cylinders containing a fluid between them. The denominator is the velocity gradient set up in the fluid as its velocity varies between that of the moving element and that of a stationary, concentric, inner element (zero).

Viscosity has been described as the reluctance of a fluid layer to slide over another one. It might be better defined as a measure of the attractive forces between molecules – forces that must be overcome if there is to be motion. It was shown to play a key role in determining the value of the Reynolds[v] number, which in turn determines the flow regime. Knowledge of viscosity helps determine the mechanical energy losses in a duct or conduit. Viscosity also influences the discharge coefficients of orifice plates and the sizes of control valves – both through the Reynolds number. It also has an influence on the velocity profile. An extreme case of this influence is the non-uniform profile of Bingham plastic flow. Viscosity is not a thermodynamic variable. It is, however, a property of fluid matter. It is a convenient mathematical coefficient that varies with pressure, temperature, composition and, sometimes, with applied stress and time.

The term "viscosity" applies to fluids. We have discussed fluids without defining them on the assumption everyone knows what a fluid is. From the point of view of fluid mechanics, a fluid is a substance, gas, vapor or liquid (or slurry) which undergoes continuous deformation when subject to shear stress. Unlike solids, liquids and gases cannot resist small shear stresses without relative motion occurring between the layers subject to shear.

The behavior of a flowing fluid depends on whether or not it is influenced by a solid boundary. If it is far removed from a solid boundary, it is not subject to shear. The parallel planes move at the same speed. Eddies cannot form (irrotational flow) and there is no compression and no dissipation of mechanical energy to incremental internal energy.

In large-scale, bulk flow, shear forces are said to be confined to the boundary layer – the layer near the solid wall. Once turbulent flow is established in pipes, the boundary layer fills the entire channel. Therefore, irreversibilities are always present across the entire section, but they are greater near the wall due to the flattened profile.

Flow of Incompressible Fluids | CHAPTER ONE

One-dimensional flow (frequently used as a model for real flow) is flow that can be represented by parallel vectors. Laminar flow is one-dimensional flow in which there is no gross lateral mixing between fluid layers. In turbulent flow, there is lateral mixing, but, in a conduit, the flow can still follow the one-dimensional model. It can still be represented by parallel vectors in the forward direction. The eddies in the normal direction all cancel one another.

Categorizing fluids based on viscous behavior

We will establish a general classification of fluids based on their viscous behavior to permit a more detailed discussion. Fluids are generally categorized according to whether or not they follow the newtonian model. If the coefficient of viscosity is independent of the shear forces and of rate of change of shear, the fluid is newtonian. If the coefficient of viscosity changes with the applied shear force or with the rate of change of shear, the fluid is non-newtonian. The newtonian model makes for fairly simple computations. These ideas will be made more concrete very shortly.

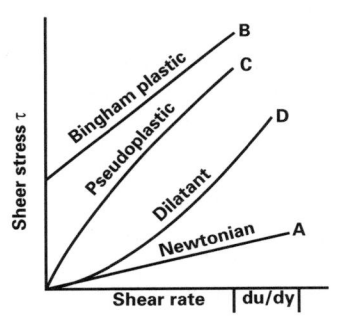

Figure I-10. Shear diagrams

Source: R. Perry and D. Green, Perry's Chemical Engineers' Handbook, 1984, reprinted by permission McGraw-Hill Companies

Non-newtonian fluids are further characterized as to whether their behavior is independent of time or is a function of time. A third characterization is whether the non-newtonian fluid displays some of the characteristics of a solid such as the ability to recover from applied stress. Non-newtonian fluids that have time-independent behavior are divided into three subcategories depending on how they deviate from the newtonian model.

Bingham plastics require a minimum stress before they begin to move. They then follow a fairly linear relationship between applied shear and rate of change of shear.

Pseudoplastics do not require a minimum stress in order to initiate movement. Therefore, their curve goes through zero similarly to a newtonian

CHAPTER ONE | Flow of Incompressible Fluids

fluid. However, their curve is not linear – it is concave downward. Less shear stress is required to produce a rate of change of shear as the stress is increased.

Dilatant fluids also have a curve that goes through zero. However, they have a curve that is concave upward, opposite to that of a pseudoplastic. It takes more shear stress to produce a rate of change of shear as the stress is increased.

Time dependent fluids are generally divided into two subcategories: those that "thin" when stress is applied over time, and those that "set up" when stress is applied over time. The first subcategory is called thixotropic and the second one is called rheopectic.

Viscoelastic fluids exhibit many of the properties of a solid. They will be discussed shortly.

Table I-1 outlines the above breakdown.

> **Table I-1. Fluid categorization by viscous behavior**
> 1. Newtonian fluids
> 2. Non-newtonian fluids
>   A. Time independent behavior
>   a. Bingham plastics
>   b. Pseudoplastics
>   c. Dilatant fluids
>   B. Time dependent behavior
>   a. Thixotropic fluids
>   b. Rheopectic fluids
>   C. Viscoelastic fluids

1. NEWTONIAN FLUIDS

Much progress in engineering has been made by establishing a model against which non-standard behavior can be compared and quantified. The newtonian model is a case in point.

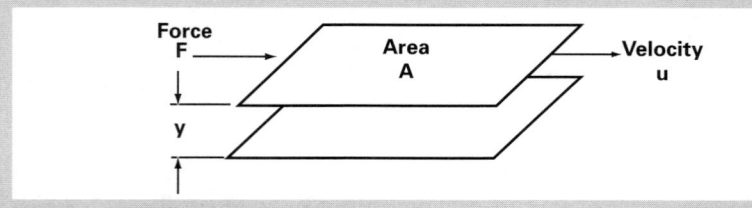

Figure I-11. Definition of viscosity

Source: R. Perry and D. Green, Perry's Chemical Engineers' Handbook, 1984, reprinted by permission McGraw-Hill Companies

Flow of Industrial Fluids—Theory and Equations

By way of example, consider two layers of fluid next to one another (Figure I-11). If no shear stress is applied, there is no motion. A shear stress can be applied by a force acting on one of two parallel plates within the fluid. The fluid is stationary relative to the plates at boundaries created by the two plates, even when one of the plates is in motion. If a force is applied to one of the two plates and the other is fixed, a velocity gradient will be created between the plates. The layer in contact with the moving plate moves at the same velocity as the plate. The layer in contact with the stationary plate is stationary. Intermediate layers have velocities depending on their relative position between the two plates. The following relationship holds:

$$F = \frac{\mu_m V A_{plate}}{g_c Z} \tag{I-6}$$

The relationship states that the force necessary to cause a velocity of a plate in the fluid is directly proportional to the velocity, to the area of the plate to which the force is applied and to a proportionality constant. It is inversely proportional to the distance between the plates, Z. The dimensional constant, g_c, is thrown in for good measure by chemical engineers using U.S. units. Various units will be discussed under the heading, Units of Viscosity, in this chapter. The units of this viscosity, μ_m, are lb_m/ft-s.

Note that equation I-6 relates a force to a velocity, not an acceleration. If the acceleration is zero, an equal and opposite force must arise in the fluid in order to prevent acceleration.

If the velocity varies linearly across the gap, V/Z may be replaced by dV/dZ, the velocity gradient. Dividing by the area gives the more usual form of the viscosity equation.

$$\frac{F}{A_{plate}} = \frac{\mu_m}{g_c} \frac{dV}{dZ} = \tau \tag{I-7}$$

The term on the left is the force per unit area applied to move the fluid layers. It is called the shear stress. The derivative term is called the shear rate. Mu is again the proportionality coefficient and g_c is the dimensional constant. The coefficient, mu, is the viscosity. It is a function of temperature, pressure and fluid composition. In a newtonian fluid, viscosity is neither a function of shear stress, nor of shear rate. The Greek letter, tau, is just a shorthand form for the shear stress.

In one practical viscometer, the parallel plates are the curved surfaces of one cylinder rotating inside another. The fluid whose viscosity is to be measured fills the volume between the two cylinders.

CHAPTER ONE | **Flow of Incompressible Fluids**

Experimentally, for newtonian fluids, it is found that the tangential force necessary to maintain a constant velocity is directly proportional to the area of the plate to which the force is applied and to the relative tangential velocity of the plates. It is inversely proportional to the distance separating the plates.

Tau is the force per unit area necessary to maintain a velocity gradient in a given fluid. It is called the shear stress. It also can be thought of as the balancing stress arising in the fluid to prevent acceleration. The derivative makes the equation more general since it covers situations when the velocity gradient is not constant.

Some insight into the term "shear rate" can be obtained by realizing that,

$$\frac{dV}{dZ} = \frac{dx/dt}{dZ} = \frac{dx/dZ}{dt} \qquad \text{(I-8)}$$

The change in the direction of motion, dx, with the change normal to the direction of motion, dZ, is called the shear. The time rate of change of this shear is equal to the change in velocity with distance normal to the flow direction.

A newtonian fluid is often defined as one in which the shear rate is linearly proportional to the shear stress – pressure and temperature being fixed. In equation I-7, the constant of proportionality between the two terms is the coefficient of viscosity divided by the dimensional constant. The use of the subscript, m, on the viscosity will be explained under the heading, Units of Viscosity. Gases, true solutions and non-colloidal liquids follow the newtonian model. Many ordinary industrial fluids follow this model. This is fortunate, since it is the simplest one.

2. NON-NEWTONIAN FLUIDS

If the coefficient of viscosity is a function of the shear stress (equivalently of the shear rate) as well as of temperature, pressure and composition, the fluid is classified as being non-newtonian. Some examples of classifications of time-independent, non-newtonian fluids are seen in Figure I-12.

To complicate the problem of viscosity, some fluids display characteristics that vary with the length of time that they are subject to stress. This means that non-newtonian fluids can be categorized as to whether they show time-independent or time-dependent behavior. The following categorization is essentially that given in Perry[x], with some commentary.

A. TIME-INDEPENDENT BEHAVIOR
a. Bingham Plastics

These fluids have a linear relationship between shear rate and shear stress. However, the line does not go through the origin. A minimum stress, τ_0, is required to initiate flow. The curve is then reasonably linear. Equation I-9 describes Bingham plastic shear stress versus rate of change of shear. Note that the symbol, η, eta, replaces the symbol, μ, mu, to indicate non-newtonian behavior.

$$\tau = \tau_0 + \frac{\eta}{g_c}\frac{dV}{dZ} \tag{I-9}$$

Water suspensions of rock (slurries), grains and sewage sludge are examples of Bingham plastics.

b. Pseudoplastics

These fluids are polymeric solutions or melts or are suspensions of paper pulp or pigments. Two gentlemen, Ostwald and de Waele[xi], established an empirical relationship that describes the behavior of pseudoplastics over fixed ranges of applied shear stress. As can be seen from the equation set I-10, this relationship is a power law one with a coefficient, K, and an index, n. The equation set describes how the basic power law can be manipulated so that the equation resembles the newtonian model. The group of terms representing a viscosity coefficient is then replaced by a single term that is an equivalent viscosity coefficient. This viscosity coefficient is not independent of applied stress, however, and it is given a different symbol.

The shear stress is related to the shear rate as follows:

$$\tau_v g_c = K\left(\frac{dV}{dZ}\right)^n, \quad n < 1 \tag{I-10}$$

$$\tau_v g_c = K\left(\frac{dV}{dZ}\right)\left|\frac{dV}{dZ}\right|^{n-1}$$

$$\tau_v g_c = K\left|\frac{dV}{dZ}\right|^{n-1}\left(\frac{dV}{dZ}\right)$$

$$\tau_v g_c = \eta\left(\frac{dV}{dZ}\right)$$

CHAPTER ONE | **Flow of Incompressible Fluids**

It can be seen that equation I-10, for power law fluids, is identical to equation I-7, for newtonian fluids, with eta replacing mu. This is a mathematical convenience under fixed circumstances. The viscosity is no longer constant at fixed pressure and temperature. It must be computed for the flowing conditions. This apparent viscosity decreases with shear rate (or with applied shear), hence the direction of curvature of the shear stress - shear rate curve (I-12).

The coefficient, K, is called an index of consistency. Hence, in the pulp and paper industry, one hears the term "consistency" more frequently than "viscosity" when talking of fiber suspensions. The index, n, is a constant that is a characteristic of the fluid. It is less than one for pseudoplastic fluids. Both the coefficient, K, and the index, n, must be established by experiment.

c. Dilatant Fluids
In dilatant fluids, the apparent viscosity coefficient increases with the shear rate. Again, the constant is no longer constant. Starch, mica suspensions, quick sand and beach sand suspensions are examples of dilatant fluids.

The same Ostwald-de Waele[xi] equation, I-10, describes dilatant fluids when the index, n, takes on values greater than one.

B. TIME DEPENDENT BEHAVIOR
a. Thixotropic Fluids
The graphic of Figure I-12 may be regarded as depicting the amount of shear stress needed to maintain a shear rate, dU/dZ (dy is used in the figure). It can be seen that if a fluid is taken fairly rapidly from point D to point A and slowly back again, the paths will be different.

Over the return path, much less stress is needed to maintain a shear rate.

If the fluid is taken fairly rapidly from point D to point A and held at that shear rate, the stress must be reduced (otherwise the rate would increase). In other words, in order to maintain the same rate of rotation of a spinning viscometer, the power to the motor must be reduced. The amount of reduction in shear stress required depends on the length of time the fluid is held at the shear rate associated with points A, B and C.

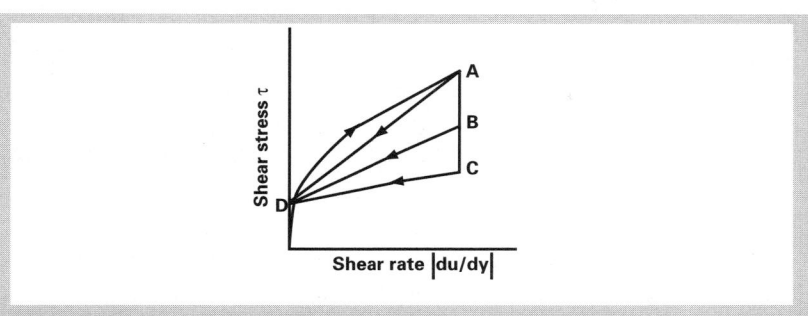

Figure I-12. Thixotropic behavior

The return paths can be from A, B, or C. They are substantially linear. Point C is a minimum for a given shear rate. It will always take some shear stress to produce a shear rate.

Mayonnaise, drilling muds, paints and inks show thixotropic behavior.

b. Rheopectic Fluids
These fluids set up. They increase their apparent viscosity on being shaken.

Bentonite sols, vanadium pentoxide sols, and gypsum suspensions in water are examples of fluids that show rheopectic behavior.

C. VISCOELASTIC FLUIDS
Viscoelastic fluids exhibit many of the properties of a solid. One of these properties is an elastic recovery from deformation. Some polymeric liquids fall into this category.

In polymer processing, mixing, extrusion, calendering, fiber spinning and sheet forming are examples of non-newtonian flow (See Bernhardt[xii], *Processing of Polymeric Materials*, for details).

This book will concentrate on fluids that can be treated as being newtonian to a close approximation. The majority of industrial problems can be solved with this approximation. Perry[x] is a good source of reference material should the reader wish more information on viscosity.

Caveat — Fluid classifications
The reader must be aware that other classifications than newtonian exist and that they must be dealt with from time to time. In fact, in some industries they must be dealt with all the time.

CHAPTER ONE | **Flow of Incompressible Fluids**

Viscosity as a fluid property

It should be evident that viscosity is a fluid property that is a function of temperature, pressure (to a minor extent for liquids), composition and sometimes time and shear. Published viscosities are usually correlations based on the first three variables. Empirical relationships are required for those fluids that are non-newtonian.

Published values for viscosities are essentially coefficients measured under laminar flow conditions; otherwise the effect of the actual flowing conditions would have to be known. Flow conditions are taken into consideration when the Reynolds[v] number is used to calculate the friction factor and the latter is inserted into the Darcy[vii] equation in order to compute the mechanical energy losses.

Viscosity's dependency on temperature and pressure

Viscosity depends primarily on temperature. It depends weakly on pressure. For gases, viscosity increases with temperature. For liquids, viscosity decreases with temperature. The viscosity of liquid water at 32°F is about six times its value at 212°F. Note that viscosity is not linear with temperature.

Other names for viscosity

Saybolt universal seconds units are just another kinematic viscosity. See Crane[xiii] for nomographs.

Generalized viscosity

When data are not available except for the viscosity of a liquid at one temperature, its viscosity at another temperature may be estimated from Figure I-13. This figure is taken from Perry[x]. Perry is a good source of information on the physical properties of liquids and gases.

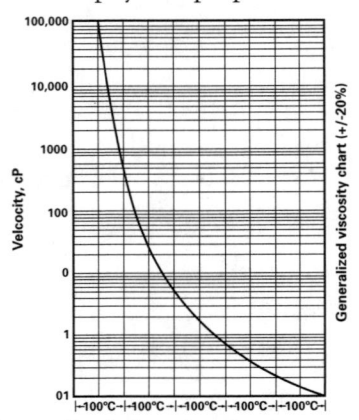

Figure I-13. Generalized viscosity chart

Source: R. Perry and D. Green, Perry's Chemical Engineers' Handbook, 1984, reprinted by permission McGraw-Hill Companies

Flow of Incompressible Fluids | CHAPTER ONE

If the viscosity in centipoise is known at one temperature for a liquid, this viscosity may be used to locate a point on the ordinate. The corresponding point on the abscissa is found. The temperature difference is then stepped off in a direction corresponding to the sign. The new viscosity corresponding to the new temperature is then read on the ordinate. The method is said to be accurate within plus or minus 20%. This means that, if information is available on a specific fluid, it should be used.

Many viscosity units

It is unfortunate there are so many different, and confusing, measures of viscosity (for instance, Saybolt Universal Seconds). It seems that every worker who has done research in the field of viscosity has developed his own units.

'Absolute' viscosity

In the English speaking engineering world, two principal measures of viscosity have been used in textbooks. The first measure has already been defined by Equation I-7 containing g_c. This equation will be repeated and the units will be derived.

$$\frac{F}{A_{plate}} = \frac{\mu_m}{g_c}\frac{dV}{dZ} = \tau \qquad \textbf{(I-11)}$$

$$[\mu_m] = \frac{g_c ZF}{AV} = \frac{lb_m \cdot ft}{lb_f \cdot s^2}\frac{ft \cdot lb_f}{ft^2 \cdot ft/s} = \frac{lb_m}{ft \cdot s}$$

The second measure of viscosity comes from simply omitting the dimensional constant in the defining equation.

$$\frac{F}{A_{plate}} = \mu_f \frac{dV}{dZ} = \tau \qquad \textbf{(I-12)}$$

$$[\mu_f] = \frac{ZF}{AV} = \frac{ft \cdot lb_f}{ft^2 \cdot ft/s} = \frac{lb_f \cdot s}{ft^2}$$

The subscripts on the viscosity terms are used simply to refer to mass and force in the defining equations. The viscosities differ only by a multiplying factor, the dimensional constant as shown in Equation I-13.

$$\mu_m = g_c \mu_f \qquad \textbf{(I-13)}$$

Since the dimensional constant is equal to 32.17, it is wise to carefully check which viscosity is being cited before using the number associated with it.

Flow of Industrial Fluids—Theory and Equations

CHAPTER ONE | Flow of Incompressible Fluids

Neither of the above viscosity coefficients has a name other than the general term, viscosity. Each is identified by its units. Both the above viscosities are referred to as being "absolute" viscosities. This just goes to prove that there is nothing absolute about the use of the word, "absolute".

In the SI system, the units of so-called "absolute" or "dynamic" viscosity are pascal-seconds, Pa•s.

(I-14)
$$\tau = \mu \frac{dV}{dZ}$$
$$[\mu] = \left[\tau \frac{dZ}{dV}\right] = \frac{N}{m^2}\frac{s}{m}m = Pa \cdot s$$

Pascals are the equivalent of newtons per meter squared. Therefore, pascal-seconds are the SI equivalent of customary U.S. pounds-force units, μ_f.

By far, the most popular unit of viscosity remains the poise or, more exactly, one derived from it, the centipoise. In other words, none of the above. The poise is based upon the older cgs (centimeter-gram-second) system. The unit of force in this system was the dyne. The dyne was a derived unit that in turn had units of gm·cm/s². When viscosity is given in poises, its units are,

(I-15)
$$[\mu_P] = \frac{FZ}{AV} = \frac{gm \cdot cm}{s^2}\frac{cm \cdot s}{cm^2 \cdot cm} = \frac{gm}{cm \cdot s}$$

It is likely that the centipoise will continue to be used, as it is a convenient decimal multiple of the official SI viscosity unit, the pascal-second. One Pa•s is equal to ten poise or 1,000 cP or 0.672 lb_m /(ft•s). So, one cP equals 0.000672 lb_m /(ft•s) or 6.72 x 10^{-4} lb_m /(ft•s).

This book will mainly use lb_m /(ft•s) mass based units and centipoise (which are anchored firmly in usage). We will name the units being used in each case.

'Kinematic' viscosity

The so-called kinematic viscosity is simply the "absolute" viscosity divided by the density of the fluid at the temperature and pressure at which the absolute viscosity was measured. It is normally given in stokes or centistokes with units of centimeter squared per second, $v = \mu / \rho$.

The units are,

$$[v] \equiv \left[\frac{\mu_p}{\rho_{cgs}}\right] = \frac{g}{cm \cdot s}\frac{cm^3}{g} = \frac{cm^2}{s} \tag{I-16}$$

A more detailed discussion of viscosity and its units can be found in Appendix AI.

I-8: 'FRICTION LOSSES'

A "friction loss" is mechanical energy which was capable of performing work (such as in moving a fluid in a pipe), but which has been transformed to internal energy by turbulence. This energy is usually seen as heat flow to the surroundings, but it can remain in the fluid as an increase in internal energy. It is considered a loss because it cannot normally be recovered as mechanical energy and external work must be supplied to make up for the loss in mechanical energy, otherwise the fluid would eventually stop flowing.

As has been pointed out, a friction loss corresponds to the concept of irreversibility in thermodynamics. The author's preference is to avoid the term friction loss wherever possible because it causes him to stop and wonder what the connection to friction is. Complete avoidance is not possible since the term is so commonly used in the literature. The term "friction factor" is one of the most common terms in fluid flow.

Irreversibilities, the conversion of mechanical energy to internal energy, are important in the calculation of flow through pipes. They are a function of the flow regime that is defined by the pipe Reynolds number. The pipe Reynolds number, in turn, is a function of velocity, density, viscosity and internal pipe diameter. Irreversibilities are also included in the coefficient used in orifice plate computations. Temperature influences most of the properties considered in the Reynolds[v] number, directly or indirectly.

Friction factor

The Moody[viii] friction factor, f_M, is the coefficient of proportionality between the irreversibility and the product of the number of pipe diameters of straight pipe and the kinetic energy (Equation I-17). The friction factor is a function only of Reynolds number for laminar flow. It is a function of Reynolds number and pipe roughness for turbulent flow.

In addition to straight pipe, a piping system consists of fittings (elbows, tees, reducers, expanders), equipment (strainers, etc.) and valves. Fittings in

CHAPTER ONE | Flow of Incompressible Fluids

a piping system are sometimes treated as so many equivalent lengths of straight pipe. Crane[xiii] gives methods of calculating equivalent lengths of various types of fittings.

Caveat — Equivalent lengths

The equivalent length method should only be used for rough approximations.

Why must irreversibilities be computed?

Some of the reasons follow:
- to establish upstream and downstream values of pressure at meters and valves;
- to calculate flow rates though piping systems;
- to establish the range of the flow regime;
- to size pumps, compressors, turbines, valves, and piping systems;
- to size relief headers and to ensure that the backpressure on relief devices will not prevent the devices from functioning adequately.

We will enter into more specific details in the next chapter.

The Darcy[vii] equation and the friction factor

If flow were ideal, there would be no permanent mechanical energy losses due to irreversibilities. All of the energy terms would be completely inter-convertible. The pressure would change with changes in fluid velocity and with elevation in the system, but the total mechanical energy, static, potential and kinetic, would remain constant. Flow, once established, would remain constant without the need for external work. In the real world, irreversibilities take their toll by converting otherwise useful mechanical energy into incremental internal energy. This incremental internal energy cannot be converted back completely to mechanical energy. It does result, however, in an increased temperature of the fluid. If the system is not adiabatic (not perfectly insulated), heat will flow to the surroundings due to the temperature difference. External work is required to maintain flow because of this conversion of mechanical energy to internal energy and its transfer as heat energy flow to the surroundings.

The potentials involved in fluid flow are mechanical ones. An irreversible conversion of a mechanical potential to an internal one means that less useful work can be performed on the fluid or by the fluid. This irreversible conversion is often described as "lost head" or "lost work". The choice of language is unfortunate and leads to confusion, but the terms are commonly used.

Flow of Incompressible Fluids | CHAPTER ONE

For liquids flowing in straight, constant diameter pipe, the irreversibilities can be predicted by the Darcy[vii] equation,

(I-17)
$$h_f = f_M \frac{L}{D}\frac{U^2}{2g_c}$$

Moody[viii] friction factor

The term, f_M, is the Moody friction factor. It is just a proportionality factor between the irreversibilities and all the other terms. It is not a constant. The subscript, M, is to remind everyone that this is the Moody friction factor. The Fanning[xiv] coefficient of friction, f_F, is smaller by a factor of four and requires that the equation be changed, that the divisor, 2, be moved to the numerator.

The L/D ratio is simply the number of pipe diameters over which the losses are being estimated. The term, U^2, is the average (across the section) fluid velocity squared. This term divided by $2g_c$ is the kinetic energy of the fluid at the section where the average velocity is U. The term, g_c, is the dimensional constant, 32.17 $lb_m \cdot ft \cdot s^{-2} \cdot lb_f^{-1}$. It is not the acceleration due to gravity, 32.17 $ft \cdot s^{-2}$. The above equation uses conventional American units, feet, seconds, pounds-force and pounds-mass. However, since the friction factor is non-dimensional, if g_c is given the value of one, without dimensions, the same equation holds for SI units.

The loss term, h_f, has units of foot-pounds-force per pound-mass, newton-meters per kilogram, or joules per kilogram depending on the choice of units. It is usually defined as the losses due to "skin friction" (primarily in a boundary layer next to the pipe wall). Remember what has been said about irreversibilities and the fact that, in turbulent flow in pipes, the boundary layer occupies the whole pipe section. We will show in the next section that the loss term is a measured quantity, from experiment.

Since the velocity is also a measured quantity, the friction factor can be correlated from experimental data. The correlations, in the form of Moody friction charts or the Churchill-Usagi[xli] equation (Appendix A-II), allow calculation of the friction factor and of mechanical energy losses due to fluid irreversibilities.

Caveats — Velocity profiles

The reader is reminded that if the wrong assumptions are made regarding the velocity profile or if a point velocity is used instead of the average velocity errors will result in the ensuing computations.

The Darcy[vii] equation must be modified somewhat for compressible flow. This will be done in the chapter on compressible flow.

The Moody[viii] friction factor in laminar flow is a function only of the Reynolds[v] number. It is,

$$f_M = \frac{64}{N_{ReD}} \qquad \text{(I-18)}$$

Fanning friction factor

The Fanning friction factor in laminar flow is,

$$f_F = \frac{16}{N_{ReD}} \qquad \text{(I-19)}$$

These equations result from substituting the Hagen-Poiseuille[xv] equation into the Darcy-Weisbach[vii] equation. These equations will be developed in Appendix AI.

This book will use the Moody friction factor exclusively. It will use the subscript, M, to identify the Moody coefficient. In the case of turbulent flow, the friction factor depends on the Reynolds number again, but not so simply as shown in Equation I-18 for laminar flow. It also depends on the pipe surface roughness. The friction factor for turbulent flow is correlated by way of friction factor charts (Crane[xiii] gives some excellent charts) or by the Churchill-Usagi[xli] equations. The Churchill-Usagi equations are an excellent tool when automating computations. The Churchill-Usagi equations will be given in Appendix AII.

It is fortunate that industrial pipe manufacturing practice allows the prediction of the surface roughness and, thus, the friction factor with a high degree of accuracy (Figure I-4).

I-9: BERNOULLI EQUATION AND THE DARCY EQUATION COMBINED

General approach to calculating flow through pipes

Most practical flow formulae are derived from the Bernoulli[i] equation. The usual integrated, steady-state form for incompressible fluids is:

$$\frac{P_1}{\rho_1}+\frac{U_1^2}{2g_c}+\frac{gZ_1}{g_c}+w_n=\frac{P_2}{\rho_2}+\frac{U_2^2}{2g_c}+\frac{gZ_2}{g_c}+h_f \qquad \text{(I-20)}$$

The Bernoulli[i] equation is derived from a force and momentum balance. The complete derivation is given in Appendix AI.

The total energy at point 1 plus the energy added (pump) or subtracted (turbine) between points 1 and 2 equals the total energy at point 2 plus energy converted (ultimately to heat flow) between points 1 and 2. Total energy is broken up into quantifiable types of energy. The first term on each side is the "static" energy or mechanical energy, due to pressure. The second term is the kinetic energy, due to average velocity in the section. The third term is the potential energy, due to position of the flowing fluid above a datum (any fixed datum). The last term on the left is the work energy per unit mass added, between points 1 and 2, by a pump, for example, or extracted (negative sign) by a turbine. The last term on the right is the "head loss" per unit of flowing mass between the same two points. It is the mechanical energy that has been converted to internal energy and that is no longer available as a mechanical driving force. It will be ultimately lost to the environment as heat energy flow. Irreversibility is a more descriptive term than head loss or friction loss.

In the Bernoulli equation, I-20, consistent units must be used. If you wish to use psia for units of pressure, simply multiply the pressure terms by 144. If you are working in SI units, treat the term, g_c, as equal to one, dimensionless. All the terms in the equation have units of energy per unit mass. If SI units are used, g_c equals 1, dimensionless, but g equals 9.81 meters/second. If you have to switch between units, it is best to use the above formula. It helps keep track of where the dimensional constants should be included. The term, h_f is added to the ideal Bernoulli equation (which did not include it) to accommodate irreversibilities between points 1 and 2. A similar looking equation, used in hydraulics practice, will be discussed shortly.

A limitation of this form of the Bernoulli equation is that it is derived for incompressible flow and it must be modified in order to be used for compressible flow. Another limitation is that average velocities across a section are used in the second term on each side. A kinetic energy correction factor is needed for more exact computations. This factor will be described in Chapter II. Another limitation is that the equation is derived

CHAPTER ONE | Flow of Incompressible Fluids

for steady state flow. This latter limitation is usually not too serious, but it must be borne in mind.

Experimental determination of the loss term, h_f

If the Bernoulli[i] equation, I-20, is applied to steady-state flow in a long, horizontal, constant diameter pipe in which a liquid is flowing, the velocity, elevation and density are constant. We can choose a section of pipe that contains no pump or turbine. The Bernoulli equation then reduces to the statement that the difference between upstream pressure and downstream pressure divided by the density of the fluid is equal to the irreversibilities (losses) in foot-pounds-force per pound-mass or in joules per kilogram.

The simple relationship just described allows irreversibilities in straight pipe to be computed from direct measurement. Once irreversibilities are computed for different flow rates in differently sized pipes flowing different fluids, these losses can be correlated with the corresponding densities, viscosities and internal pipe characteristics (roughness, diameter). The Darcy[vii] equation and the Moody[viii] friction factor are the result of these correlations. It is fortunate for us that so much effort has been spent establishing what, on the surface, seem to be such simple relationships.

The Darcy equation

The Darcy equation is,

$$h_f = f_M \frac{L}{D} \frac{U^2}{2g_c} = K \frac{U^2}{2g_c}$$

(I-21)

The equation allows estimating irreversibilities in a pipe of given linear length and diameter for a given flow (average velocity) and a given friction factor. The last term allows irreversibilities to be estimated across fittings with a given K factor. The K factor must be associated with the pipe diameter in which the velocity was estimated.

When the irreversibilities are estimated between two points, the values can be plugged into the Bernoulli equation. The Bernoulli equation can then be solved for the remaining quantities. The Darcy equation applies to laminar and turbulent flow of incompressible fluids. Its modification for compressible fluid computations will be discussed later. Both the friction factor and the K factor are established by correlation after the irreversibilities have been computed for a known situation. They can then be used in hypothetical situations.

The Moody[viii] friction factor, f_M, is a function of the Reynolds[v] number (and through it of viscosity) and the pipe roughness. The friction factor has been established by correlation and experiment and plotted on log/log paper for various sized pipes and various relative roughnesses. Figure I-4 is an example. The friction factor may also be estimated with greater accuracy using the Churchill-Usagi [xli]equations given in Appendix AII.

Loss coefficients

For constant diameter pipe, the Darcy[vii] equation shows the lost mechanical energy is proportional to the Moody friction factor, the pipe length to diameter ratio, and the square of the average velocity across the pipe section. Various fittings, valves and restrictions may be characterized by making use of equivalent pipe lengths or by using an equivalent K (or $f_M L/D$) in the Darcy equation.

The loss coefficient, K, is often called the number of velocity heads lost through a fitting. This terminology is pure jargon. The factor, K, is best thought of as the fraction of kinetic energy that is converted to internal energy and thermal energy by flow by the fitting. The term, $U^2/2g_c$, is the velocity energy or the kinetic energy per unit mass. Remember the energy is only "lost" to mechanical use. It is still present as internal energy or it passes to the environment as thermal energy flow.

The factor, K, is associated with the diameter in which the velocity is computed. For fittings, it is treated as being independent of Reynolds number (therefore, of velocity, viscosity and density) and of the friction factor. For such short lengths, it can be correlated by geometrical similarity among fitting types. We will discuss this fact in greater detail in the next chapter. For straight pipe, K is not independent of Reynolds number or the friction factor,

$$K = f_M \frac{L}{D} \qquad \text{(I-22)}$$

Experimentally, for fittings, K varies as does $f_M L/D$ for clean, commercial pipe. Crane[xiii] makes use of this fact to establish some simple correlations between the coefficient K and the types of fittings. We will discuss this in Chapter II. Table I-3 gives values of K for liquid flow in various types of fittings and valves. The figure is taken from Perry[x]. The lower case reference letters refer to the original source of the coefficients. Perry should be consulted if you wish further information.

CHAPTER ONE | Flow of Incompressible Fluids

I-10: CONSERVATION OF ENERGY IN HYDRAULICS PRACTICE

Energy is the same entity regardless of the discipline analyzing it. It is sometimes called by different names and different groupings of variables are used to describe it. We will summarize the treatment of energy and include a description of how hydraulics practice treats energy.

It has already been stated that energy cannot be destroyed, but it can be converted to various forms, one of which is not useful mechanically. This less useful form is often described as being "lost" energy. The mechanically useful forms are to be found in Table I-2.

TABLE I-2. Mechanically useful forms of specific energy		
Energy	J/kg (SI)	ft-lb$_f$/lb$_m$ (U.S.)
static energy	P/ρ	P/ρ
potential energy	gZ	gZ/g_c
kinetic energy	$U^2/2$	$U^2/2g_c$

The total mechanical energy per unit mass, e, is the sum of the above three terms. The computed value of e at a section will be the same as the computed value at any other section in the ideal case. The ideal case gives a starting point for making corrections to suit the real world.

The SI units of specific energy are newton-meters per kilogram, N-m/kg, or the equivalent joules per kilogram, J/kg. The U.S. units are foot-pounds-force per pound-mass, ft-lb$_f$/lb$_m$.

In hydraulics engineering it is said to be more convenient to express energy in terms of unit weight instead of unit mass. In the author's opinion, this is an approach that leads to confusion. Starting with the total mechanical energy balance for the ideal case (no losses), energy per unit mass is,

Flow of Incompressible Fluids | CHAPTER ONE

$$e = gZ + \frac{P}{\rho} + \frac{U^2}{2} \quad (SI) \tag{I-23}$$

$$\frac{e}{g} \equiv H = Z + \frac{P}{\rho g} + \frac{U^2}{2g}$$

$$e = \frac{g}{g_c} Z + \frac{P}{\rho} + \frac{U^2}{2g_c} \quad (US)$$

$$\frac{g_c e}{g} \equiv H = Z + \frac{P}{\rho \frac{g}{g_c}} + \frac{U^2}{2g}$$

$$H = Z + \frac{P}{\gamma} + \frac{U^2}{2g} \quad (hydraulic)$$

The weight density, γ, has different values and different formulae in metric and in U.S. practice.

$$\gamma \equiv \rho g \quad or \quad \gamma \equiv \rho \frac{g}{g_c} \tag{I-24}$$

(*metric*)   (US)

CHAPTER ONE | Flow of Incompressible Fluids

Additional Frictional Loss for Turbulent Flow through Fittings and Valves[a]

Type of fitting or valve	Additional friction loss, equivalent no. of velocity heads, K
45° ell, standard[b,c,d,e,f]	0.35
45° ell, long radius[c]	0.2
90° ell, standard[b,c,e,f,g,h]	0.75
Long radius[b,c,d,e]	0.45
Square or miter[h]	1.3
180° bend, close return[b,c,e]	1.5
Tee, standard, along run, branch blanked off[e]	0.4
Used as ell, entering run[g,i]	1.0
Used as ell, entering branch[c,g,i]	1.0
Branching flow[i,j,k]	1[l]
Coupling[c,e]	0.04
Union[e]	0.04
Gate valve,[b,e,m] open	0.17
¾ open[n]	0.9
½ open[n]	4.5
¼ open[n]	24.0
Diaphragm valve,[o] open	2.3
¾ open[n]	2.6
½ open[n]	4.3
¼ open[n]	21.0
Globe valve,[e,m] bevel seat, open	6.0
½ open[n]	9.5
Composition seat, open	6.0
½ open[n]	8.5
Plug disk, open	9.0
¾ open[n]	13.0
½ open[n]	36.0
¼ open[n]	112.0
Angle valve,[b,e] open	2.0
Y or blowoff valve,[b,m] open	3.0
Plug cock[p] (Fig. 5-42) $\theta = 5°$	0.05
10°	0.29
20°	1.56
40°	17.3
60°	206.0
Butterfly valve[p] (Fig. 5-43) $\theta = 5°$	0.24
10°	0.52
20°	1.54
40°	10.8
60°	118.0
Check valve,[b,e,m] swing	2.0[q]
Disk	10.0[q]
Ball	70.0[q]
Foot valve[e]	15.0
Water meter,[h] disk	7.0[r]
Piston	15.0[r]
Rotary (star-shaped disk)	10.0[r]
Turbine-wheel	6.0[r]

Table I-3. Irreversibilities in turbulent flow

Source: R. Perry and D. Green, Perry's Chemical Engineers' Handbook, 1984, reprinted by permission McGraw-Hill Companies

The units of the weight density as defined by gamma are force per unit volume. The SI units are newtons per meter cubed. The customary U.S. units are pounds-force per foot cubed.

The use of weight density allows the hydraulic equation developed above to have all of its terms expressed in meters or feet of fluid. This is another source of the term "head". Physically, the hydraulic equation (last equation in the set I-23) may be interpreted as shown in Figure I-14.

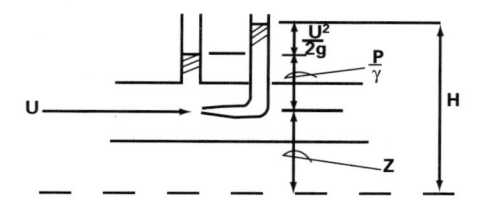

Figure I-14. Interpretation of hydraulic 'head'

'Hydraulic head' or 'piezometric head'

In hydraulics practice, it is usual to give the name "hydraulic head" or "piezometric head" (h) to the sum of the elevation above the datum (Z) and the static pressure energy term (P/γ).

$$h = Z + \frac{P}{\gamma} \qquad \textbf{(I-25)}$$

So, the total mechanical energy in terms of feet or meters of fluid becomes,

$$H = h + \frac{U^2}{2g} \qquad \textbf{(I-26)}$$

The term derived from the kinetic energy term in the set of equations I-23 is so similar to it (g replacing g_c) as to lead to confusion. This is particularly true in metric practice when g equals 9.81 and g_c equals one.

Low-pressure gases can often be treated as incompressible fluids. In hydraulics practice, the pressures and differential pressures are often measured with manometers. The manometers are referenced to atmospheric pressure as shown in Figure I-15. It can be seen from the above equation that the velocity head is equal to the difference between the total

CHAPTER ONE | Flow of Incompressible Fluids

mechanical energy head and the hydraulic (piezometric) head. Notice that, if the manometers are all topped up to the same levels initially, each one will have a uniform displacement about the same center line. The total mechanical energy will equal the sum of the sum of the energy due to velocity ("head") and the piezometric ("head").

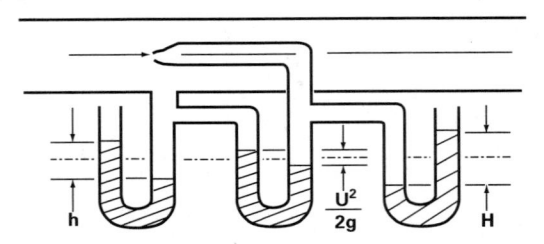

Figure I-15. Manometers in hydraulic practice

When the manometers that measure these two variables are referred to atmospheric pressure, it makes no difference to the equation whether the variables are absolute or relative. The pressure equivalent to the atmospheric pressure would have to be added to both terms. Hence, psig units are often used in place of psia units. This is a practice that can lead to error when dealing with compressible fluid computations.

Because the general engineering approach is less restrictive than the hydraulic engineering one, we will use the general approach almost exclusively. The general approach also gives a better understanding of physical problems.

I-11: WORKED EXAMPLES

We will give some very basic computations in this chapter. Examples of more extensive computations may be found in Chapter II.

Examples of hydraulic practice

Example I-1: Pressure corresponding to 50 mm WC
We can assume that 50 mm WC is the measured head, h, on a manometer with no flow. The datum, Z, is zero. Since pressure equals the head times the weight density or the head times the mass density times the acceleration of gravity, we can write the set I-27.

Flow of Incompressible Fluids | CHAPTER ONE

$$H = h = \frac{50}{1000} m = \frac{P}{\gamma} = \frac{P}{\rho g} \tag{I-27}$$

$$P = \frac{50}{1000} m 1000 \frac{kg}{m^3} 9.81 \frac{m}{s^2} = 490.5 \frac{kg}{ms^2} \left(\frac{N}{m^2} = Pa \right)$$

The conversion from kg/m-s² was performed by making use of Newton's force-acceleration law: N = kg-m/s². A pressure of 490.5 pascals is close to two inches of water column. Since this pressure was measured with a manometer, it is assumed that we are dealing with gauge pressure, not absolute pressure.

Example I-2: Pressure corresponding to 100 inches WC

The same assumptions apply as in the first example, only the units and their values change. An additional subtlety is that we are dealing with weight density that is numerically equal to mass density only in the U.S. customary system.

$$H = h = \frac{100}{12} ft = \frac{P}{\gamma} = \frac{P}{\rho g / g_c} \tag{I-28}$$

$$P = \frac{100}{12} ft \cdot 62.371 \frac{lb_f}{ft^3} = 519.8 \frac{lb_f}{ft^2} = \frac{519.8}{144} \frac{lb_f}{in^2} = 3.609 \, psig$$

We can write psig because of the assumptions that were accepted.

Conversion difficulties

Difficulties in conversion lie not so much in finding the right conversion factors as in understanding what it really is that one is converting. Appendix AIII will discuss in detail the different uses of the word "head" and the fact that it sometimes means a pressure difference, sometimes a height difference and sometimes an energy per unit mass difference. In this chapter, let us assume we have identified what it is we are talking about and simply want to convert pressure units to head units. More exactly, we are interested in converting differential pressure into differential manometric head (equivalent height of a column of liquid) or vice versa.

The easiest way to effect the conversion is to change all units to standard SI or customary U.S. units. The next step is to visualize a column of liquid

CHAPTER ONE | Flow of Incompressible Fluids

in tube and to ask what the height of the column would be which would exactly balance the pressure at the bottom. It is very important to remember the difference between pressure and force and to use the appropriate SI or U.S. customary form of Newton's law to solve the problem.

Example I-3: Height of WC under maximum vacuum, SI

Suppose we want to know the absolute maximum height that we can raise a column of water at 60°F or 15.6°C under the best vacuum we can achieve. The first point to consider is that water has a vapor pressure, so we cannot achieve a perfect vacuum. At the temperature in question, the vapor pressure is 0.258 psia or 1.779 kPa. We will assume the atmospheric pressure forcing the liquid up the column is 14.700 psia or 101.350 kPa. The vapor pressure is not insignificant; it is 1.76% atmospheric pressure. The differential pressure holding up the column of water will be 14.442 psi or 99.571 kPa. These are the numbers with which we must work.

The next step is to make use of the appropriate form of Newton's law — but not blindly. Logically, we know that the mass of fluid represented by the column of water would accelerate if released from an elevation equal to that of the bottom of the column. We also know from elementary physics that this acceleration is uniform over normal distances and is equal for all bodies when not subject to any other forces than the force of gravity. We now reason that in order to stop the mass from accelerating an equal and opposite force must be applied at the bottom of the column and that this force is given by Newton's law.

We know the pressure and we know that pressure is force per unit area. The equation set I-29 gives us the solution for the SI case.

$$F = mg$$ (I-29)
$$F = \rho V g = \rho A Z g$$
$$\frac{F}{A}\frac{1}{\rho g} = Z$$
$$\frac{P}{\rho g} = Z$$
$$\frac{99{,}571}{999(9.805)} = 10.165 m$$

Hydraulic practice makes the direct connection between pressure and manometric head by saying that a pressure (differential) of 99.574 kPa is equal to a head of 10.186 m.

Flow of Incompressible Fluids | CHAPTER ONE

U.S. customary units

Example 1-4: Height in customary U.S. units

If we wish to solve the same problem in customary U.S. units, we transform pounds per square inch to pounds per square foot and include the dimensional constant in Newton's law. The equation set I-29 gives the solution.

$$F = \frac{mg}{g_c} \tag{I-30}$$

$$F = \frac{\rho V g}{g_c} = \frac{\rho A Z g}{g_c}$$

$$\frac{F}{A}\frac{1}{\rho}\frac{g_c}{g} = Z$$

$$\frac{P}{\rho}\frac{g_c}{g} = Z$$

$$\frac{14.442(144)}{62.35}\frac{32.17}{32.17} = 33.354 \; ft$$

It is common in hydraulics practice to say that 14.442 pounds per square inch is equal to 33.354 feet of head. What is more, as has been discussed previously, since at fixed places on the globe, the acceleration due to gravity and the dimensional constant are equal numerically, it is common to lump them into the density and change the name of mass density to weight density.

In U.S. hydraulic units, the second from last line in I-29 would read:

$$\frac{P}{\gamma} = Z \tag{I-31}$$

Weight density is equal to mass density times the acceleration of gravity divided by the dimensional constant. In U.S. practice, the numbers are the same, but the units change.

In European metric hydraulic practice mass density and weight density is not numerically the same. The denominator in the second from the last line in I-29, ρg, would be replaced by γ.

$$\gamma = \rho g \tag{I-32}$$

CHAPTER ONE | Flow of Incompressible Fluids

The weight density in European metric practice differs from the mass density by the multiplying factor g (9.805 m/s^2). In U.S. hydraulic practice, the weight and mass densities are numerically identical, but not dimensionally.

Converting energy per unit mass units to manometric head units

When using the U.S. systems, the conversion from customary U.S. units to U.S. hydraulic units is easy. Only the names of the dimensions change, the numbers remain the same. Ten ft-lb$_f$/lb$_m$ become ten feet of fluid.

When converting SI units to metric hydraulic units, joules per kilogram or newton-meters per kilogram have to be divided by the acceleration of gravity, g, which equals 9.805 m/s^2.

Frequently, even those of us who dislike using hydraulic units are forced to use them – usually because a particular pump data sheet requires their use. We will therefore give two examples of the conversion process. Suppose that we have computed the kinetic energy, the velocity head, in a pipe and need the equivalent feet or meters of head. How is it done?

Example I-5: customary U.S. units

For the customary U.S. unit example we will assume an average velocity across the pipe of 10 feet per second. The kinetic energy (still called "head" by chemical engineers) is given by,

$$h_v = \frac{U^2}{2g_c} = \frac{10^2}{2(32.17)} = 1.554 \; ft - lb_f / lb_m \quad \text{(I-33)}$$

Example I-6: U.S. hydraulic units

In U.S. hydraulics practice this would be written,

$$h_v = \frac{U^2}{2g} = \frac{10^2}{2(32.17)} = 1.554 \; ft \quad \text{(I-34)}$$

Note the subtle change of the dimensional constant to the acceleration of gravity and the change in units, but not in their numerical value.

Example I-7: SI units

In SI units, the equivalent velocity would be 3.048 meters per second. The kinetic energy in energy per unit mass would be,

$$h_v = \frac{U^2}{2} = \frac{3.048^2}{2} = 4.645 \, J/kg \, (N-m/kg) \qquad \text{(I-35)}$$

The above conversion can be checked by using the dimensions associated with Newton's acceleration law, N=kg-m/s² to replace seconds squared. To convert the SI units to meters of manometric head, one has to divide by the acceleration of gravity, g, which equals 9.805 meters per second squared. The formula would be,

$$h_v = \frac{U^2}{2g} = \frac{3.048^2}{2(9.805)} = 0.474 \, \text{meters of fluid} \qquad \text{(I-36)}$$

The two hydraulic computations give approximately equivalent quantities, 1.544 feet equals 0.474 meters. We have used, however, four similar but subtly different formulae to arrive at results in different units. We can reduce the problem substantially if we use only SI or customary U.S. units.

I-12: CHAPTER SUMMARY

We have now defined most of the basic concepts necessary to solve problems involving circular liquid lines flowing full. The principal concepts to be retained from this chapter (and even memorized) are:

- total energy is constant, it can be transformed to different forms: potential, static, kinetic and internal. The transformation of incremental internal energy has a unidirectional component that prevents it from being transformed to one of the other, more mechanically useful forms, in a flowing system. This explains friction losses or irreversibilities;
- flow of liquids in circular pipes flowing full can be characterized as laminar, transitional or turbulent;
- laminar flow is characterized by concentric laminae sliding past one another, with very little lateral mixing;
- turbulent flow is characterized by very small eddies and vortices with much lateral mixing;
- transitional flow is an unstable mixture of the two other flow regimes;
- in a long straight pipe, at steady state, each regime develops a characteristic flow profile or velocity distribution across the pipe;
- the profiles are parabolic for laminar flow and tend to be flattened for turbulent flow;

CHAPTER ONE | **Flow of Incompressible Fluids**

- control over or knowledge of the profile is important to flow measurements in many cases. For instance, a Pitot tube measures the point velocity in the profile. If the sensor is located at a point where the velocity profile is unknown or distorted, there are no guarantees as to accuracy;
- some measuring instruments are very sensitive to distorted profiles, others less so;
- an index exists, the Reynolds[v] number, which can predict the flow regime and which can be used to calculate the irreversibilities (losses) in pipe;
- to calculate irreversibilities in laminar flow, a simple formula exists for finding the friction factor;
- to calculate losses in turbulent flow, graphical methods, or the Churchill-Usagi[xli] equations must be used to find the friction factor;
- although the concept of viscosity being the proportionality factor between shear stress and shear rate is quite simple, the plethora of units is very confusing and frequently leads to error. Always carefully check which units are being used and convert them if necessary before using them in any formula;
- there are two commonly used friction factors, the Moody[viii] and the Fanning[xiv] factors;
- the Moody coefficient is larger than the Fanning coefficient by a factor of four. The Darcy[vii] equation as used in this book employs the Moody friction factor. If the Fanning factor is used, the equation has to be adjusted accordingly;
- the Bernoulli[i] equation is probably the most used equation in the field of fluid mechanics. It is a force-momentum balance that equates driving potentials in energy per unit mass to changes in kinetic and potential energies and to lost work;
- lost work is a term for irreversibilities or energy that has been "degraded" into incremental internal energy;
- the Bernoulli equation can also be derived from the first and second laws of thermodynamics — a fact that establishes its generality;
- in hydraulics practice, a deceptively similar equation to the Bernoulli equation is used.

CHAPTER TWO

Incompressible Fluid Flow – Losses Of Mechanical Energy

II-1: SCOPE OF CHAPTER — APPLYING BASIC CONCEPTS

This chapter will apply the basic concepts developed in Chapter I and Appendix AI to the solution of some typical fluid flow problems. Many of the problems relate to measurement and control applications. The reader might be tempted to ask why he or she should invest time in learning of things that are somewhat remote from his or her field of interest. The answer is a general knowledge of flow phenomena helps avoid errors due to incorrect assumptions having been made. This general knowledge may be obtained only through broad study.

We will show how to apply the Bernoulli[i] equation to specific problems. These problems require an understanding of the following:

- how to estimate irreversibilities and the associated pressure drop in straight pipe;
- how to estimate irreversibilities and pressure drop in various types of fittings;
- what the difference is between pressure drop and irreversibilities;
- how the irreversibilities of in-line components are estimated;
- what pressure recovery is.

We will take a systematic approach to the solution of flow problems in an attempt to show it is impossible to work in isolation (or in ignorance of other people's problems). This statement is intended to mean that one cannot isolate "measurement" problems from system flow problems such as those associated with pumps, compressors, pipes and piping components.

CHAPTER TWO: Incompressible Fluid Flow – Losses of Mechanical Energy

The chapter will concentrate on the passive elements in a piping system. However, the chapter will prepare the way for the understanding and analysis of complete systems.

The differences between units of energy per unit mass and "feet" or "meters" of head will be discussed in detail in Appendix AII.

This chapter is intended to be of practical utility. Tables of data and appropriate references will be given. When the reader has understood this chapter, it is hoped he or she will be able to approach typical incompressible fluid loss estimation problems with confidence.

II-2: REASONED APPROACH TO DESIGN – A LITTLE PERSONAL PHILOSOPHY

Experience teaches that, in any engineering effort (a project) involving different disciplines over time, work must be divided among the various participants. Schedules have to be met. Plants are built only after considerable engineering co-ordination has taken place.

The opportunities for error are enormous, and new plants frequently require considerable modification before they function to the satisfaction of the owners. Much time is spent by managers devising quality control methods and programs. Slogans such as, "Do it right the first time!" abound. The impossibility of doing something right without a few iterations is not often addressed.

The pressure of schedules usually means process engineers estimate line sizes conservatively at an early date. The sizes sometimes are not corrected to match the final plant design. Mechanical engineers take early data from the process engineers and add their own safety factors for the specification of the prime movers such as pumps. Control systems engineers size their valves and other instruments based on the same conservative data and also add safety factors.

One piece of evidence that supports the above statements is the frequent discovery of oversized (and expensive) control valves during plant start-up. A control valve is meant to minimize flow problems; an oversized valve can create them.

Incompressible Fluid Flow – Losses of Mechanical Energy — CHAPTER TWO

A more reasoned approach, from the point of view of a control systems engineer, would be to delay final specification of control valve sizes until the piping layout were reasonably fixed. The control systems personnel would then have the responsibility of performing a final hydraulic study on a more realistic system. Real data could be used for pipe lengths, numbers and types of fittings and for pump heads. Design changes could be accommodated more readily. Oversize pumps at least could be identified. There might even be time to trim pump impellers.

Unfortunately, schedules sometimes get in the way of reason. It is not always permissible to do things right the first time. They must be fixed afterwards.

Regardless of whether you are trying to do it right the first time or to fix it afterwards, you must follow the same analytic process. In the first case you may be acting on assumed (hypothetical) data. In the second case you will be acting on real (as-built) data. The same engineering knowledge is required in both cases.

II-3: THE BERNOULLI EQUATION REVISITED

$$\frac{P_1}{\rho_1} + \frac{\alpha_1 U_1^2}{2g_c} + \frac{gZ_1}{g_c} + w_n = \frac{P_2}{\rho_2} + \frac{\alpha_2 U_2^2}{2g_c} + \frac{gZ_2}{g_c} + h_L \quad \text{(II–1)}$$

The Bernoulli[i] equation is so important to all aspects of fluid flow that it is worth revisiting in greater detail. The complete, integrated form of the Bernoulli equation contains eight terms, two of which have correction factors (alpha). It is a steady-state balance of the energy per unit mass at one point in a piping system against that at another point. Two of the terms, w_n and h_L, refer to energy per unit mass added or converted to thermal energy, between sections 1 and 2. Not all the terms or factors are necessary to solve all problems. For instance, in the case of incompressible flow:

- if there is no pump or turbine, the fourth term on the left disappears;
- if the elevations at the two points are equal, the third term on each side cancels;
- in most cases of completely turbulent flow, the kinetic energy correction factors, alpha 1 and 2, each can be treated as being equal to one;
- if the fluid is incompressible, and the pipe at both points has the same diameter, the second term on each side is equal and cancels (the alphas each being made equal to one).

CHAPTER TWO | Incompressible Fluid Flow – Losses of Mechanical Energy

It is worth taking a close look at each term in the Bernoulli[i] equation to see how it is applied in practice.

The variables of the first three terms on each side have numbered subscripts. The terms represent different forms of mechanical energy per unit mass of the flowing fluid at the sections labeled 1 and 2. The last terms on each side do not have numbered subscripts. They are the mechanical energy per unit mass of flowing fluid added or subtracted (on left), or "lost" (on right) between points 1 and 2.

The word "lost" means mechanical energy has been converted to internal energy. If the system is adiabatic, the fluid gets hotter. If it is diabatic or isothermal, there is a transfer of heat across the system boundaries (the pipe wall).

The complete Bernoulli equation represents a total energy balance between two points. The word "balance" suggests each term may have its value altered and the changes may be balanced by changes in all or some of the other terms. There are constraints, however. The last term on the right is always a positive increment. It is the term that accounts for lost mechanical energy or, better stated, the conversion of mechanical energy to internal energy and, ultimately, to heat flow to the environment.

We now will examine each term of the Bernoulli equation individually. The word "specific" in the headings refers to the fact that the energy is on a per unit mass basis.

Specific pressure-volume energy

The first term on each side of the Bernoulli equation is the energy per unit mass associated with the static absolute pressure. Static pressure is the pressure exerted normal to the flow. It can be measured by a pressure gauge on the pipe wall if the pipe tap is flush with the pipe and has no burrs. It is also the pressure that a hypothetical device flowing with the fluid would measure. If the pressure gauge measures "gauge" pressure, the atmospheric pressure must be added in order to obtain absolute pressure.

Caveat — Hydraulic engineering practice

In hydraulic engineering practice, when dealing with liquids, gauge pressure is often used in the above equation in place of absolute pressure. The result is usually correct since, in the Bernoulli equation, differences are being computed. It is best always to use absolute pressures in order to avoid forgetting them when they are needed (mainly for compressible fluids).

Incompressible Fluid Flow – Losses of Mechanical Energy | CHAPTER TWO

The density in the denominator of the first term on each side of the Bernoulli[1] equation carries a subscript because density can change in compressible flow. In the case of incompressible flow, the densities are equal on either side only if the temperatures, pressures and compositions remain the same.

Consistent units must be used. If the density is in lb_m/ft^3, the pressure must be in lb_f/ft^2. If you wish to use psia, then multiply pressure in psia by 144 in the equation. If the density is in kg/m^3, the pressure must be in Pa. In this case the energy units are joules per kilogram. If kilopascals are used, the specific energy units for that term will be kilojoules/kg. The first term on each side of the equation must be multiplied by 1,000.

Specific kinetic energy

The second term on each side is the kinetic energy. This term becomes important when dealing with changes in pipe diameter and when there is acceleration due to expansion of compressible fluids. The dimensional constant, g_c, is important in the customary U.S. system. In the SI system, it is not needed. However, it is worth being kept in the equation and simply being treated as having a value of one, dimensionless, when SI units are used. This habit makes it easier for those who have to switch back and forth between systems to remember its location in the equation.

The velocity, U, is the average velocity across the pipe at the numbered section. It is not the point velocity. In the U.S. system the units are feet per second. The SI system uses meters per second.

Alpha is the kinetic energy correction factor that is close to (but greater than) one for turbulent flow and is exactly two for laminar flow. Alpha is necessary because the average velocity is used in most equations whereas, in truth, the integrated point velocity should be used. The point velocity varies across the section studied. In most industrial engineering problems involving turbulent flow, alpha is treated as being equal to one without too much loss of accuracy, but the assumption should always be checked.

Specific potential energy

The third term on each side represents the energy per unit mass due to position (potential energy) at sections, 1 and 2. The units of Z are feet or meters. The dimensional constant is again 32.17 $lb_m \cdot ft \cdot lb_f^{-1} \cdot s^{-2}$ in the customary U.S. system. It can be treated as one, dimensionless, in the SI system. Acceleration due to gravity, g, is 32.17 $ft \cdot s^{-2}$ in the U.S. system.

CHAPTER TWO | Incompressible Fluid Flow – Losses of Mechanical Energy

It is 9.81 m•s^{-2} in the SI system. These last two numbers are approximations. They are true only at fixed latitude and at sea level. Their use is adequate for most engineering computations in industrial practice.

When using real numbers for Z, it is best the select the lowest point in the system as a datum line and to measure all elevations from it. This avoids the use of negative numbers when the pipe dips below its starting point. The base line is selected for convenience in the computation. Again, because we are dealing with differences, there is no absolute height.

Caveat — Gravity trap

Be aware of the trap of treating the gravitational constant and the dimensional constant as being equal and capable of being cancelled. This leads to simplified equations, but conceptual difficulties, and sometimes to errors.

Specific energy added or subtracted

The fourth term on the left is the energy added per unit mass by a pump or subtracted (negative sign) by a turbine. This term does not represent the energy added or subtracted at a particular section. It is the energy added or subtracted at any location between the two sections.

The energy added or subtracted per unit mass flowing is not all of the energy delivered to the motor or removed from the shaft of a turbine per unit mass flowing. There will be irreversibilities in both cases. More energy is delivered to a pump motor than is transferred to the fluid and more energy is extracted from a fluid than is delivered to a turbine shaft. Efficiency factors have to be applied. The Bernoulli[i] equation applies only to energy changes within the fluid unless it is modified to contain an efficiency factor. Note efficiency factors are specific to a physical transmission means – a shaft, a pump impeller or motor, for instance. Watch the definitions!

Specific energy converted to internal energy

The fourth term on the right is the mechanical energy converted to internal energy by "wall (skin) friction" or by "form friction". These terms will be explained later. The fourth term is often given in terms of pressure drop. Equating the loss term to pressure drop is a source of confusion, because of the phenomena of pressure recovery and irreversibility. It is best to use specific energy units so as to avoid mistakes when performing estimations.

Incompressible Fluid Flow — Losses of Mechanical Energy | CHAPTER TWO

Caveat — Bernoulli balance

The complete Bernoulli[i] equation with the so-called loss term is often called a mechanical energy balance. It is not. It is a total energy balance when the loss term is included. The "loss" is the conversion of mechanical to internal energy.

II-4: IRREVERSIBILITIES DUE TO PIPE AND FITTINGS

Considerable effort has been expended to establish correlations for fluid flow through various types and sizes of pipe, valves and fittings and the resulting permanent losses of mechanical energy (the fourth term on the right of the Bernoulli equation). Mechanical energy losses can be categorized as being those associated with:

- straight pipe of uniform diameter;
- sudden contractions;
- sudden expansions;
- gradual expansions or contractions;
- valves treated as fittings;
- orifices;
- miscellaneous fittings;
- manifolds with dividing or combining flows;
- the total system (all of the above).

We will discuss each of the above categories and give the basic equations for estimating the irreversibilities. Following this discussion we will give a series of examples of each case. The last example will treat several devices as a system.

There are two main approaches to estimating irreversibilities. One approach, mainly applicable to straight conduits, uses the Darcy[vii] equation directly. The other approach makes use of the component loss coefficient, the "K" factor. We will first discuss straight pipes of uniform diameter and then explain the component loss coefficients.

Straight pipe of uniform diameter

For incompressible fluids flowing through straight, horizontal pipe of uniform diameter with no pump or turbine in the segment, the Bernoulli

CHAPTER TWO | Incompressible Fluid Flow – Losses of Mechanical Energy

equation (II-1) reduces to,

$$\frac{P_1}{\rho_1} = \frac{P_2}{\rho_2} + h_L$$ (II–2)

$$h_L = \frac{P_1 - P_2}{\rho}$$

The last equation gives the means of measuring head "losses" in constant diameter pipe. This is about the only situation in which the irreversibilities equal the pressure drop divided by density. The pressure drop divided by the fluid density, in this case, gives the energy per unit mass converted to internal energy. Most often, we do not have to measure this loss. We have to predict it.

Prediction of the loss term is based on many correlations that have been performed over the years. For uniform diameter pipe, the loss term is given by the Darcy[vii] equation,

$$h_L = f_M \frac{L}{D} \frac{U^2}{2g_c}$$ (II–3)

It seems fairly logical that the conversion of mechanical energy to internal energy should be proportional to the kinetic energy ($U^2/2g_c$), the length (L), to a friction factor (f_M) and inversely proportional to the pipe diameter (D).

The Darcy equation applies to laminar and turbulent flow. It applies also to incompressible and compressible flow. However, for compressible flow, the equation must be used in differential form because velocity and the friction factor are not constant. The compressible flow form of the equation will be developed in the chapter on compressible flow.

For pipes, the friction factor has been correlated with fluid viscosity and density, pipe diameter, wall roughness and fluid velocity. The correlation is presented as two charts; one for any type of commercial pipe, Figure I-4, and the other for clean commercial pipe, Figure II-1. It may also be computed from the Churchill-Usagi[xli] equations given in Appendix AII.

The correlation for any type of commercial pipe gives the Moody[viii] friction factor versus the Reynolds[v] number and the pipe relative roughness. The relative roughness is the average height of the wall irregularities, ε, divided by the diameter, D, in the same units. In fully turbulent flow, the friction factor increases with relative roughness. In laminar flow it does not.

Incompressible Fluid Flow – Losses of Mechanical Energy | CHAPTER TWO

Note that, for the same type of commercially available pipe, the absolute roughness will be independent of diameter. The relative roughness, therefore, will vary inversely with diameter.

To use Figure I-4, you have to know the type of pipe material. The average absolute roughness of several pipe materials is given in Table II-1 Simply divide the absolute roughness in feet or millimeters by the pipe internal diameter in feet or millimeters to obtain the relative roughness. The correlation for clean commercial pipe, Figure II-1, gives the friction factor as a function of the Reynolds[v] number, the pipe diameter and schedule number. This chart is the one most commonly used in design engineering. Once the Moody[viii] friction factor has been obtained, the mechanical energy losses in ft-lb$_f$/lb$_m$ or J/kg for uniform diameter pipe may be found by plugging the appropriate data (length, diameter, velocity and Moody friction factor) into the above Darcy[vii] equation.

Table II-2 is extracted from, Internal Flow Systems, D.S. Miller, 2nd Edition[xvi]. It is intended for general hydraulic practice. The minor differences with Crane[xiii] may be due to differences in European and North American pipe fabrication practices. Miller states that experience with similar systems is the best guide to selecting roughness values and deterioration tolerances. Pipes conveying water are particularly prone to deposits, erosion, corrosion, bacterial slimes and growths and to marine and fresh water fowling. For water, he recommends using an absolute roughness ranging from 0.5 mm to 2.0 mm. The smaller value is for chlorinated, filtered, clear, unaggressive and non scale-forming water. The larger value is for worst-case unchlorinated water. Both values are for water pipes or conduits after several years of service. Industrial fluids tend to be more predictable than water over the long term.

TABLE II-1. Absolute roughness of various materials		Data from Crane
Material	**Absolute roughness, ε, feet**	**mm**
Drawn tubing	0.000005	0.0015
Commercial steel	0.00015	0.046
Asphalted cast iron	0.0004	0.122
Galvanized iron	0.0005	0.152
Cast iron	0.00085	0.259
Wood stave	0.0006 - 0.003	0.183-0.914
Concrete	0.001 - 0.01	0.305-3.05
Riveted steel	0.003 - 0.03	0.914-9.14

CHAPTER TWO | Incompressible Fluid Flow – Losses of Mechanical Energy

Friction Factors for Clean Commercial Steel Pipe[10]

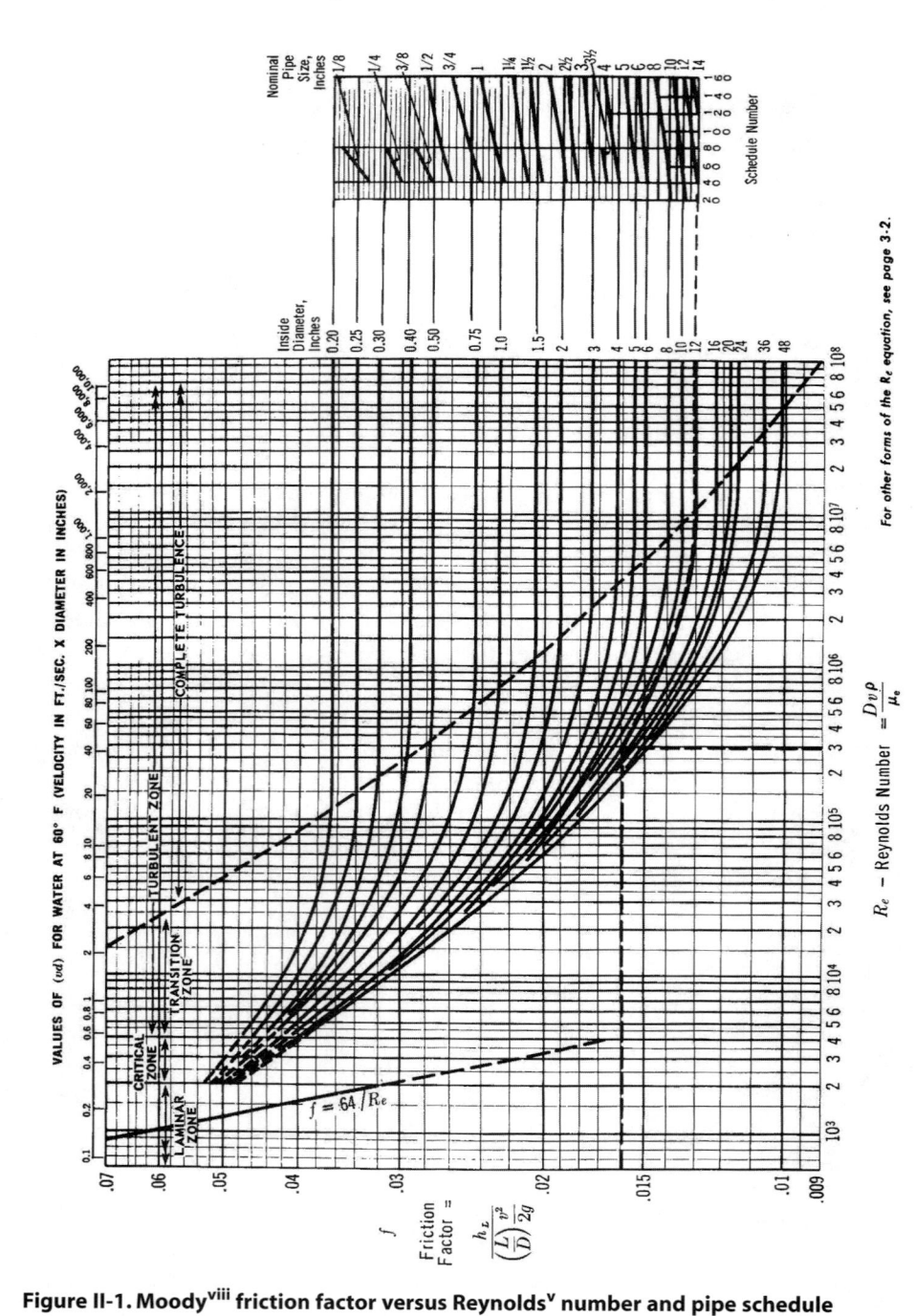

R_e – Reynolds Number $= \dfrac{D v \rho}{\mu_e}$

Friction Factor $= f = \dfrac{h_L}{\left(\dfrac{L}{D}\right) \dfrac{v^2}{2g}}$

For other forms of the R_e equation, see page 3-2.

Problem: Determine the friction factor for 12-inch Schedule 40 pipe at a flow having a Reynolds number of 300,000.

Solution: The friction factor (f) equals 0.016.

Source: ©Crane Co. All rights reserved.

Figure II-1. Moody[viii] friction factor versus Reynolds[v] number and pipe schedule

Flow of Industrial Fluids—Theory and Equations

Incompressible Fluid Flow – Losses of Mechanical Energy | CHAPTER TWO

TABLE II-2. Absolute roughness of various materials		Data from D.S. Miller
Material	**Absolute roughness, ε feet**	**mm**
Smooth pipes		
Drawn brass, copper, aluminum, etc.	0.0000082	0.0025
Glass, plastic, Perspex, Fiberglas, etc.	0.0000082	0.0025
Steel pipes		
New smooth pipes	0.000082	0.025
Centrifugally applied enamels	0.000082	0.025
Mortar lined, good finish	0.00016	0.05
Mortar lined, average finish	0.00033	0.1
Light rust	0.00082	0.25
Heavy brush enamels and tars	0.0016	0.5
Heavy rust	0.0033	1.0
Water mains with general tuberculations	0.0039	1.2
Concrete pipes		
New, unusually smooth concrete with smooth joints	0.000082	0.025
Steel forms, first class workmanship with smooth joints	0.000082	0.025
New, or fairly new, smooth concrete and joints	0.00033	0.1
Steel forms, average workmanship, smooth joints	0.00033	0.1
Wood floated or brushed surface in good condition with good joints	0.00082	0.25
Eroded by sharp material in transit, marks visible from wooden forms	0.0016	0.5
Precast pipes, good surface finish, average joints	0.00082	0.25
Segmental lined conduits in good ground conditions with expanded wedge block linings	0.0033	1.0
Segmental lined conduits with other conditions	0.0066	2.0
Other pipes		
Sheet metal ducts with smooth joints	0.0000082	0.0025
Galvanized metals, normal finish	0.00049	0.15
Galvanized metals, smooth finish	0.000082	0.025
Cast iron, uncoated and coated	0.00049	0.15
Asbestos cement	0.000082	0.025
Flexible straight rubber pipe with a smooth bore	0.000082	0.025
Mature foul sewers	0.0098	3.0

CHAPTER TWO | Incompressible Fluid Flow – Losses of Mechanical Energy

Accuracy of loss estimates for pipes – the use of safety factors

Accuracy of loss estimates depends on knowing and using the actual pipe internal diameter, not the nominal diameter. For metering devices that make use of piping to form part of their "run", the diameter is of great consequence. This is why meter runs are often prefabricated so they may be swaged into larger diameter piping. Permanent mechanical energy losses vary inversely with approximately the fourth power of the diameter. Miller[xvi] states that a 100% error in the roughness causes a 10% error in the friction coefficient.

Caveat — Safety factors

Do not add "safety factors" indiscriminately with each computation. It is best to perform all computations without them and then to logically decide where they must be applied in light of the total problem. For instance, in choosing a pipe size one must choose a commercially available pipe. The diameter will be greater than the required diameter. This in itself constitutes a safety factor.

Component loss coefficients, 'K' factors and 'skin friction'

When an incompressible fluid is flowing in a long, straight, horizontal pipe, the Darcy[vii] equation gives the mechanical energy losses. These irreversibilities are often described as "skin friction". This unfortunate term is another of those loose and misleading terms used to describe natural phenomena. It arises from fact that when studying relative flow past profiles (not in pipes), turbulent flow develops next to a solid boundary and then spreads in the normal direction. In confined flow, turbulence may be more intense near the solid boundary, but it certainly does not constitute a "skin". In fact, this term really only applies to structures such as aeroplane wings. In pipes, the boundary layer fills the entire cross section.

When fluid passes through a valve or a fitting, the flow profile is radically disturbed and there are large eddies. Boundary layer separation occurs and we are dealing with what is known as "form friction". Form friction is actually the irreversibilities associated with the vortices that arise from major changes in direction – obstructions, for instance. Friction factor correlations were developed for skin friction, not form friction. K factor correlations were developed to measure the irreversibilities associated with form friction.

The flow paths through some valves and some fittings change drastically and it is not possible to derive theoretical equations for each case. This does not stop engineers from establishing empirical relationships.

Incompressible Fluid Flow – Losses of Mechanical Energy | CHAPTER TWO

The Darcy[vii] equation and the 'K' factor equation compared

We can compare the Darcy equation for straight pipe with the theoretic equation for a sudden expansion and the empirical one for a sudden contraction. We can establish a relationship between the loss coefficient, K, and the product of the friction factor and the length to diameter ratio.

(II–4)

$$h_L = f_M \frac{L}{D} \frac{U^2}{2g_c}$$

$$h_L = K \frac{U^2}{2g_c}$$

$$K = f_M \frac{L}{D}$$

We can say a given fitting would cause the same mechanical head loss as an equivalent length of straight pipe with a diameter ratio, L/D, whose friction factor, under specific flowing conditions, would be given by f_M. The velocities and therefore the diameters in which the velocities occurred would be the same.

The loss coefficient, K, for a fitting with one inlet and one outlet can be seen to be the ratio of the permanent mechanical energy losses across a fitting and the kinetic energy in one of its legs (inlet or outlet). It is much more a function of form friction than skin (path) friction. Therefore, it frequently is considered to be independent of the friction factor. It is frequently considered to be dependent only on the geometry of a particular valve or fitting. This is not quite true, so a correction must be made to the assumption.

Caveat — Velocity and the K factor

The numerical value of K depends upon the leg in which the velocity is measured. The leg in which the velocity is measured must always be specified and the appropriate K factor must be used.

Geometric similarity and dissimilarity

In theory, geometrically similar valves and fittings would have the same loss coefficient, no matter what the size. This turns out to be the case for sudden contractions and expansions, which are only dependent on the beta ratio, for pipe entrances, which are only dependent on the geometry (rounding), and for pipe exits.

CHAPTER TWO | Incompressible Fluid Flow – Losses of Mechanical Energy

For pipe exits (to a tank or to atmosphere) the loss coefficient is one (1.0), regardless of pipe size. This means the permanent mechanical energy losses are numerically equal to what is called one velocity head ($U^2/2g_c$) across any pipe exit, based on the velocity within the pipe. The loss across any pipe exit in energy per unit mass units can be found by multiplying by one the average cross sectional velocity in the pipe exit squared divided by two and the dimensional constant. The customary U.S. units for the lost mechanical energy would be foot-pounds force per pound mass. The SI units would be joules per kilogram (N•m/kg).

Pressure drop not always equal to head loss

It is worthwhile pausing here to drive a point home. The Bernoulli[i] equation is often manipulated to change specific energy units into head units or even to pressure units. When pressure units are used it is particularly difficult to keep in mind which pressure is being discussed. For instance, the head loss across a square edged pipe exit, when expressed in pressure units, is given as,

$$\Delta P = \rho \frac{U^2}{2g_c} \quad \text{(II–5)}$$

It is very easy to assume this pressure drop means the pressure inside the pipe at the discharge point is greater than that within the tank at the same level. Such is not always the case. For instance, if water is flowing at 15 fps through a pipe outlet to a tank, expressed in pounds force per square inch, the head loss across the square edged exit would be approximately,

$$\Delta P = \frac{62.4(15)^2}{(144)2g_c} = 1.5\,psi \quad \text{(II–6)}$$

Expressing the loss in pressure units leads very easily to the assumption that this is the measured pressure drop across the exit. This is not true. The following development will prove this statement.

In this type of equation (II-6), the differential pressure term is positive

If the Bernoulli equation is left in its specific energy form, it is easier to see the real relationship. We can state that the losses (the non recoverable mechanical energy), across a square edged pipe outlet are,

Incompressible Fluid Flow – Losses of Mechanical Energy | CHAPTER TWO

$$h_L = K \frac{U^2}{2g_c} = 1\frac{(15)^2}{2g_c} = 3.5 \frac{ft-lb_f}{lb_m} \quad \text{(II–7)}$$

or

$$h_L = 1\frac{(4.572)^2}{2(1)} = 10.45 \frac{m^2}{s^2} = 10.45 \frac{J}{kg}$$

There is no confusion with either of these units. We know we are talking about the loss term in the Bernoulli[i] equation, II-1.

In Equation II-8, section 1 is just inside the exit from a horizontal pipe; section 2 is in the tank. The Bernoulli equation, the first line of set II-8, can now be manipulated as follows (the alpha corrections each equal one).

$$\text{(II–8)}$$

$$h_L = K_1 \frac{U_1^2}{2g_c} = \left(\frac{P_1}{\rho_1} + \frac{U_1^2}{2g_c} + \frac{gZ_1}{g_c}\right) + w_n - \left(\frac{P_2}{\rho_2} + \frac{U_2^2}{2g_c} + \frac{gZ_2}{g_c}\right)$$

$$Z_1 = Z_2, U_2 \approx 0, K_1 \approx 1.0, \rho_1 = \rho_2 = \rho, w_n = 0$$

$$\frac{P_1}{\rho} + \frac{U_1^2}{2g_c} = \frac{P_2}{\rho} + K_1 \frac{U_1^2}{2g_c} = \frac{P_2}{\rho} + 1.0 \frac{U_1^2}{2g_c}$$

$$P_1 = P_2$$

The density of the flowing fluid will have decreased negligibly due to the increased temperature.

The pressure inside the pipe, P_1, equals the pressure, P_2, at the same elevation within the tank at the point where the velocity is negligible. There is no measurable difference between the pressure immediately inside the pipe exit and that at the same elevation in the tank at a point where the fluid is practically stationary. The kinetic energy has been completely converted to internal energy and the downstream fluid is warmer than the upstream fluid.

This becomes clear if it is remembered the term associated with the K factor represents irreversibilities. In the second from last equation of the set II-8, the first two terms on the left, static energy and kinetic energy, represent the driving potential that is exactly balanced by the static pressure potential and the irreversibilities on the right. If there were no irreversibilities, the static pressure potential downstream would balance both

CHAPTER TWO | Incompressible Fluid Flow – Losses of Mechanical Energy

driving potentials. For the same velocity, the pressure upstream would be less. Recovery would have taken place. In the above example, if K is not equal to one, the pressure will not be equal.

We will see later that recovery, a higher pressure downstream than upstream, is a common occurrence with certain types of fittings and with valves.

Caveat — Mechanical energy 'losses' versus pressure

Do not fall into the temptation of equating real mechanical energy losses to pressure drop, as is very frequently done in the literature.

The loss coefficient, K_2, based on the velocity within the pipe, for a sharp edged pipe entrance is 0.5, regardless of pipe size. The irreversibilities for the same velocity (kinetic energy) in the same sized pipe would be exactly half those of a pipe exit.

In practice, with the exceptions given above, similar fittings are not geometrically similar over all sizes. Curvatures within fittings and valves change as sizes change. The relative roughnesses of different sizes are not equivalent. The value of the absolute roughness, ε, generally remains the same as the diameter increases, so the relative roughness decreases with the increase in diameter. In any case, the effects of flow separation in fittings and valves usually outweigh the effects of normal pipe turbulence.

Useful practical correlations from Crane[xiii]

Crane gives a useful empirical correlation among the type and size of a valve or fitting and its loss coefficient. The equivalent K, $f_M L/D$, at fully developed turbulent flow for 30 diameters of the same size of pipe is the basis for the correlation. The chart is reproduced as Figure II-2 (Fig. 2-14, Crane). It shows that, for a given type of valve or fitting, the slopes of the K versus valve or fitting size are very similar to the slope of the $f_M L/D$, versus the same size of pipe 30 diameters long. The pipe chosen is clean, commercial Sch. 40 pipe.

Crane states that it is probably a coincidence the effect of geometric dissimilarity between different sizes of the same line of valves or fittings upon the resistance coefficient, K, is similar to that of relative roughness upon the friction factor.

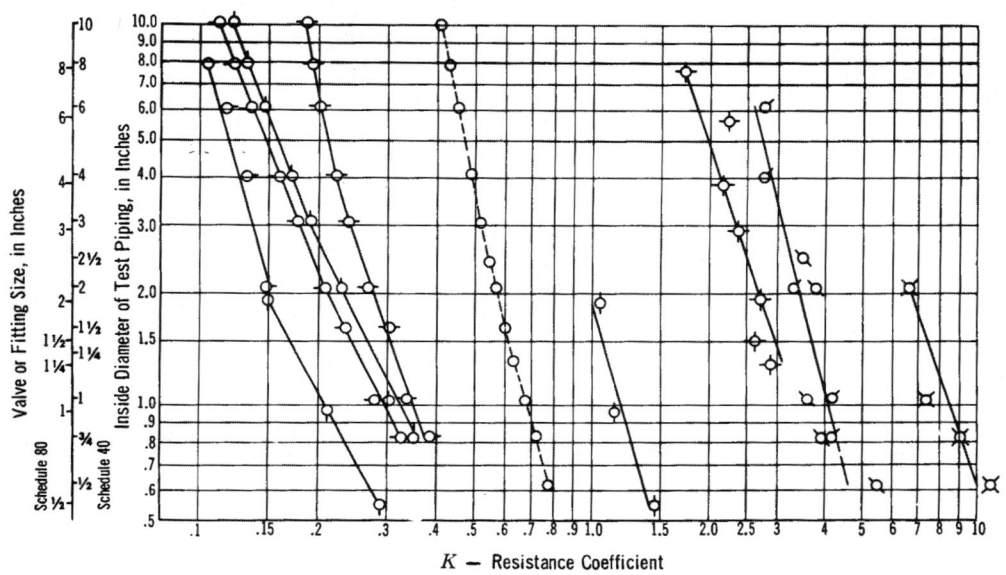

Incompressible Fluid Flow – Losses of Mechanical Energy | CHAPTER TWO

Resistance Coefficient K, Equivalent Length L/D, And Flow Coefficient C_v — continued

Figure 2-14, Variation of Resistance Coefficient K ($=f$ L/D) with Size

Symbol	Product Tested	Authority
○	Schedule 40 Pipe, 30 Diameters Long ($K = 30\ f_T$)*	Moody A.S.M.E. Trans., Nov.-1944[18]
⊙	Class 125 Iron Body Wedge Gate Valves	Univ. of Wisc. Exp. Sta. Bull., Vol. 9, No. 1, 1922[16]
⊘	Class 600 Steel Wedge Gate Valves	Crane Tests
⊖	90 Degree Pipe Bends, $R/D = 2$	Pigott A.S.M.E. Trans., 1950[6]
◇	90 Degree Pipe Bends, $R/D = 3$	Pigott A.S.M.E. Trans., 1950[6]
⊖	90 Degree Pipe Bends, $R/D = 1$	Pigott A.S.M.E. Trans., 1950[6]
◊	Class 600 Steel Wedge Gate Valves, Seat Reduced	Crane Tests
✧	Class 300 Steel Venturi Ball-Cage Gate Valves	Crane-Armour Tests
⊘	Class 125 Iron Body Y-Pattern Globe Valves	Crane-Armour Tests
⋈	Class 125 Brass Angle Valves, Composition Disc	Crane Tests
⋈	Class 125 Brass Globe Valves, Composition Disc	Crane Tests

*f_T = friction factor for flow in the zone of complete turbulence; see page A-26.

Source: ©Crane Co. All rights reserved.

Figure II-2. Component loss coefficient correlation

In other words, the happy coincidence is that, as the pipe diameter increases, the relative roughness, ε/D, for the same material in pipes decreases. This decrease influences the friction factor at complete turbulence in the same direction and at a similar rate as the geometric dissimilarity of commercially available valves and fittings influences the K factor.

Flow of Industrial Fluids—Theory and Equations

CHAPTER TWO | Incompressible Fluid Flow – Losses of Mechanical Energy

The correlation is really between the friction factor, which is relative to "skin" turbulence, and the K factor, which is relative to "form" turbulence. This is a good place to point out the location of the cause of the turbulence does not necessarily coincide with the location of the irreversibilities. Irreversibilities occur downstream where the turbulence is destroyed.

The above correlation allows Crane[xiii] to establish simple relationships among the type of valve or fitting, the size and the friction factor at complete turbulence of the same sized schedule 40 pipe. These relationships may be found in Crane Technical Paper No. 410. They are essentially all statements that the K factor we are interested in is equal to a friction factor times a constant. This constant is the equivalent L/D of the fitting. The friction factor is that which would be associated with fully developed turbulent flow in schedule 40 pipe of the same diameter.

Note this methodology is not applicable to all types of fittings. We will discuss these cases shortly. The method is useful when there is no better one or for quick estimates.

Component loss coefficients

The component loss coefficient, or the K factor, is a performance parameter for a component within a system. It gives a measure of the permanent mechanical energy conversion to thermal energy due to the presence of a specific component within the system.

Figure II-3 is an example. It depicts an orifice plate that can be removed from a pipe without disturbing the pipe's characteristics. Total energy is shown on the ordinate and length is depicted on the abscissa. The pipe is horizontal in order to simplify the analysis. Although an orifice plate is used in the example, any component could have been used.

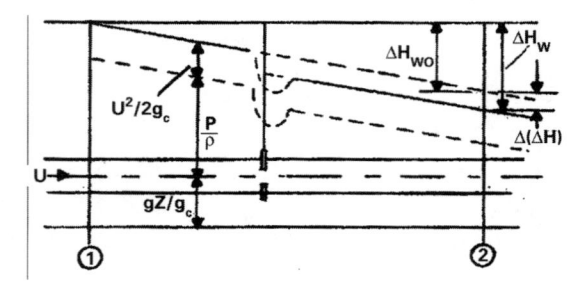

Figure II-3. Component loss coefficients

68   Flow of Industrial Fluids—Theory and Equations

Incompressible Fluid Flow — Losses of Mechanical Energy | CHAPTER TWO

The sections numbered 1 and 2 are sufficiently removed from the orifice location so the flow profiles at these two points are substantially undisturbed when the orifice is in place. The flow profile will be disturbed on either side of the orifice; but at the numbered locations, it is normal. Section 1 is the reference location. Properties downstream of this location may be compared to its properties.

The figure shows the changes when the orifice is in place. The flow rate is the same in both cases, with or without the orifice plate. The datum line is the bottom horizontal line. It could have been taken as the center line of the pipe. The total mechanical energy is given by the sum of the elevation head, the static pressure head and the velocity head.

Upstream of the orifice, the profile and the head loss will be the same with or without the orifice except in the immediate vicinity of the plate. Downstream, with the orifice in place, there will be a more radical change that will become of constant slope after thirty to fifty pipe diameters.

Caveat — Pressure recovery

When measuring irreversibilities, care must be taken to allow sufficient distance upstream and downstream so the flow profile becomes re-established. Measuring pressure drop across an orifice, for instance, will include both permanent pressure losses and recoverable pressure drop as can be seen from Figure II-3. The quantity of interest is really the irreversibilities that occur due to the presence of the plate. These irreversibilities may occur away from the plate as the profile recovers.

The total head or mechanical energy loss can be computed in both cases, with and without the orifice plate. Subtracting the loss without the plate, ΔH_{WO}, from the loss with the plate, ΔH_W, gives the loss attributable to the plate alone, $\Delta(\Delta H)$. This method of arriving at the irreversibilities accounts for upstream and downstream disturbances due to the presence of the component only.

Tightening up terminology

It is to be noted the difference between irreversibilities computed for two sections is normally symbolised as ΔH. The loss attributed to the component is given the same symbol, but this habit lumps the upstream and downstream irreversibilities into those due to the component. In the figure, we have used $\Delta(\Delta H)$ to try to be more specific.

CHAPTER TWO | Incompressible Fluid Flow – Losses of Mechanical Energy

Looking at the figure, it can be seen that, with fully developed flow profiles in straight pipe, the total head lines can be projected to the orifice plate location. The difference between the points of intersection with the plate will be the same as the difference referred to previously as $\Delta(\Delta H)$. The permanent irreversibilities due to the plate or any other component are given by $\Delta(\Delta H)$. These irreversibilities will vary with the component but will be found to be proportional to the kinetic energy, $U^2/2g_c$. The average velocity must be associated with a specific section (of pipe, not of the orifice) when the K factor is computed.

Note we have an equation in the form of,

$$\Delta(\Delta H) = K_i \frac{U_i^2}{2g_c}$$

(II–9)

The term on the left is the difference between the head losses with and without the component. We are talking about changes in total mechanical energy that occur slightly upstream and within 30 to 50 pipe diameters downstream of the component. The velocity, U, must be measured or estimated at the same time. The component loss coefficient, K_i, is then taken as the ratio of the difference in projected mechanical energy across the component to the kinetic energy at a particular point.

The coefficient must always be associated with the point at which the velocity was measured. K_i is dimensionless. It has the same value in SI units as it does in customary U.S. units.

Figure II-4 gives an example of how the loss coefficient is established. The example is that of a 90 degree mitered pipe. The example is modified from D.S. Miller[xvi]. The pipe is horizontal. The pipe is of uniform diameter (velocity at point 1 is equal to velocity at point 3 because the fluid is incompressible). Point 1 is upstream of the miter and the manometer at this point gives the static pressure at the inlet. Point 3 is sufficiently downstream as to represent the point at which pressure recovery is complete. Point 2 represents a vena contracta that is formed due to the sudden change in direction at the miter, the wall separation and reverse flow that takes place immediately behind the bend.

The average velocity at point 2 will be greater than at points 1 and 3. The series of manometers depict the interchange of static pressure energy and kinetic energy and the partial reverse process as pressure energy is recov-

Incompressible Fluid Flow – Losses of Mechanical Energy | CHAPTER TWO

ered. The permanent irreversibilities are seen to be approximately numerically equal to $U^2/2g_c$ or what is termed one velocity head. If the velocity were measured at the vena contracta, it could be inferred an additional two velocity heads had been lost. If the velocity were measured at point 3 a more realistic picture of the non-recoverable losses emerges. Point 3 should be at least 30 pipe diameters downstream of the miter.

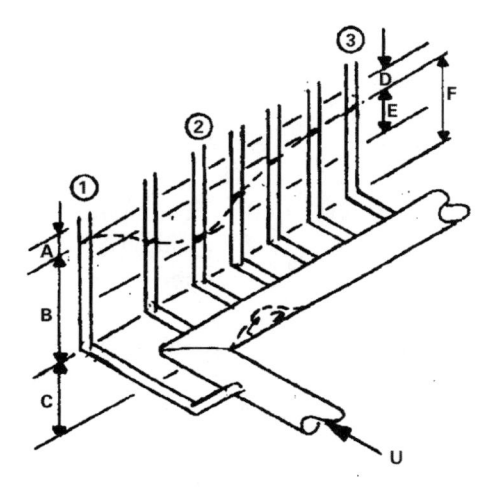

Figure II-4. Establishment of loss coefficient for 90 degree mitered elbow

The letters identify the terms in the Bernoulli[1] equation.

A equals $U_1^2/2g_c$, the "velocity head" at section 1, upstream of the miter. Without flow, the liquid would reach the upper line of the A interval. With flow, its height is indicated by the dashed curve.

B equals P_1/ρ, the "static head" at point 1.

C equals gZ/g_c, the "elevation (potential) head".

D equals h_L or $KU_1^2/2g_c$, the permanent losses or irreversibilities.

E equals approximately $2U_1^2/2g_c$. It represents the static head, P/ρ that will be recovered over and above the head drop that occurs at the vena contracta.

F equals P_3/ρ, the static head at section 3.

Note that, to obtain the true measure of the influence of the miter bend, identical data would have to be obtained on a straight run of identical piping. The difference between the head loss with the miter in place and that without it gives the loss attributable to the miter alone. It is this concept that is emphasised by the symbol $\Delta(\Delta H)$.

Flow of Industrial Fluids—Theory and Equations

CHAPTER TWO | Incompressible Fluid Flow - Losses of Mechanical Energy

To quantify the error that would occur if pressure measurements taken at points 1 and 2 were used to compute a head loss, the following development can be made:

$$K_{12} = \frac{\Delta(\Delta H_{12})}{U_1^2/2g_c} = \frac{\Delta(\Delta P/\rho)_{12}}{U_1^2/2g_c} \qquad \text{(II–10)}$$

$$\left(\frac{\Delta(\Delta P)}{\rho}\right)_{12} = \frac{U_1^2}{2g_c} + \frac{2U_1^2}{2g_c} = \frac{3U_1^2}{2g_c}$$

$$K_{12} = \frac{3U_1^2/2g_c}{U_1^2/2g_c} = 3$$

If differential pressures between points 1 and 2 were transformed directly to permanent losses by dividing by the density, there would be about a three-fold error. Pressure recovery would have been neglected.

Note carefully the subtle distinctions drawn by the subscripts. The K factors and the permanent losses are between two points. The average velocity (kinetic energy term) is measured at a fixed section whose location must be clearly stated.

If a similar development is followed for the loss coefficient between points 1 and 3, it will be seen that K_{13} equals approximately 1. This is much closer to the true value. The above example shows clearly why the differential across the normal flange taps of an orifice plate cannot be used to represent permanent losses. Pipe taps would give a closer approximation.

Basic loss coefficients

It is convenient to establish loss coefficients at a particular Reynolds[v] number and to make adjustments to the coefficient for other values of the Reynolds numbers. D.S. Miller[xvi] gives the following reasoning in picking the standard Reynolds number:
1. most major industrial flows operate between $10^6 < N_{Re} < 10^8$;
2. variations in K are quite small for $N_{Re} > 0.5 \times 10^6$;
3. commercial laboratories cannot economically run tests at $N_{Re} > 10^6$, therefore:
4. N_{Re} equals 1×10^6 is an acceptable standard Reynolds number upon which to base modifications (as a starting point for computation).

The use of the basic loss coefficients allows tabulation of standard K factors that are then modified for actual conditions. We will demonstrate this by examples. It should be noted other authors use 1 x 10^5 as a standard Reynolds[v] number.

Sudden contraction

The worst case loss for a reduction in dimension is a sudden contraction. It occurs when a smaller pipe is flanged directly downstream of a larger pipe (Figure II-5).

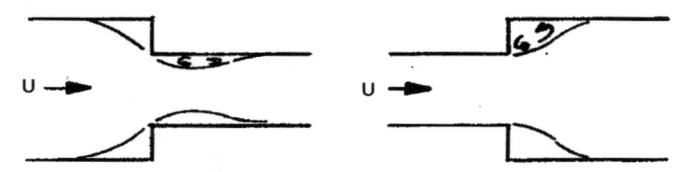

Figure II-5. Sudden contraction   **Figure II-6. Sudden expansion**

It can be seen that a vena contracta occurs downstream of the entrance to the smaller pipe with the usual turbulence and losses in mechanical energy. When the loss is taken as proportional to the velocity head (kinetic energy term) in the *smaller pipe*, the loss in mechanical energy is established empirically as,

$$h_L = \Delta(\Delta H) = K_c \frac{U_2^2}{2g_c} \qquad \text{(II–11)}$$

In other words the loss in mechanical energy per unit mass due to the contraction is proportional to the square of the average velocity at section 2 (the point of full recovery of profile) divided by two and the dimensional constant. The factor of proportionality, K_c, is the contraction loss coefficient. It is initially established by measurement. The measurements are then correlated with pipe dimensions to establish a semi-empirical formula. The use of the formula then obviates the necessity of making further measurements. The irreversibilities are negligibly small for laminar flow.

For turbulent flow the loss coefficient is given by,

$$K_c = 0.5\left(1 - \frac{A_2}{A_1}\right) = 0.5\left(1 - \frac{D_2^2}{D_1^2}\right) = 0.5\left(1 - \beta^2\right) \qquad \text{(II–12)}$$

The beta ratio is seen to be simply the ratio of the smaller to the larger diameter.

CHAPTER TWO | Incompressible Fluid Flow – Losses of Mechanical Energy

Due to the forming of guided flow streams resembling those of a venturi, the irreversibilities from a sudden contraction are less than those from a sudden expansion.

To estimate losses in mechanical energy across a sudden contraction, the average velocity in the smaller pipe at a point where the profile is fully developed and the beta ratio are needed. These numbers are plugged into Equations II-11 and II-12 to obtain the loss in mechanical energy. In addition, if it is necessary to estimate the pressure drop across the same contraction, the Bernoulli[1] equation is necessary. *The permanent head loss and the pressure drop divided by density will not be the same.* This is a point that should be kept in mind.

With all piping components, external forces must be exerted by the piping anchors to counteract the forces developed by the flowing fluid. Otherwise, the piping will move. These forces are easily estimated from pressure drop and impulse-momentum data. This exercise is required of piping engineers. Often they require input from process and control systems personnel. We will consider the estimation of these forces to be beyond the scope of this book.

Sudden expansion

The worst case for an increase in dimension is a sudden expansion. The gradual tapers associated with expanders have lower head losses than those of a sudden expansion. Figure II-6 represents a larger pipe flanged directly to a smaller one.

The losses in mechanical energy associated with a sudden expansion are proportional to the kinetic energy, the velocity head, at a section where the velocity profile is fully developed in the smaller (upstream) pipe. The loss equation is,

$$h_L = \Delta(\Delta H) = K_e \frac{U_1^2}{2g_c} \quad \text{(II–13)}$$

This equation is almost identical to that for a sudden contraction. The loss coefficient for a sudden expansion has a different value than that for a sudden contraction. It is,

$$K_e = \left(1 - \frac{A_1}{A_2}\right)^2 = \left(1 - \frac{D_1^2}{D_2^2}\right)^2 = \left(1 - \beta^2\right)^2 \quad \text{(II–14)}$$

The above equation is not entirely empirical; it can be derived from theoretical considerations (see Appendix AII).

Incompressible Fluid Flow – Losses of Mechanical Energy | CHAPTER TWO

It is instructive to compare the relative magnitudes of a sudden expansion and a sudden contraction with the same beta ratios – which we will take as 0.5.

$$\frac{K_e}{K_c} = \frac{(1-\beta^2)^2}{0.5(1-\beta^2)} = 2(1-\beta^2) = 2(1-0.5^2) = 1.5 \tag{II–15}$$

The mechanical energy conversion at a sudden expansion is greater by a factor of one and a half than that at a sudden contraction when the beta ratio of each is 0.5.

Valves as fittings

Valves are treated in two ways in fluid flow engineering depending on the purpose of the computation. They are regarded as control valves when they are being sized or when reasonably exact estimations of overall irreversibilities and pressure drops must be performed. When the total system is being analyzed, they are frequently regarded as just another fitting — just another obstruction. When they are regarded as control valves, equations such as those of ISA-75.01.01 are used. Factors are used to adjust the equations to the specific case and type of valve.

When a valve is a block valve or an on-off valve, it is usually fully open when fluid is flowing. In this case it frequently is treated as a fitting and standard K factors are used. In truth, treating the valve as a control valve in all cases would give more accurate results, since the valve sizing equations apply to all valves. In one of the examples that follow we will treat the valve as a fitting. The more accurate treatment will be given later.

Caveat — Source of K factor correlation

It should be stated that valves from different manufacturers would have different flow patterns. They will therefore have different K factor correlations. This means that, if the K factor method is used to estimate irreversibilities, the factors should be obtained from the specific manufacturer's literature.

Rather than use one manufacturer's K factor for another's valve, a better method is to use L.R. Driskell's Table of Representative Valve Factors, *ISA Handbook of Control Valves*[xvii], to obtain a typical valve flow coefficient, C_v, for a specific valve type and then to transform it to a K value using the method given in Appendix AII.

CHAPTER TWO | Incompressible Fluid Flow — Losses of Mechanical Energy

Orifice plates — recovery and permanent losses across an orifice plate

There are two distinct components of pressure drop across an orifice plate and the ratio of the two changes with the location of the pressure taps. The components are combined and appear as one when measured directly. However, they must be computed separately because one is recoverable and the other is not. The pressure profile across an orifice plate is shown in Figure II-7.

The pressure drop associated with metered flow includes both components. If vena contracta taps are used, the measured pressure drop is considered the maximum pressure drop available across the restriction. It gives the largest signal for a given flow rate. This pressure drop is useful to infer flow; the greater the differential pressure, the easier it is to read the signal. It tells us nothing about the irreversibilities. (Note the maximum available pressure drop is really the drop between an upstream corner tap and the vena contracta, but these locations are not standard ones.)

Figure II-7 shows the longitudinal flow profile and pressure profile over a length of piping corresponding to the length between pipe taps (specified by the distances from the upstream edge of the plate). The reason for the slight pressure build-up upstream of the plate is that kinetic energy is converted to static energy as the fluid is decelerated in the corner. Note this extra pressure is necessary to accelerate the fluid centripetally toward the vena contracta. After again decelerating during the expansion, the fluid is once again at unaccelerated steady-state.

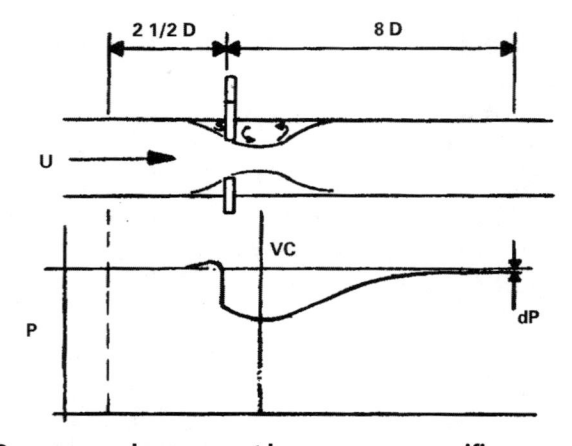

Figure II-7. Recovery and permanent losses across an orifice

Incompressible Fluid Flow — Losses of Mechanical Energy | CHAPTER TWO

For pipe taps, the term dP is the pressure drop ten and a half pipe diameters downstream of the upstream pressure tap. This pressure drop divided by the density is often used to estimate the permanent losses even though this term still contains some recoverable energy. The quantity $\Delta P/\rho$ estimated from pressure drop over these taps is not strictly equivalent to the permanent losses, $\Delta(\Delta H)$ or h_L. It is within two or three percent of the permanent losses in mechanical energy of an orifice plate.

It is necessary to know the irreversibilities if we wish not to compute a flow for metering purposes, but to establish what the flow will be under various circumstances. In other words, we wish to consider the orifice as just another device that creates irreversiblities.

There are two ways to estimate permanent losses across an orifice. The first is to make use of the fact that the pressure drop across pipe taps, 2 1/2 diameters upstream and 8 diameters downstream of the upstream face of the plate, approximate the permanent losses. In this case the permanent losses (converted to pressure drop) and the pressure drop measured across the taps coincide to within a few percent. It is assumed all the recovery has taken place within eight diameters (remember a minimum of 30 diameters is recommended for complete profile recovery).

The permanent pressure drop can be estimated from the formula for pipe taps. The second method is to make use of a head loss formula similar to those using the K factor. We will discuss this method here.

Head loss (mechanical energy) formulae

A head loss (mechanical energy) formula for square edged, concentric orifice plates (Simpson, L.L., *Chemical Engineering*, July 17, 1968[xviii]) is:

(II–16)
$$h_o = K_o \frac{U_D^2}{2g_c}$$

$$K_o = 2.8(1-\beta^2)\left[\left(\frac{1}{\beta}\right)^4 - 1\right]$$

Note that in Equation II-16 the velocity is that in the pipe, not the orifice. If the beta ratio is known, Equation II-16 can be used to estimate K_o. This factor then can be used with the average velocity *in the pipe* at fully developed profile to estimate the head losses due to the presence of the orifice.

CHAPTER TWO | Incompressible Fluid Flow – Losses of Mechanical Energy

An empirical formula for permanent loss across an orifice is given in R.W. Miller[xix]. It is curve-fitted to the ASME head loss curve (*ASME Research Report on Fluid Meters*). The formula is,

$$\frac{h}{h_w} = 1.0 - 0.24\beta - 0.52\beta^2 - 0.16\beta^3 \qquad \text{(II–17)}$$

The loss is given in terms of the ratio of the permanent pressure drop to the full-scale differential pressure across the orifice plate. Since the terms on the left constitute a ratio, any units may be used as long as they are the same for both terms.

It is sometimes necessary to use an orifice as a restriction orifice in order to fix the maximum flow rate through a system. In this case, the total irreversibilities necessary to fix the flow can be estimated. The irreversibilities available because of actual system configuration may then be estimated. The difference between the two gives the irreversibilities that must be generated by the restriction orifice. Example II-6 treats this case.

Miscellaneous fittings and manifolds

So far we have discussed rather simple fittings such as sudden enlargements or contractions. These fittings are rarely encountered. They are used as worst-cases for conservative estimates of irreversibilities.

Since the detail involved in estimating the irreversibilities of some of the more complex fittings goes beyond the intent of this chapter, we will consider them more fully in Appendix AII. We will describe some of the different cases here to remind the reader of their existence.

Gradual enlargements

These are swages from a smaller to a larger diameter (expanders). The K factor is less than that of a sudden enlargement. A formula is given in Appendix AII for the case where the included angle is equal to or less than 45 degrees. When the included angle is larger the 45 degrees, the fitting can be treated as a sudden enlargement without too much loss of accuracy.

Gradual contractions

These are swages from larger to smaller diameters (reducers). Two formulae are given in Appendix AII. The break point for the application of one or the other formula is 45 degrees of the included angle between the walls of the swage.

Incompressible Fluid Flow – Losses of Mechanical Energy | CHAPTER TWO

Inward projecting pipe entrance

This involves a pipe fixed to a tank wall in such a manner that it projects through it. The entrance has geometric similarity for all sizes. Therefore, K based on average fully-developed flow in the pipe is constant (0.78).

Square edged and rounded entrances

These entrances to a pipe from a tank are flush with the tank wall and have different amounts of rounding on the internal edge. The square edged entrance has a K value of 0.5. If the edge is rounded the associated K value is inversely proportional to the degree of rounding.

Manifold flows (dividing and combining flows)

This term is a general description for either a single stream that is split into two or more streams or for two or more streams that are combined into a single one. The loss coefficient, K, depends not only on the fitting configuration and dimensions, but also on the relative flows through the branches and on their proximity to one another. Some of these cases are the most difficult to estimate with any degree of accuracy.

These and other cases will be considered in more detail in Appendix AII. A list of common K factors and their formulae will be given. Their sources will also be given where they are known.

Total (system) mechanical energy losses

If we have a series of fittings, pipe sizes, valves and other obstructions, *sufficiently far apart from one another*, we can simply sum all of the mechanical energy losses per unit mass of flowing fluid to get the total loss per unit mass between two points.

$$h_T = \sum_i h_{Li} \qquad \text{(II–18)}$$

The subscript, Li, stands for all the various irreversibilities such as those due to sudden expansions and contractions, various types and sizes of fittings, valves and other forms of obstructions. It is necessary to have a correlation as a formula to obtain the loss over a certain type of obstruction. Many of these formulae can be found in Crane[xiii]. We have already presented some of them. D.S. Miller[xvi] gives extensive charts for combining and dividing fittings of various configurations. He also discusses the problem of combining closely connected fittings. This is a problem that is often overlooked in the general literature on fluid flow.

CHAPTER TWO | Incompressible Fluid Flow – Losses of Mechanical Energy

Caveat — Proximity of disturbances
When fittings, bends and devices are close to one another, their effects on total system irreversibilities are not necessarily additive. See D.S. Miller[xvi] for more details.

Once the total mechanical energy losses are obtained for a particular flow rate, the sum may be substituted into the Bernoulli[i] equation and various manipulations may be performed, depending on the problem to be solved.

Total energy, mechanical energy and hydraulic energy grade lines
A useful tool to visualize the energy relationships of a flowing fluid is shown in Figure II-8, associated with Example II-5. It represents the specific energy content of a fluid and the mechanical energy converted to internal energy.

Example II-5, System Losses, makes use of these lines and attempts to integrate most of the cases we have discussed.

II-5: EXAMPLES OF ESTIMATIONS OF IRREVERSIBILITIES

Example II-1: Straight pipe of uniform diameter
Suppose we have 100 feet of horizontal pipe, 6" Sch. 40, with 60°F water flowing at 900 gallons (U.S.) per minute. What is the value of the loss term and what is the pressure drop? The pipe is clean, commercial steel pipe.

All the data are readily available from Crane, Technical Paper No. 410.[xiii]

L = 100 feet
D = 6.065/12 = 0.505 feet
U = 9.99 feet per second
g_c = 32.17 $lb_m \cdot ft \cdot lb_f^{-1} \cdot s^{-2}$

The use of Figure II-1 for a 6"- Sch. 40 pipe to obtain a friction factor requires only computation of the Reynolds[v] number and the size and schedule of pipe. Remember that, although the symbol for viscosity is similar in all the terms of Equation II-19 the units of the last two terms are different. The last two terms use centipoise. Because of the multiplicity of units for viscosity, this is not an uncommon occurrence. Similarly, the fact that mixed units (gpm, lbm/h, cP, fps) must be manipulated is not uncommon.

Incompressible Fluid Flow – Losses of Mechanical Energy | CHAPTER TWO

Caveat — Viscosity units

Pay particular attention to the viscosity units when performing computations. They are one of the most common sources of error.

The test that must always be performed is to check that the Reynolds[v] number computed with the units chosen is indeed dimensionless.

$$N_{Re} = \frac{DU\rho}{\mu} = \frac{DG}{\mu} = 6.31\frac{W}{d\mu_{cP}} = 6.31U\frac{\pi D^2}{4}\rho\frac{3600}{d\mu_{cP}} \quad \textbf{(II–19)}$$

The pipe internal diameter and the fluid velocity have already been given. The density of 60°F water is 62.365 lb_m/ft^3. The viscosity is 1.14 cP. So, the Reynolds number is 410,699 or approximately $4.1(10^5)$.

If centipoise is the unit of viscosity the third equation of II-19 may be used with W in pounds-mass per hour and d in inches. Alternatively, centipoise can be converted to pounds-mass per foot-second units by multiplying it by 6.72 (10^{-4}) and the first two terms can be used. In these cases, D is in feet, U is in feet per second, rho is in pounds-mass per foot cubed, and mu is in $lb_m ft^{-1} \cdot s^{-1}$.

If SI units are required, centipoise must be converted to $N \cdot s/m^2$ or, what is the same thing, $Pa \cdot s$, by multiplying cP by 10^{-3}. In this case the diameter, D, is in meters, average velocity, U, is in m/s, rho is in kg/m^3.

To compute the Reynolds number using SI units, but starting from mixed units, ANSI piping and with viscosity given in centipoise (which is not an SI unit), the following exercise may be performed:

D = 6.065 inches = 0.505 feet = 0.1539 meters
U = 9.99 fps = 3.0449 m/s
ρ = 62.365 lb_m/ft^3 = 998.98 kg/m^3
μ = 1.14 cP x 10^{-3} = 1.14 x 10^{-3} $N \cdot s/m^2$ (or $Pa \cdot s$)
$N_{Re} = DU\rho/\mu$ = (0.1539m x 3.0449 m/s x 998.98 Kg/m^3)/1.14 x 10^{-3} $N \cdot s/m^2$ = 410,642 ~ 4.1 x 10^{-5}

The formula for the Reynolds number and the units must always be carefully checked. It is a good idea to do the computation twice with two different formulae. It is also a good idea to check the dimensions. If it is remembered that, in SI units, force (newtons) is a derived quantity equal

CHAPTER TWO | Incompressible Fluid Flow – Losses of Mechanical Energy

to mass times acceleration, then kg•m/s^2 may be substituted for N and it will be found all the units cancel. A pure number results.

Figures I-4 and II-1 give a value for the Moody[viii] friction factor of 0.0165. The same value is obtained from either chart. The energy converted to internal energy can be computed from the lost mechanical energy formula as found in II-20.

$$h_f = f_M (L/D)U^2/2g_c = 0.0165(100/0.505)\, 9.99^2/2g_c = 5.07 \text{ ft-lb}_f/\text{lb}_m \qquad \text{(II–20)}$$

To obtain the pressure drop, simply multiply the energy term by the density.

$$P_1 - P_2 = 5.07 \times 62.365 = 316.2 \text{ psf} = 2.2 \text{ psi} \qquad \text{(II–21)}$$

Note in this case the loss term can be converted directly to pressure drop. In most other cases the pipe diameter changes and the elevation changes. These factors must be included in the Bernoulli[i] equation. The advantage of keeping the complete Bernoulli equation on hand is that details such as this are not forgotten.

Example II-2: Sudden contraction

Take the previous example (II-1) and attach a sudden contraction to the six-inch pipe. The contraction is constituted of a flanged connection and four-inch diameter, schedule 40 pipe downstream. What are the head losses and what is the pressure drop across the flange?

If we look at the terms of the Bernoulli equation, we see there is no pump and the pipe is still horizontal; so the elevation terms cancel. We do, however, have a change in diameter. Therefore, there is a change in velocity and the kinetic energy terms must be left in the equation. We are still dealing with an incompressible fluid, so the density is constant. The Bernoulli equation reduces to,

$$\frac{P_1}{\rho} + \frac{U_1^2}{2g_c} = \frac{P_2}{\rho} + \frac{U_2^2}{2g_c} + h_L \qquad \text{(II–22)}$$

The upstream velocity is known from the previous problem. It is 9.99 fps. The downstream velocity can be immediately computed from the continuity equation,

Incompressible Fluid Flow — Losses of Mechanical Energy | CHAPTER TWO

$$\rho A_1 U_1 = \rho A_2 U_2 \qquad \text{(II–23)}$$

$$U_2 = \frac{A_1}{A_2} U_1 = \frac{U_1}{\beta^2}$$

We have simply replaced the area ratio by the square of the ratio of the diameters. The beta ratio is in the denominator because the downstream diameter is smaller.

The downstream velocity, U_2, equals 9.99 fps divided by the square of 0.664, the beta ratio. It is 22.67 fps. It has gone up by a factor of 2.3 because the pipe diameter decreased by a factor of 1.5. Note this velocity is much greater than what is normally considered reasonable, but it can occur and, therefore, it will occur.

The downstream kinetic energy or velocity head term is,

$$\frac{U_2^2}{2g_c} = \frac{22.67^2}{2g_c} = 7.99 \, \frac{ft-lb_f}{lb_m} \qquad \text{(II–24)}$$

The number of velocity heads lost is given by the loss coefficient, K_c.

$$K_c = 0.5\left(1 - \beta^2\right) = 0.5(1 - 0.664^2) = 0.280 \qquad \text{(II–25)}$$

The permanent losses of mechanical energy are given by the product of the loss coefficient and the velocity head based on the smaller pipe. The number is 2.24 ft-lb_f/lb_m

Notice loss in mechanical energy from a single fitting, in this case, is almost half that from one hundred feet of straight, six-inch pipe that was previously computed as 5.07 ft-lb_f/lb_m. This is typical of fittings. The bulk of the permanent losses in a piping system normally is due to fittings and bends.

To show pressure drop is different from mechanical energy loss, we will take the applicable form of the Bernoulli[i] equation and transform it. Sections 1 and 2 are at the ends of the length over which the permanent losses were computed.

CHAPTER TWO | Incompressible Fluid Flow – Losses of Mechanical Energy

(II-26)
$$\frac{P_1}{\rho} - \frac{P_2}{\rho} = \frac{U_2^2}{2g_c} - \frac{U_1^2}{2g_c} + h_L$$

$$P_1 - P_2 = \rho\left(\frac{U_2^2}{2g_c} - \frac{U_1^2}{2g_c} + h_L\right)$$

$$P_1 - P_2 = 62.34\left(\frac{22.67^2}{2g_c} - \frac{9.99^2}{2g_c} + 2.24\right)$$

$$P_1 - P_2 = 540.9 \; psf = 3.76 \; psi$$

The value of the irreversibilities (2.24 ft-lb$_f$/lb$_m$) times the density divided by 144 is 0.97 psi. The two numbers are not the same and should not be confused even when they are expressed in the same pressure units. The difference between the pressure drop, 3.76 psi and the permanent losses, 0.97 psi is equal to 2.79 psi. This number represents pressure drop that is recoverable. The difference will become clear when we introduce the energy line and hydraulic grade line graphs.

Example II-3: Sudden expansion
Building on the previous example, we will add a sudden expansion by flanging a six-inch schedule 40 pipe to the four-inch pipe. The flow rates and velocities will be those of the previous example.

The expansion loss coefficient is computed from,

$$K_e = \left(1 - \beta^2\right)^2 = \left(1 - 0.664^2\right)^2 = 0.313 \quad \text{(II–27)}$$

The mechanical energy head loss is,

(II–28)
$$h_f = K_e \frac{U_1^2}{2g_c} = 0.313 \frac{22.67^2}{2g_c} = 2.5 \; ft-lb_f/lb_m$$

The change in static pressure can be found by applying the Bernoulli[1] equation over the sudden expansion. The elevations are the same and cancel.

(II–29)
$$P_1 - P_2 = \rho\left(\frac{U_2^2}{2g_c} - \frac{U_1^2}{2g_c} + h_L\right)$$

$$P_1 - P_2 = 62.34\left(\frac{9.99^2}{2g_c} - \frac{22.67^2}{2g_c} + 2.5\right)$$

$$P_1 - P_2 = -245.4 \; psf = -1.7 \; psi$$

Incompressible Fluid Flow – Losses of Mechanical Energy | CHAPTER TWO

Notice the negative sign. The downstream pressure is greater than the upstream pressure by 1.7 psi! This result can be disconcerting to people who are used to thinking that flow is due to a pressure difference. Here, we have met the phenomenon of pressure recovery. The pressure recovery phenomenon arises frequently in piping systems (especially safety vent headers), in orifice plate installations, and in valve bodies.

From the above equation pressure recovery can be seen to occur when the downstream kinetic energy is much less than the upstream kinetic energy and their difference numerically exceeds the mechanical energy loss term. Obviously, the assumption that pressure difference is the only source of motive force is wrong. The rate of change of momentum must be considered, as is done by the application of the Bernoulli[i] equation.

Example II-4: Valves

A four-inch globe valve is placed in the four-inch line of the previous example with the same flow rates. What is the equivalent L/D ratio and how much equivalent additional pipe does the valve add? What are the permanent mechanical energy losses due to the presence of the valve?

We will use two different methods to solve the problem. The first uses the general correlation from Crane[xiii]. The second uses a more accurate method involving the valve C_v, which is the manufacturer's valve flow coefficient obtained from test data.

From Crane, on the assumption the seat diameter is the same as that of the pipe, K equals $340f_T$. The constant, 340, is the equivalent L/D of a standard globe valve with a beta ratio of one. It is obtained from Crane Technical Paper 410. The factor, f_T, is the friction factor at fully developed turbulent flow in a four inch schedule 40 pipe. It can be obtained from a convenient tabulation in the same publication or from the flat part of the curve for four-inch schedule 40 pipe in Figure II-2. More details on this methodology will be given in Appendix AII.

The friction factor at complete turbulence is 0.017. Therefore,
K = 0.017(340) = 5.78.

The first part of the question is answered above. The equivalent L/D is 340. The equivalent length is this number multiplied by the pipe inside diameter in feet. So, the equivalent length is 114 feet for one obstruction, a globe valve.

CHAPTER TWO | Incompressible Fluid Flow - Losses of Mechanical Energy

The permanent mechanical energy losses are found by applying the Darcy[vii] equation with the above friction factor and L/D ratio and the rated flow through the line.

$$h_L = f_M \frac{L}{D} \frac{U^2}{2g_c} = 0.017(340)\frac{22.67^2}{2g_c} = 46.2 \; \frac{ft-lb_f}{lb_m} \quad \text{(II-30)}$$

Using data from a specific valve, we can obtain a more accurate estimation. The valve flow coefficient, C_v, can be converted to a loss coefficient, K, by formula. The derivation for the conversion will be given in Appendix AII.

The formula in U.S. customary units is,

$$K = \frac{891 d^4}{(C_v)^2} \quad \text{(II-31)}$$

The internal diameter, d, of a four-inch schedule 40 pipe is 4.026 inches. If we have a four-inch Masoneilan globe valve, 20000 series, ANSI Class 150, full port, the catalogue data gives the C_v as 195.

$$K = \frac{891(4.026)^4}{195^2} = 6.16 \quad \text{(II-32)}$$

This more accurate value for K contrasts with 5.78 from the more general method.

Since K is equal to f_M (L/D), the L/D ratio is obtained by dividing K by the friction factor for fully developed flow in four inch schedule 40 pipe, 0.017. The ratio is 362.4. The equivalent length is 362.4(4.026/12) = 121.6 feet. This contrasts with 114 feet from the more general method.

The permanent mechanical energy losses are given by,

$$h_f = f_M \frac{L}{D} \frac{U^2}{2g_c} = 0.017(362.4)\frac{22.67^2}{2g_c} = 49.2 \; \frac{ft-lb_f}{lb_m} \quad \text{(II-33)}$$

These irreversibilities contrast with 46.2 foot-pounds force per pound mass computed by the more general method.

The pressure drop across the valve can be considered to be equivalent to the permanent losses in this case because the valve is horizontal and the upstream and downstream diameters are equal. It is 49.2(62.365/144) = 21.3 psi. The less accurate method gives 20 psi. The pressure drop to the vena contracta somewhere within the valve body will be greater than 21.3 psi.

Incompressible Fluid Flow — Losses of Mechanical Energy | CHAPTER TWO

It should be noted the vapor pressure of 60°F water is approximately 0.26 psia. If the upstream pressure at the valve inlet is less than about 21 psia, the vena contracta pressure will dip below the bubble point of the water. The water will vaporise to some extent. With any pressure recovery in the valve body, the vapor bubbles will collapse and cavitation can result. Cavitation will be dealt with more thoroughly later in the book.

Example II-5: System irreversibilities

Figure II-8 represents a length of piping which contains a flowing fluid, water. The conditions at section 1 are known. We are going to explore the changes in velocity, pressure and specific energy as the flow passes through the following: a sudden contraction, a fully open globe valve, a 90 degree elbow, a vertical segment, another elbow, a horizontal segment, a sudden expansion and another horizontal segment.

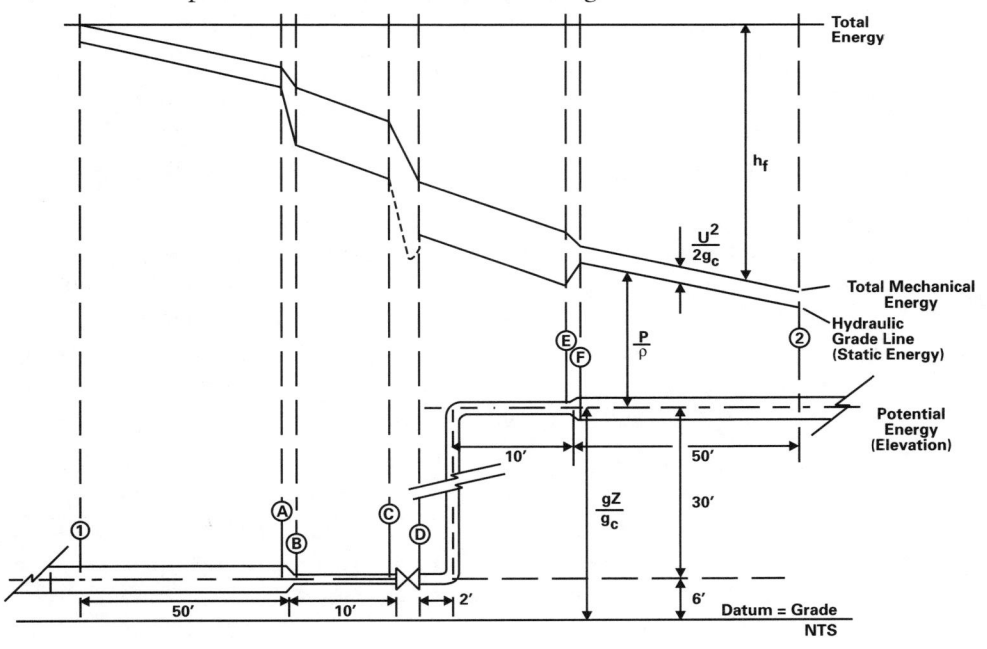

Figure II-8. Example of system irreversibilities

For the sake of continuity, we have used the data from the previous examples, but we have changed the elevations in order to show their effect. In addition, a datum line has been added and the pressure at section 1 is assumed to be 100 psig. We could have performed the computations using gauge pressures since we are dealing with water, an incompressible fluid, but we chose to use absolute pressures, for reasons given previously.

CHAPTER TWO — Incompressible Fluid Flow – Losses of Mechanical Energy

The figure is not to scale. Five lines are shown on the figure. They are defined as follows:

1. **The datum line.** This is an arbitrarily established horizontal line from which all elevation changes will be measured. Since the Bernoulli[i] equation is established in terms of the differences between two points, we can choose any horizontal level we wish. It is usually best to choose the lowest elevation in the flowing system in order to avoid dealing with negative numbers. Since the end points of the pipe are not shown, we have assumed, arbitrarily, that the lowest elevation is 6 feet below the pipe at section 1.

2. **The elevation line.** This line is used to measure elevations above the datum line for flow in the pipe. If we have a tank with a liquid surface, the surface will often correspond to the elevation. It also corresponds to the center line of the pipe. The line represents the potential energy of the fluid due to its position above the datum.

3. **The hydraulic grade line.** This line represents the energy per unit mass attributable to static pressure plus the potential energy. With liquids, it corresponds to the height the liquid would reach if a suitable manometer could be attached at successive points. For gases, it simply represents a computed value.

4. **The total mechanical energy line.** This line represents the sum of the static (pressure) energy, the potential (elevation) energy, and the kinetic energy. In the case of liquids flowing in uniform diameter pipe, the line is above the hydraulic grade line by a constant factor equal to the velocity head, the kinetic energy, in the pipe. In the case of gases, the line will deviate from the hydraulic grade line as the velocity increases with expansion.

5. **The total energy line.** This line is a horizontal line parallel to the datum line. It represents the fact that, by the first law of thermodynamics, total (mechanical and internal) energy is conserved. The difference between the total energy line and the total mechanical energy line represents the irreversibilities. It is the mechanical energy converted to internal energy. Notice it corresponds to the total mechanical energy line at section 1. We are only concerned with differences. Irreversibilities occurring upstream of the starting point of the analysis are not included in the solution to a particular problem.

On the sketch are shown the differences between the various lines and what these differences signify. The first difference is numerically (but not dimensionally) equal to the elevation above the datum. The second is equal to the absolute pressure divided by the density. The third is the kinetic energy. The fourth difference represents irreversibilities.

Incompressible Fluid Flow — Losses of Mechanical Energy | CHAPTER TWO

Points to bear in mind

The following points may not be obvious at first glance. At the sudden expansion between sections E and F, there is some pressure recovery. This is demonstrated in the computations and in the convergence of the hydraulic grade line and the total mechanical energy line across the expansion. The phenomenon is quite common, but is usually unexpected.

What is even less apparent is the same phenomenon occurs in the globe valve. Since we adopted the short cut of using an equivalent K factor, it is possible to lose sight of the fact a valve represents a tortuous path and a restriction. There will be a vena contracta at which the pressure will be much less than valve discharge pressure. Under circumstances of low inlet pressures and fairly pure liquids with high vapor pressures, this low pressure can cause cavitation, slugging and possibly damage to the valve and piping.

Plan of attack

For any problem solving exercise, it is best to have a plan of attack. This problem is fairly straightforward. We are given the conditions at section 1 and we have to compute the conditions sequentially downstream.

The general approach, in this case, is to compute the total energy at the starting section by summing all the terms on the left of the Bernoulli[i] equation. At the first section the total mechanical energy equals the total energy. The mechanical energy losses for a length of pipe, a valve or a fitting are computed by the Darcy[vii] equation or by using the appropriate K factor equation. Each loss is subtracted from the value of the total mechanical energy of the previous segment to obtain the total mechanical energy at the next section.

Given the irreversibilities along each length and across each valve and fitting, the total mechanical energy line can be plotted. Given the velocity heads, the hydraulic grade line can be plotted below the total mechanical energy line. Given the irreversibilities, velocity heads and elevations, the Bernoulli equation may be used to estimate the downstream pressures.

Frequently, engineers will convert all valves and fittings and different sized pipes to an equivalent (theoretical) length of constant diameter pipe and then perform the estimations. This habit is satisfactory for most simple cases for liquids although it is not accurate as has already been demonstrated. It cannot be used for gases and vapors. It also has the disadvantage of causing a loss of insight into what is actually taking place. We will

CHAPTER TWO | Incompressible Fluid Flow – Losses of Mechanical Energy

follow the more general approach, but we will convert the two elbows to equivalent pipe lengths to show what is involved.

Table II-3 gives the results of the computations at each section. Although the computations are made in terms of energy per unit mass, gauge pressure is frequently of immediate interest. A column for gauge pressure has been included in the table.

Most of the computations have already been given in previously. We will consolidate the results here for convenience.

Conditions at section 1

We can immediately fill in the first column of Table II-3 from the given data.

Elevation	6 feet
Pressure	100 psig, 114.7 psia
Flow	900 gpm, 450,150 lb_m/h
Temperature	60°F
Velocity	9.99 fps

At section 1:

Potential energy		6.00 ft•lb_f/lb_m
Static energy	114.7(144)/62.365 =	264.80 ft•lb_f/lb_m
Kinetic energy	9.99^2/2(32.17) =	1.55 ft•lb_f/lb_m
Irreversibilities		0.00
Total energy		272.35 ft•lb_f/lb_m
Hydraulic grade line	272.35 - 1.55 =	270.80 ft•lb_f/lb_m

Pipe segment 1 to A, 50 feet of six inch, Sch. 40 clean commercial pipe

This problem has already been solved in Example II-1 for 100 feet of pipe. Since we are dealing with horizontal pipe of uniform diameter, the irreversibilities are exactly one half, 2.53 ft•lb_f/lb_m.

As a reminder, the procedure is to compute the Reynolds[v] number based on known viscosity, diameter and flow rate. The Moody[viii] friction factor is then obtained from charts or from the Churchill-Usagi[xli] relationship. This factor is used in the Darcy[vii] equation to obtain the irreversibilities, h_L.

Incompressible Fluid Flow – Losses of Mechanical Energy   CHAPTER TWO

The irreversibilities for liquids flowing in straight, horizontal pipe can be converted directly to pressure difference by multiplying by the density. They are 1.1 psi in this case.

Conditions at section A
At section A, the total mechanical energy line has dropped below the total energy line by the amount of the irreversibilities, 2.53-ft•lb$_f$/lb$_m$. The total mechanical energy is now 269.8 ft•lb$_f$/lb$_m$. The hydraulic grade line is below this line by the kinetic energy in that line. This was computed above as 1.55 ft•lb$_f$/lb$_m$. The fluid is incompressible and the pipe is uniform, so the two lines are parallel. The elevation of the pipe stayed at 6 feet, so the potential energy is still 6 ft•lb$_f$/lb$_m$.

Conditions at section B
The irreversibilities across the sudden contraction were given in Example II-2 as 2.24 ft•lb$_f$/lb$_m$. This means the total mechanical energy line dips by this much across the fitting. The hydraulic grade line dips below the total mechanical energy line by the amount of the kinetic energy in the smaller downstream pipe, 7.99 ft•lb$_f$/lb$_m$.

Pipe segment B to C, 10 feet of four inch, Sch. 40 clean commercial pipe
For the four-inch pipe, the mass flow rate remains the same as for the six-inch pipe. The Reynolds[v] number can be computed as,

$$N_{Re} = 6.31 \frac{450{,}150}{4.026(1.1)} = 641{,}387 \approx 6.4 \times 10^5 \qquad \text{(II–34)}$$

The friction factor is obtained from Figure II-2. It is 0.017.

The velocity in the four-inch pipe has already been computed in Example II-2 as 22.67 fps. We can obtain the irreversibilities from the Darcy[vii] equation, as,

$$h_L = 0.017 \frac{10}{4.026/12} \frac{22.67^2}{2g_c} = 4.047 \frac{ft \cdot lb_f}{lb_m} \qquad \text{(II–35)}$$

The total mechanical energy line will dip below its starting point at section B by this amount. The hydraulic grade line will be below this inclined line by the amount of the velocity head in the four-inch line, 7.99 ft•lb$_f$/lb$_m$.

CHAPTER TWO | Incompressible Fluid Flow — Losses of Mechanical Energy

Globe valve
The irreversibilities across the valve were given in Example II-4 as 49.2 ft·lb_f/lb_m. This is the amount by which the total mechanical energy line dips between section C and section D. Normally the hydraulic grade line is shown parallel to this total mechanical energy line by the amount of the kinetic energy at the inlet and discharge of the valve. We have chosen to represent the hydraulic grade line by a dashed line to show that somewhere in the body of the valve is a vena contracta. It will serve to remind us to look for potential problems.

Pipe segment D to E
In this segment we have two elbows and a change in elevation. We will convert the elbows to equivalent lengths of straight pipe and then compute the irreversibilities for the sum of the straight pipe and the equivalent lengths.

If we assume standard 90 degree elbows, from Crane[xiii], K equals 2(30)f_T. It is to be noted the friction factor is that at completely turbulent flow, not the actual friction factor. From Crane, this is also 0.017 so, in our case, they are the same. K = 1.02 for both elbows. The equivalent L/D would be 60 x 4.026/12, which equals 20 feet. We must add 20 feet to the 42 feet of straight pipe to obtain an equivalent length of 62 feet.

To obtain the permanent losses, we can use the Darcy[vii] equation, as before. Alternatively, since we already have the irreversibilities for ten feet of straight four-inch pipe, we can simply ratio the lengths and multiply by the previously computed irreversibilities. This method gives
4.047 x 62/10 ~ 25.1 ft·lb_f/lb_m.

The total mechanical energy line will drop by this much to section E.

The hydraulic grade line will be below the total mechanical energy line by the kinetic energy term in the four-inch line, 7.99 ft·lb_f/lb_m.

The potential energy line follows the elevation changes of the pipe.

The difference between the potential energy line and the hydraulic grade line is equal to P/ρ. So the pressure in psia at section E can be found by multiplying the differences by the density and dividing by 144. To obtain the pressure seen by a pressure gauge, subtract the atmospheric pressure, 14.7 psia to obtain psig.

Incompressible Fluid Flow – Losses of Mechanical Energy | CHAPTER TWO

Sudden expansion, E to F

Example II-3 gave the irreversibilities across this segment E to F as being 2.5 ft•lb$_f$/lb$_m$. The pressure drop across this segment was given as minus 1.7 psi so we talk of pressure recovery. This is not only a theoretical consideration. Pressure readings taken across such an expansion will show a difference in gauge pressures. The upstream pressure will be lower than the downstream pressure. This phenomenon is not what we expect from common sense alone. Common sense tells us that flow should be in the direction of decreasing pressure, not increasing pressure.

Application of the Bernoulli[i] equation from any point upstream to section E and then to section F will confirm what the pressure gauge tells us.

The convergence of the total mechanical energy line and the hydraulic grade line in the figure, not only shows the phenomenon, but also shows its limits. The two lines can only get as close as the amount of the kinetic energy, the velocity head, downstream. In addition, although pressure has been recovered, the sum of the total irreversibilities (cumulative irreversibilities) keeps on increasing in the direction of flow. These irreversibilities are represented by h$_L$, the distance from the total energy line to the mechanical energy line.

Pipe segment F to 2

This segment of pipe is identical to the segment from 1 to A. The only difference is the elevation. The irreversibilities will be identical to those from section 1 to A: 2.53 ft•lb$_f$/lb$_m$. The total mechanical energy line will dip by this much. The hydraulic grade line will be below this line by exactly 1.55 ft•lb$_f$/lb$_m$.

It is noteworthy the pressure at section 2 is higher than the pressure upstream at section E.

All of these numbers are tabulated in Table II-3.

Table II-3. Results of computations
Energy distribution : pipe, 60°F water, 900 gpm. Upstream pressure at Section 1 :114.7 psia

Section	1	A	B	C	D	E	F	2
Sch. 40, in	6.065	6.065	4.026	4.026	4.026	4.026	6.065	6.065
Elevation, ft	6.00	6.00	6.00	6.00	6.00	36.00	36.00	36.00
Velocity, fps	9.99	9.99	22.67	22.67	22.67	22.67	9.99	9.99
P.E. ft-lb_f/lb_m	6.0	6.0	6.0	6.0	6.0	36.0	36.0	36.0
K.E. ft-b_f/lb_m	1.550	1.550	7.990	7.990	7.990	7.990	1.550	1.550
S.E. ft-lb_f/lb_m	264.8	262.3	253.6	249.5	200.3	145.2	149.2	146.7
Total M.E. ft-lb_f/lb_m	272.3	269.8	267.6	263.5	214.3	189.2	186.7	184.2
M.E. losses, ft-lb_f/lb_m	0.000	2.530	2.240	4.047	49.200	25.10	2.500	2.530
Cumulative M.E. losses, ft-lb_f/lb_m	0.000	2.530	4.770	8.817	58.017	83.117	85.617	88.147
Total energy, ft-lb_f/lb_m	272.3	272.3	272.3	272.3	272.3	272.3	272.3	272.3
P from Bernoulli¹ psia	114.7	113.6	109.8	108.1	86.7	62.9	64.6	63.5

Conditions at section 1

Density, lb_m/ft^3	62.365
Pressure, psia	114.7
Temperature, F	60.0
Volumetric flow, gpm	900.0
Mass flow, lb_m/h	45015

Example II-6: Restriction orifice

Figure II-9 represents an extension to the system of Figure II-8. It is a hypothetical, but fairly frequent, case. Suppose water is to be transferred from tank, T1, to reactor, R1, by pressure difference. The maximum flow rate must be limited to 900 gpm (U.S.). The minimum flow rate is not important. We will also ignore the need for check valves for the purpose of the exercise.

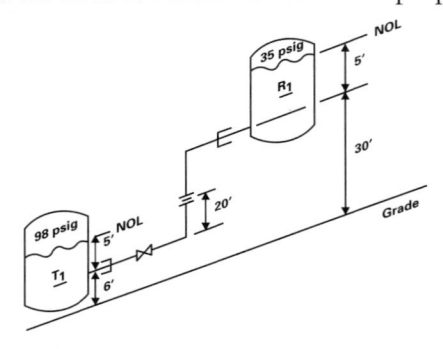

Figure II-9. Extension to system irreversibility example

Incompressible Fluid Flow – Losses of Mechanical Energy | CHAPTER TWO

The maximum pressure in T1 is 98 psig. The minimum pressure in R1 is 35 psig. These pressures have been observed on (calibrated) pressure gauges on the two vessels. They correspond to the operating intent. The specific gravity of the reactor fluid is very close to 1.0, so the density of the fluid in the reactor will be assumed equal to that of the water, 62.365 lb_m/ft^3.

The problem is the operator has no means of knowing what the actual flow rate is. He has the habit of opening the globe valve wide open. This is detrimental to the reaction. How can we be sure to limit the maximum flow rate to 900 gpm with a limited budget?

A restriction orifice could be of use if there were a convenient location for it. It would not be wise to place the orifice directly at the valve or too close to an upstream fitting. The disturbance to the flow profile caused by the valve or fitting would cause an uncertainty in the estimation. Suppose a flange is located in the vertical four-inch pipe about elevation 26 feet. If a restriction orifice were placed here, the globe valve could be fully opened as part of the operating procedure and the flow would never exceed 900 gpm. How do we go about sizing the restriction orifice?

Plan of attack

- Use the Bernoulli[i] equation between one unit mass of fluid at the surface of the water in tank, T1, and one unit mass of fluid at the surface of the reaction mixture in reactor, R1.
- Compute the system irreversibilities *required* to limit the flow rate to 900 gpm.
- Compute the system irreversibilities *available* without the restriction orifice.
- Subtract the irreversibilities available from the irreversibilities required. This gives the irreversibilities the restriction orifice must create.
- Use the orifice loss formula to compute beta iteratively.
- Multiply beta by the pipe internal diameter to obtain the required diameter of the orifice.

CHAPTER TWO | Incompressible Fluid Flow – Losses of Mechanical Energy

Required irreversibilities to limit the maximum flow to 900 gpm

$$\frac{P_1}{\rho_1} + \frac{\alpha_1 U_1^2}{2g_c} + \frac{gZ_1}{g_c} + w_n = \frac{P_2}{\rho_2} + \frac{\alpha_2 U_2^2}{2g_c} + \frac{gZ_2}{g_c} + h_L \quad \text{(II–36)}$$

$$\alpha_1 = \alpha_2 \sim 1,\ w_n = 0,\ \rho_1 = \rho_2$$

$$\frac{P_1}{\rho} + \frac{U_1^2}{2g_c} + \frac{gZ_1}{g_c} = \frac{P_2}{\rho} + \frac{U_2^2}{2g_c} + \frac{gZ_2}{g_c} + h_L$$

$$\frac{144(98+14.7)}{62.365} + 0 + \frac{g(11)}{g_c} = \frac{144(35+14.7)}{62.365} + 0 + \frac{g(35)}{g_c} + h_L$$

$$h_L = 121.5$$

Customary U.S. units have been used. The velocities at the surfaces of the liquids are zero. The factor, g/g_c, equals one numerically, but is left in as a reminder. The units of h_L are ft·lb$_f$/lb$_m$ in this case.

Computation of available irreversibilities

IRREVERSIBILITIES FROM A PIPE ENTRANCE FROM TANK T1

$$K = 0.5 \quad \text{(II–37)}$$

$$h_f = K\frac{U_2^2}{2g_c} = 0.5\frac{9.99^2}{2g_c} = 0.7756$$

The value of K is that value for a square edged entrance. See Appendix AII for details.

IRREVERSIBILITIES FROM THE PIPE EXIT TO REACTOR, R1

$$K = 1.0 \quad \text{(II–38)}$$

$$h_f = 1.0\frac{9.99^2}{2g_c} = 1.551$$

The value of K is that value for a square edged exit. See Appendix AII.

AVAILABLE SYSTEM IRREVERSIBILITIES

$$0.7756 + 1.551 + 88.15 = 90.47 \quad \text{(II–39)}$$

The last number on the right is from the previous computations. It is the cumulative irreversibilities at section 2 of the previous system. See Table II-3.

IRREVERSIBILITIES REQUIRED OF THE RESTRICTION ORIFICE

$$121.5 - 90.47 = 31.03 \qquad \textbf{(II–40)}$$

These mechanical energy losses have units of ft-lb$_f$/lb$_m$. The irreversibilities to be created by the restriction orifice must equal the total required irreversibilities less the total available irreversibilities.

Restriction orifice sizing

The last number computed, 31.03 ft-lb$_f$/lb$_m$, can be used in the restriction orifice K factor equation and a value for beta may be found by iteration, as follows:

$$h_o = K_o \frac{U_D^2}{2g_c} \qquad \textbf{(II–41)}$$

$$K_o = 2.8(1-\beta^2)\left[\left(\frac{1}{\beta}\right)^4 - 1\right]$$

$$31.03 = 2.8(1-\beta^2)\left[\left(\frac{1}{\beta}\right)^4 - 1\right]\frac{9.99^2}{2g_c}$$

$$7.145 = (1-\beta^2)\left[\left(\frac{1}{\beta}\right)^4 - 1\right]$$

Iteration	Trial Beta	Computed Loss Value	
1	0.5	11.25	TOO HIGH
2	0.6	4.30	TOO LOW
3	0.559	6.35	TOO LOW
4	0.5467	7.148	CLOSE ENOUGH

To obtain the bore dimension, the internal diameter, 4.025 inches, is multiplied by the d/D (beta) ratio, 0.5467, to obtain 2.200 inches (to the nearest thousandth of an inch).

If an orifice is bored to this dimension and to this tolerance and it is placed in the flange, the flow will be limited to 900 gpm under the given specifications.

CHAPTER TWO | Incompressible Fluid Flow – Losses of Mechanical Energy

II-6: CHAPTER SUMMARY

This chapter has taken the theoretical concepts of Chapter I and Appendix AI and has given them practical form. In particular, the problems arising from the use of loose terminology have been addressed and the author has had the opportunity to edit some of his own loose terminology. The reader should now have obtained some insight into the conversion of mechanical energy into less useful internal energy. He should be able to tackle with some confidence the problems associated with incompressible flow through pipes when fluid machinery (pumps and turbines) is not considered.

The chapter has:
- analyzed the Bernoulli[i] equation in some detail. It has discussed the importance of its terms one by one. The importance of the last term dealing with permanent losses was discussed in detail;
- described the main causes of irreversibilities in piping systems;
- discussed geometric similarity;
- compared the Darcy[vii] equation with the generalized K factor equation;
- given a general equation for permanent losses across square edged, concentric orifice plates;
- discussed the use of grade lines: total energy, mechanical energy and hydraulic;
- given examples of loss computations for the more common cases;
- presented a plan of attack to be applied to systems.

Appendix AII will give additional related information. In particular, it will cover:
- the relationship between energy/mass units and head units;
- the Churchill-Usagi[xli] friction factor equation;
- manifold flows (combining and dividing flows);
- permanent losses due to various instruments.

CHAPTER THREE III

Pumps: Theory and Equations

III-1: SCOPE OF CHAPTER — PUMPS AND THEIR PERFORMANCE CAPABILITIES

Chapter III's intent is to provide enough information about pumps to permit the reader to understand their performance capabilities and their interactions with the other components of a complete hydraulic system. Very little mathematical description is given in this chapter. A few worked examples of head computations may be found in Section III-9. Mathematical detail can be found in Appendix AIII. Mechanical design considerations, such as flange and casing ratings, will not be considered unless the practical utility of the information to the reader might not seem obvious.

There are many excellent texts that deal with pump and turbine engineering, and there are many texts, such as *Chemical Engineers' Handbook*, and *Unit Operations*, intended to help chemical engineers specify fluid machinery. This chapter and its associated appendix are intended for those seeking an understanding of pumps because they must work with systems that include such fluid machinery within a larger system. The latter personnel normally do not specify fluid machinery. They are, however, faced with the task of dealing with the characteristics of such machinery and controlling flow through it, or measuring that flow. The related task of machinery monitoring and control is beyond the scope of this book. It will only be mentioned superficially, where it seems appropriate to the author.

Since different types of pumps have different installed characteristics and different control requirements, a general classification will be given. The classification will be used as the framework for further discussion.

More time will be spent on centrifugal pumps because that type of pump is most common in industrial practice. Their connection in parallel and series configurations will be considered. Turbines will be discussed incidentally to form a more complete picture of fluid machinery.

CHAPTER THREE | **Pumps — Theory and Equations**

III-2: FUNCTIONS OF PUMPS

Pumps move liquids through a system of conduits and equipment by increasing the total mechanical energy of the liquid at the pump location. Sometimes the intent is to move a liquid from one place to another – transportation of the liquid. Sometimes it is simply to increase the pressure so the higher pressure can be made use of in a hydraulic machine. Sometimes the intent is to produce kinetic energy in a stream of liquid – water issuing from a fire hose, for example.

Liquid moves through conduits when the total mechanical energy upstream is greater than the total mechanical energy downstream. Note thermal energy does not contribute to bulk motion of a fluid, only to molecular motion. In other words, liquid moves through a conduit from A to B when the sum total of static, kinetic and potential energies is greater at A than at B. If the total mechanical energies are equal, there is no bulk movement. If the total mechanical energy at B is greater than at A, the flow will be in the opposite direction, from B to A (if there are no check valves). In a flowing system without deliberate thermal input or extraction, the difference between the total mechanical energies at A and B is the amount of mechanical energy converted to internal energy, and ultimately to heat energy flow, by viscous drag, turbulence and shock. This difference is what is termed "irreversibility" in this book.

To get liquid to move through a conduit (if we discount buoyancy due to temperature differences), mechanical energy must be transferred to the liquid. There must also be some physical control over the direction of liquid flow. Increasing the energy level at a point does no good if the flow is allowed to go in all directions.

Means of energy transfer to liquids

The common means of transferring mechanical energy to a liquid, discounting buoyancy, and of giving the liquid a direction in which to flow are by the use of:
- gravity;
- centrifugal force;
- volumetric displacement;
- momentum transfer;
- mechanical impulse;
- electromagnetic force.

Pumps — Theory and Equations | CHAPTER THREE

Energy transfer by gravity

Gravity is the natural means of moving liquid from a high point to a low point – in conduit or in open channels. A head tank is a common means of maintaining a constant pressure when flow control is critical. Head tanks and water towers are also used to maintain reserve capacity for peak periods and to permit smaller, less expensive pumps to be used for the service. The direction of flow is governed by the direction of the applied force, gravity, and the restraining walls of the conduit. The pressure at a point is a function of the elevation difference, the hydrostatic head, and the fluid density.

The characteristics of transfer by gravity are:
- Gravity is free, unless the liquid has to be pumped to a head tank before gravity can be used;
- in closed conduits, flow control, including shutoff, is easy;
- the pressure is limited by the available hydrostatic head. If the conduit size is fixed, a booster pump may be necessary to obtain a desired pressure or flow rate.

Energy transfer by centrifugal force

Centrifugal force is a quite sophisticated way of transferring mechanical energy to liquids (and gases). By performing work on a shaft (by muscle power, water power, wind power or electrical power), the shaft can be made to turn. A suitable impeller causes liquid to be forced to travel radially while it gathers velocity. The mechanical energy input to the shaft has been transformed to kinetic energy. Some of the mechanical energy will be transformed to thermal (internal) energy and will be lost to practical use. This transformation is a mechanical inefficiency.

If sufficient suction pressure is maintained, the liquid flow will be continuous. Liquid will move into the eye of the impeller at the same rate it is propelled to the exterior. Once the liquid reaches the tip of the impeller, its kinetic energy must be transformed into static (pressure) energy in such a fashion as to minimize conversion of mechanical energy to internal energy. This is usually done by carefully guiding the liquid with diffusers or vanes. The purpose of the guidance is to smoothly slow down the stream and to allow the kinetic energy to be converted to pressure energy.

The physical limitations imposed upon the flow rates of centrifugal pumps are related to maximum tip speed of the impeller. A given

CHAPTER THREE | **Pumps — Theory and Equations**

impeller design is limited to a maximum tip speed (to avoid cavitation). A single stage is, therefore, limited to a maximum differential pressure. Typical designs cover the range from 70 to 100 ft-lb_f/lb_m (30 to 43 psi or 207 to 296 kPa with water). By carefully guiding the liquid leaving one stage, and by carefully increasing the channel diameter, the liquid can be made to enter another stage at its original velocity but at a higher pressure and a higher total energy. Additional mechanical energy can be added in a second, a third or even a seventh stage. It should be obvious the overall mechanical efficiency diminishes with an increased number of stages as some of the impeller shaft energy is converted to internal energy at each stage.

Characteristics of centrifugal pumps

Centrifugal pumps have the following general characteristics:
- they are relatively pulsation free;
- their mechanical design allows higher capacity at (generally) lower heads than positive displacement pumps;
- they have fairly efficient performance over a wide range of pressures and capacities (at least, single stage pumps do);
- control can be by discharge throttling when consideration is given to overheating and surging at very low flow rates;
- discharge pressure is a function of density at fixed angular velocities (rotational speeds);
- they are capable of high rotational speeds (up to the speed, which produces cavitation);
- they can be made to handle slurries and corrosive or erosive liquids by suitable coating or alloys;
- they can be relatively inexpensive.

Energy transfer by volumetric displacement

Volumetric displacement was the first method invented for pumping liquid. The flow direction was controlled by the use of check valves or by automatically operated valves. Volume can be displaced by mechanical means, as by a piston, or by another fluid directly or indirectly (behind a diaphragm), or by the same fluid directly, but in a different phase.

Some condensate pumps use direct steam as the motive force. Acid eggs use air. In all of these devices, energy transfer is by the movement of a force through a distance. It is easily recognizable as work.

Pumps — Theory and Equations | CHAPTER THREE

With rotary pumps, a volume of liquid is trapped by a rotating member (rotating vane or gear pumps). The design produces relative smooth, continuous flow.

Some of the general characteristics of pumps using the volumetric displacement principle are:
- flow is often pulsating, requiring some external damping such as by a pulsation-damping bottle (in both the suction and the discharge piping);
- flow cannot be throttled at the discharge without stalling the driver, bursting the equipment, relieving a pressure safety valve, or actuating a diversion valve;
- flow can be varied by varying the speed of the driver or by diverting flow to the source or elsewhere;
- volumetric pumps are often associated with low capacity, high-pressure service;
- positive displacement pumps can be designed for extremely low flow rates.

In spite of the generalities just mentioned, positive displacement pumps are used for "frac" jobs. These pumps pump sand slurries and emulsions at high pressures (5,000 psig) into oil bearing formations for purposes of oil well stimulation.

Energy transfer by momentum transfer

Jets and eductors are examples of momentum transfer devices. The deceleration of a motive fluid causes acceleration of a pumped fluid.

The characteristics of jets are:
- they can pump liquids and gases;
- they are usually very inexpensive to purchase;
- they are usually expensive to operate as they are relatively inefficient;
- being small, they can be made of exotic materials without incurring too much first cost;
- the pumped stream can contain solids.

That being said, many evaporators or crystallizers could not be without them. They do have characteristics that cannot be reproduced readily by centrifugal pumps – the ability to produce vacuum, low pressure opera-

CHAPTER THREE | **Pumps — Theory and Equations**

tion and the ease with which they handle solids and corrosive or erosive fluids.

Energy transfer by mechanical impulse

Regenerative pumps (also called turbine pumps, but they are different from a vertical turbine pump) use the principle of imparting a mechanical impulse in an axial as opposed to a centrifugal direction.

Energy transfer by electromagnetic force

If the fluid is a conductor, such as a molten metal, it can be made to follow a moving electromagnetic field. This technique is highly specialized. It is used in nuclear reactors to pump heat transfer fluids.

III-3: A BRIEF HISTORY OF PUMPS

Man has always had a need to raise liquids (water) through heights. If we consider a bucket or a skin bag dropped into a well on the end of a rope as a pump, then it is difficult to give credit to the inventor of the first pump.

The ancient Greeks had a fairly well developed hydraulic technology. They used water power to open temple doors, for instance. Hero[xx], in the second century BC, developed a fire pump with two cylinders and a hand-operated rocking beam to supply motive force. The early pumps were all hand-operated, obviously, but it did not take long for people to realize donkeys, camels, wind and water could supply the necessary input power.

One ingenious example of early technology is the use of a hollow rod (bamboo?) into which was threaded a rope. The rope was connected at its ends to make a continuous length. Rags were tied to the rope at convenient lengths. The hollow rod was placed vertically in a well and fixed in place. The rope was then pulled through it. Each section of the rope between the rags would trap a quantity of water that could be made to flow at the well head in a more-or-less continuous manner. This was a positive displacement pump.

The Greeks used piston and cylinder arrangements, made of wood, to force water from wells. The motivation for the development of the early pumps was the need to supply water for drinking, irrigation and laundry.

The Romans took over and developed Greek technology. They developed bronze pumps. All of the early pumps were positive displacement pumps.

Pumps — Theory and Equations | CHAPTER THREE

In the 16th century the German Agricola[xxi] described the extensive use of pumps in the mining industry for mine dewatering purposes. This was probably the first large-scale industrial use of pumps. He also described the coupling of multiple pumps to a water wheel. This represented a jump in technology from the use of muscle power (human or animal) to the use of an external force (gravity).

In the Middle Ages, bilge pumps were used aboard ships. The cylinders were often square in section — a fact we tend to forget. Another fact we tend to forget is pumps were frequently used in a suction mode. That is, suitable check valves allowed water to enter the cylinder on the outward stroke, but not on the inward stroke. Water was allowed to leave on the inward stroke either via another check valve and discharge port or simply by leaking past the cylinder packing. In this case, the packing was devised to fold in one direction (chevron packing). The physical limitation on suction lift, about 33 feet or 10 meters for water was soon discovered — but it was not explained too well.

With the industrial revolution, the increase in industrial activity created a demand for pumps to move water from ever deeper coal mines and to supply motive power to hydraulic rams. One of the first uses of the steam engine was as the motive power to dewatering pumps. The invention of the centrifugal pump is generally credited to Denis Papin[xxii], a French inventor, in 1689. Its use had spread around the world by the mid 1800's. Today, the centrifugal pump is an omnipresent part of our existence.

III-4: CLASSIFICATION OF PUMPS

Although positive displacement pumps were the first to be used historically, in present day use, centrifugal pumps are more prevalent. Therefore these pumps are listed first.

A lot of engineering thought has gone into developing reliable pumps for different services. The purpose of classification is to group common characteristics so these characteristics can be discussed with some economy of effort. The following grouping is to allow people interested in fluid systems involving pumps to seize the essentials — it is not meant to be all-inclusive.

CHAPTER THREE | **Pumps — Theory and Equations**

1. Centrifugal pumps

Single stage pumps

a. Radial pumps. (See Figure III-1). These are true centrifugal pumps. Subcategories are volute, diffuser, and turbine pumps.

b. Axial pumps. (See Figure III-2). These pumps have a lifting or propeller action. This is not really a centrifugal pump in the strictest sense of the word "centrifugal". The liquid does not flee the center.

c. Mixed flow pumps. These pumps combine both radial and axial actions. These pumps are usually vertical and are also called "turbine" pumps but should not be confused with the pump that combines mechanical impulse with centrifugal force and that is called a "regenerative" pump.

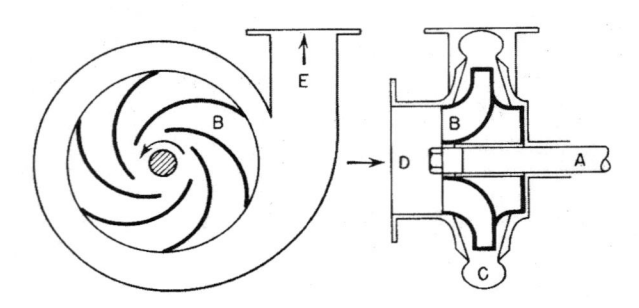

Figure III-1. Simple centrifugal pump

Source: R. Perry and D. Green, Perry's Chemical Engineers' Handbook, 1984, reprinted by permission McGraw-Hill Companies

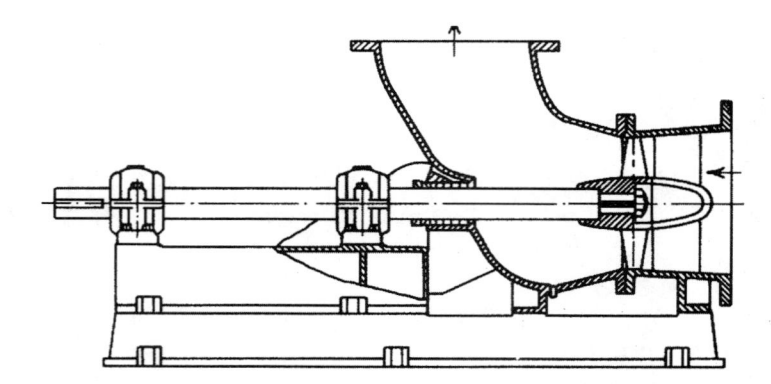

Figure III-2. Axial-flow elbow-type propeller pump *(Courtesy of Lawrenece Pumps, Inc.)*

Source: R. Perry and D. Green, Perry's Chemical Engineers' Handbook, 1984, reprinted by permission McGraw-Hill Companies

Multistage pumps (See Figure III-3)
These pumps overcome the tip velocity limitations of a single stage by combining more than one stage. They can have seven or more stages on the same shaft.

Pumps — Theory and Equations | CHAPTER THREE

2. Positive displacement pumps

a. Piston and cylinder pumps. These pumps can be direct acting, steam-driven pumps. They are classified further as simplex, duplex, single or double acting.

b. Rotating member pumps. This is not really a single category. It includes cam and piston, gear, lobular, screw, vane and peristaltic pumps.

c. Diaphragm, metering pumps. Again, this is not a single category of pump.

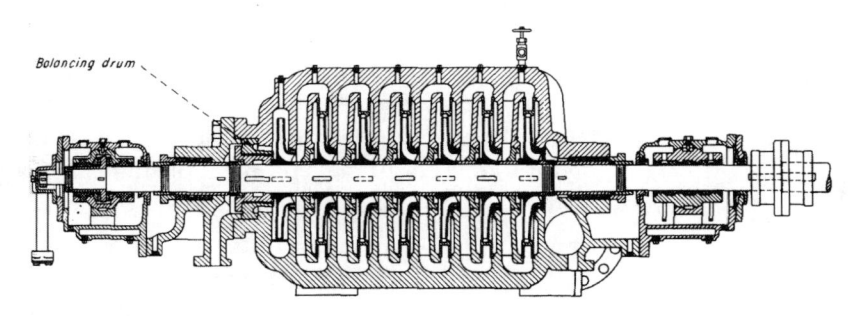

Figure III-3. Seven-stage diffuser-type pump

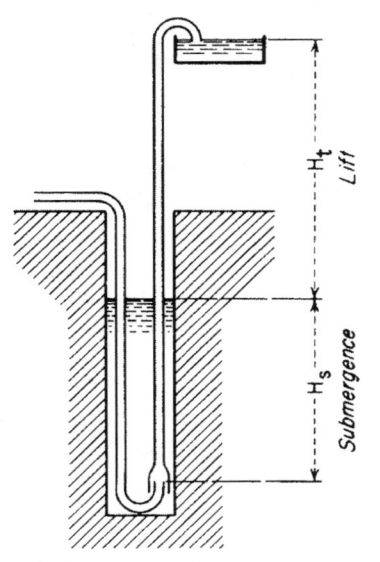

Figure III-4. Simplified sketch of an air lift, showing submergence and total head

Flow of Industrial Fluids — Theory and Equations   **107**

CHAPTER THREE | **Pumps — Theory and Equations**

3. Miscellaneous pumping devices

a. Gas, or air, lifts. (See Figure III-4)

b. Jet pumps. (See Figure III-5) Jet pumps are sometimes called ejectors. An ejector is a jet device that uses a motive fluid at high pressure to entrain another fluid at low pressure. It discharges the mixed fluids at an intermediate pressure. Ejectors can use liquids or gases as motive fluids and can pump (compress and move) liquids or gases. Figure III-5 from Schutte and Koerting[xxiii] demonstrates the large variety and the many names given these pumps.

c. Liquid metal pumps. These pumps make use of the motor principle to move molten metal heat exchange media through piping. If a magnetic field and an electric current are mutually perpendicular and each is perpendicular to the axis of the pipe, the molten metal will be subject to an axial force that will propel it in the direction of the force and the pipe.

The grouping is rather arbitrary as can be seen by the inclusion of axial pumps with radial pumps in the "centrifugal" category.

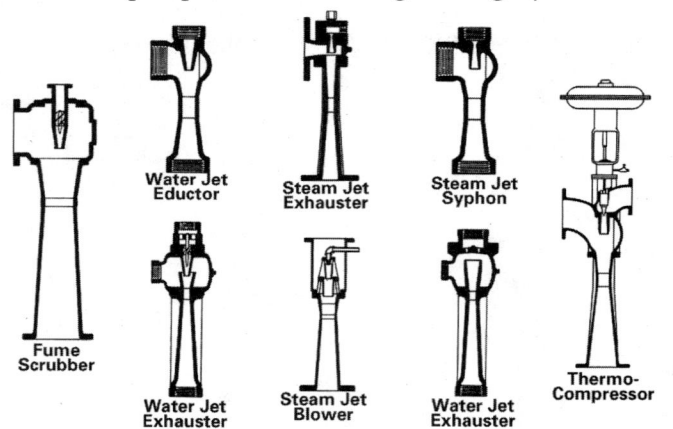

Figure III-5. Ejectors

Notes: Eight ejectors with different designs and different names
The fume scrubber is a water jet air pump for low differential pressures and large capacities.
The water jet eductor is a liquid jet pump for liquids.
The water jet exhauster is a liquid jet pump for gases.
The steam jet siphon is a steam jet pump for liquids.
The steam jet exhauster (also called a vacuum pump) is a gas jet gas pump (two shown).
The steam jet blower is a steam jet pump for air for low differential pressures and large capacities.
The thermocompressor is a steam jet steam compressor.

III-5: CHARACTERISTICS OF PUMPS

In this section we will generally describe the pumps classified above and will briefly describe their characteristics. Again, the emphasis will be on characteristics from the point of view of understanding their interaction with the flowing system. Pump selection is the job of the specialist. Nevertheless, a more technical description will be given in Appendix AIII.

By far, the most common pump application is one involving some type of single stage centrifugal pump. This fact is due to their relatively simple construction, low first cost and ease of maintenance. They can handle a wide variety of corrosive and erosive materials (by proper choice of coatings or alloys). They are offered in many different sizes and operating speeds in off-the-shelf designs. Generally, a single stage centrifugal pump is the first choice unless some process requirement demands otherwise.

The engineer or technician interested in a measurement and control problem is usually faced with an already designed system. His or her problem is not to select a pump for the application. Rather, it is to make it work efficiently within the system. He or she must still understand the basic principles behind the selection and must have an overall grasp of the operating characteristics of various pumps.

1. General characteristics of centrifugal pumps

Centrifugal pumps are usually specified on the basis of volumetric flow rate in cubic meters per hour or U.S. gallons per minute. The differential pressure across the pump is usually given in kilopascals or pounds per square inch. Pressure is usually converted to "head" in meters or feet of the fluid to be pumped. This "head" is the number of meters or feet of fluid equal to the measured differential pressure plus the differential kinetic energy.

A given impeller rotating at a given speed in a given casing will produce a given head, independently of the fluid involved. This will be shown mathematically in Appendix AIII. The actual differential pressure has to be computed from the head and the known density. This permits the issuance of general performance curves without specifying fluids.

The use of the concept of head has already been discussed in Chapter I, Section 10. It is much easier to keep energy relations straight if we think in terms of energy per unit mass and not "head". This applies in partic-

CHAPTER THREE | **Pumps — Theory and Equations**

ular when the term "head" is not clearly defined. However, the use of the term is so entrenched in engineering documents, such as pump calculation sheets, that it must be clearly understood. It is worth taking the time to establish which head is being discussed. The term should be defined clearly on any document employing it. We will give examples of the use of the head concept in Appendix AIII.

Flow control is frequently by throttling the discharge of a centrifugal pump (not the suction, because of the risk of lowering the suction pressure below the NPSH requirements). Many centrifugal pumps are capable of wide turndowns, but it is wise not to assume that all of them are. Consult the pump expert when in doubt.

Priming centrifugal pumps operating in a suction mode

Liquid pumps are designed to operate on liquids, not gases. To fix ideas, let us refer back to the statement that the total dynamic head of a single stage is independent of the fluid being pumped. The average maximum head of a single stage is about 85 ft-lb_f/lb_m in customary U.S. units. Multiplying the head by the density will give the pressure in pounds-force per square foot. Dividing by 144 will give the pressure in pounds-force per square inch. The differential pressure equivalent to the total dynamic head of 60°F water will be 85 x 62.34/144 = 36.8 pounds per square inch. This will not be exactly the pressure read on a differential gauge across the same pump, although it will not be far from it. The difference between the two will be due to kinetic energy and elevation differences.

As long as water is available to its suction, such a pump can more than adequately cause a negative pressure so as to ensure it remains flooded at all times. In other words, if the pump discharge is at atmospheric pressure, the pump can cause a negative pressure at the suction that will ensure water flow from a receiver that is also at atmospheric pressure, provided the pump is initially primed.

If the same pump is started with 60°F air in its casing and in the suction, the differential pressure equivalent to the same 85 ft-lb_f/lb_m would be 85 x 0.0764/144 = 0.045 pounds per square inch or 1.2 inches of water column or about 32 mm of water column. By lowering the pressure in the suction piping, the pump might be able to raise water to this height, but no more.

The term "prime" refers to ensuring the presence of liquid within the pump by means of pressure or vacuum. Pumps located above their source can be primed by using a source of vacuum (an eductor, for example) on the pump casing in order to raise (usually, non-hazardous or non-corrosive liquid) from the normal source. If foot valves are present, pumps can be primed from another source while being vented.

Characteristic curves

The characteristic curve of a centrifugal pump is a useful and enlightening tool – Figure III-6 is an example. The figure has capacity in gallons per minute or it may have some other volumetric rate as the abscissa and a variety of variables on the ordinate. Units used on the ordinate can be head, efficiency, brake horsepower or shaft power and net positive suction head required. All of these terms will be defined more carefully shortly.

The principal curves of interest for the present are the head-capacity curve and the efficiency capacity curve. The head curve shown is relative smooth and it decreases continuously from shutoff to the maximum flow rate. The particular efficiency curve shown starts at zero, curves up toward 70% efficiency and then dips back down rather sharply. The head curve represents the energy per unit mass the pump is capable of transferring to the fluid at a given volumetric flow rate.

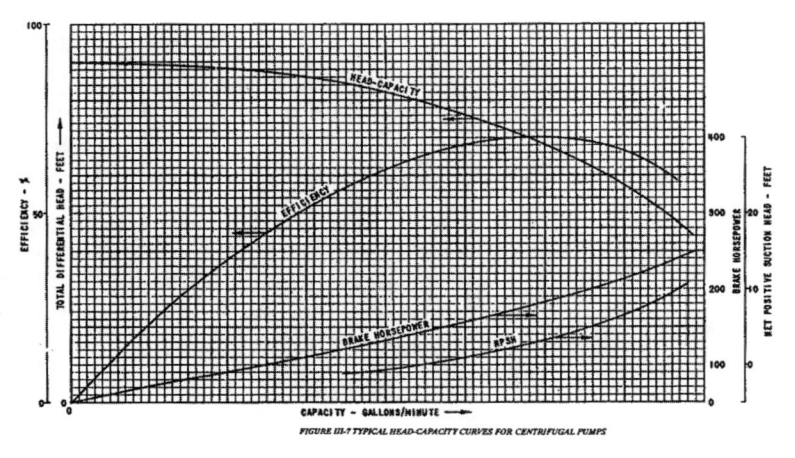

Figure III-6. Typical head-capacity curves for centrifugal pumps

The shape of the efficiency curve explains why, even though a centrifugal pump is capable of wide turndown, it is not necessarily economical to operate it for long periods of time away from the optimum point. Some pumps, notably regenerative turbine pumps, are designed for low flows with high heads per stage. The efficiency curve of regenerative pumps is very peaked, so the useful operating range is even more restricted.

CHAPTER THREE | **Pumps — Theory and Equations**

A plot of average design efficiencies of commercially available single-stage pumps versus their nominal capacities will show low volumetric flow pumps are generally less efficient than the larger pumps. The smaller pumps have efficiencies between 10% and 37% at 20 gpm. The efficiencies increase with capacity to lie between 78% and 86% for larger pumps at 10,000 gpm. The efficiency curve can be used to compute the extra energy per unit mass that must be transferred to the shaft by the driver over and above that transferred to the fluid. Dividing the head at a given flow rate by the fractional efficiency at the same flow rate gives the energy the driver must supply to the pump. The difference between the two energy numbers represents the irreversibilities (within the pump only) – the amount of mechanical energy converted to internal energy and then to heat energy flow.

Another factor to consider is that, frequently, the head curve also has a hump near the shutoff point. This characteristic is frequently associated with a steep head curve and a high-speed machine. The characteristic is similar to that of a centrifugal compressor. There are two equilibrium points, two flow rates, for the same discharge pressure. Surging can result at low turndowns as the pump "hunts" between the two points.

A third factor is that, even though a centrifugal pump might accept shutoff conditions without surging, the low efficiency close to shutoff means probable overheating of the fluid. When overheating or surging may be a problem, minimum flow control systems that recirculate flow to suction, sometimes via a cooler, are used.

The brake horsepower curve is simply the power, expressed in equivalent horsepower, that will have to be applied to the shaft to maintain a given flow rate and head. The horsepower requirement increases with increasing flow rate (and diminishing head). A portion of the shaft power will be transmitted to the fluid as fluid power. The rest will heat up the fluid, any cooling or flushing liquid and, ultimately, the environment. The power transmitted to the fluid is equal to the shaft power ("brake" power) times the pump efficiency. The pump has its own efficiency rating which is separate and distinct from that of the driver. The two efficiencies (driver and pump) must be multiplied to obtain the overall efficiency of the pump and driver.

The net positive suction head (required), NPSH(R), curve is of interest mainly to the person who specifies the pump, but it is also important when troubleshooting. In fact, if the pump is not operated within the limitations of the minimum NPSH required, the pump may self-destruct. The net positive suction head required is established by the pump vendor. Basically it is the minimum head in excess of the vapor pressure that must be available at the suction flange to prevent the pump from cavitating. It is to be noted this required head increases with increasing flow rate. NPSH(A) is a term used for a computed head available to the pump. This head is dependent on the piping configuration and layout. It must be greater than NPSH(R) over the total turndown flow range of the pump. More details will be given later in this chapter and in Appendix AIII.

Measures of performance of real centrifugal pumps

It is common to describe the performance of a centrifugal pump in terms of its "capacity" and its "head". Both of these terms are rather loosely used and they should be defined each time they are used. The conversions between head and pressure in both SI and customary U.S. systems were introduced in Chapter I, Section I-10.

Capacity can mean mass flow per unit time or it can mean volumetric flow per unit time, both in various units. Head can mean differential pressure, differential energy per unit mass, discharge pressure or the height in feet or meters a column of liquid would be raised if it were subject to an equivalent pressure at the base (but which pressure?). We also talk of velocity head, static head, total head, and so on. Given the potential for misinterpretation, it behoves all of us to pay particular attention to definitions.

Capacity is easily dealt with. It is flow per unit time. It can be mass flow or volumetric flow. The mass or volumetric units and the time units should be clearly stated so they can be translated to appropriate units by the person using the information. Difficulties within the English speaking world come when the volume units are gallons and it is not specified whether U.S. gallons or imperial gallons are being discussed. Imperial gallons are only about 20% larger than U.S. gallons!

Head is not so easily dealt with given the general looseness with which the term is used. We shall begin by given the definitions of various pump "heads". The reader is advised to pay attention and to learn to discriminate amongst them. In addition, words like "total" and "absolute" and "perfect" are used rather loosely in engineering. They can be misleading.

CHAPTER THREE | **Pumps — Theory and Equations**

Fundamental definition of head

Head ultimately means energy per unit mass at a section of conduit. The fundamental units are foot-pounds force per pound mass or newton-meters per kilogram (joules per kilogram). Because it is not easy to measure energy directly, units such as feet of fluid and meters of fluid have sprung up as a substitute measure. However, ultimately in the Bernoulli[1] equation and in other energy balances all of these *convenience units* must be translated into energy per unit mass.

It is wise to fix the concept of head as being one or more of the different forms of mechanical energy per unit mass of a fluid available at a section of conduit. Computed at a cross section of conduit, it will consist of three terms: static (pressure) energy, kinetic (velocity) energy and potential (by virtue of elevation) energy. Their sum is usually given the name "total" head. Each of the individual terms is also called a head. In addition, each of the terms used to compute the fundamental quantities in the head equation is also called "head": static discharge, static suction, etc. It is worthwhile spending a little time fixing the concepts.

Total discharge head

The total discharge head, h_d, is the gauge reading at the discharge flange *"corrected to the pump center line"* (or to the eye of the inlet impeller in vertical pumps), plus the barometer reading, plus the velocity head. All of the individual terms must be summed, so they must have identical units. In hydraulics work, the units are usually feet or meters of fluid. Note the gauge reading could be negative and the sign must be obeyed. We will use energy per unit mass units for computations and convert to feet or meters where necessary.

Explanation of the term 'corrected to the pump center line'

The term, *"corrected to the pump center line"*, can be misleading, but it is commonly used, so an explanation is warranted. When sizing pumps, and frequently when analyzing their performance, the pump center line is chosen for the datum plane. The elevations of interest for potential and static energy measurement are the center lines of the inlet and discharge piping of the pump. Pressure gauges are frequently located at a different level than the center lines. Therefore, if we apply the pressure gauge reading in the appropriate units directly into the Bernoulli equation, we will be in error.

Pumps — Theory and Equations | CHAPTER THREE

The correction is quite simple. If the gauge were on the center line of the pipe, the reading plus the atmospheric pressure would be the correct one to use. If the gauge were higher, the pressure reading would be less and, if it were lower, the reading would be greater. The error would be directly proportional to the distance between the gauge center line and the pipe center line.

Since, under steady state conditions, the total discharge head and the total suction head are fixed, we can correct for a different location of the pressure gauge by changing the potential head (the elevation) by the negative of the error in the static head (pressure). This is done automatically if the elevation, Z, we use in the Bernoulli[i] equation is taken, not at the pipe center line, but at the gauge center line. Doing this automatically corrects the total head to the pipe center line. What was lost in pressure energy is gained in potential energy. This statement applies only to an isolated pump. When performing computations on systems, the appropriate elevation must be used.

Since the velocity term is fixed, if we were to raise or lower the gauge to the center line of the pump, we would have the same total energy as long as we took the elevation term at the gauge center line. So, it makes no difference where the gauge is as long as the elevation term is changed to match the location of the gauge. The term, "corrected to the pump center line", arises out of this correspondence. It would be better to state that the reading is corrected to the pipe center line. The three basic energy terms are still present, but one of them is hidden as a "correction to the pump center line".

Total suction head

The total suction head, h_s, is the gauge reading at the suction flange corrected to the pump center line (or to the eye of the inlet impeller in vertical pumps) plus the barometer reading plus the velocity head. All of the individual terms must be summed, so they must have identical units. In hydraulics work, the units are usually feet or meters of fluid. Note the gauge reading could be, and frequently is, negative and the sign must be obeyed.

Total dynamic head

The total dynamic head is the difference between the total discharge head and the total suction head, $H = h_d - h_s$. This is a potential source of error. We have defined the discharge and suction head terms as the sums of the absolute pressure heads, the velocity heads and the potential heads. We have explained in Chapter I, Section I-10 the conversion between

CHAPTER THREE | **Pumps — Theory and Equations**

pressure and head. If the reader is not a hydraulician and if he or she does hydraulic computations infrequently, it is best to use energy per unit mass units — they are easier to handle logically.

The total dynamic head is important in computing the energy requirements of a pump. It represents the total mechanical energy transmitted to the fluid (not the pump) per unit mass of flowing fluid. Mechanical energy converted to thermal energy is not directly considered. It appears in the efficiency computation.

Static discharge head

The static discharge head, h_{sd}, is the vertical distance to the free surface of the liquid in the discharge receiver *or in the discharge piping system* plus the absolute pressure on this liquid surface (gauge plus atmospheric) in consistent units. This head, h_{sd}, is also corrected to the pump center line.

We have emphasised the fact that the static discharge head could be measured within the piping because this is exactly what occurs when the pump is started. It is also what occurs when the pipe discharges to the top of the vapor space of a receiver. The static discharge head varies from a minimum to a maximum and then, usually, drops to a lower level. This phenomenon occurs because the piping usually runs overhead to the receiver. Once the pipe is filled, the siphon effect reduces the head requirements of the pump. If a pump is sized based only on the receiver elevation, without considering the maximum piping elevation, it may not be able to get liquid to the receiver. Such a pump could operate if the pipe were filled from a different source. It is always wise to check the actual routing of piping and its maximum elevation.

Static suction head

The static suction head, h_{ss}, is the vertical distance to the free surface of the liquid source plus the absolute (gauge plus atmospheric) pressure corrected to the pump center line. All terms in the sum must be converted to the same units.

Total static head

The total static head, h_{ts}, is defined as the difference between the static discharge head and the static suction heads, $h_{ts} = h_{sd} - h_{ss}$.

Pumps — Theory and Equations | CHAPTER THREE

Velocity

We have already discussed the differences between average velocities (all fluid assumed traveling at the same velocity) and point velocities. These differences do not seem too important until we start to measure flow. It is wise, however, always to remember the difference so as not to fall into the trap of neglecting it when it is important.

Most data is given in volumetric units, meters cubed per hour, liters per minute or gallons per minute, to name but a few sets of units. The time base for flow rates when performing most computations is seconds and the flow frequently has to be converted to mass units per second. It is usually best to do all conversions on the initial statement and the final statement of all computations and to maintain a consistent set of units. This is particularly important for automatic calculations using computers.

To obtain average velocity from volumetric flow, we first convert all volumetric and time units to cubic meters or cubic feet and to seconds. We then use the equality, q = AU, where the flow rate, q, is equated to the cross sectional area, A, of the flow path times the *average* velocity, U.

Mixed metric units are sometimes quoted as being SI units. For instance, if the flow rate, Q, is given in cubic meters per hour and the pipe diameter, d, in centimeters, the following expression results:

$$U = \frac{q}{A} = \frac{Q/3600}{\pi(d/100)^2/4} = 3.537\frac{Q}{d^2} \, m/s \quad \textbf{(III-1)}$$

This is an example of mixed metric units, not SI units. The coefficient, 3.537, converts the units (hours and centimeters) to the appropriate ones and includes the constant, π.

A similar expression in mixed U.S. units of Q gpm and d inches would be:

$$U = \frac{q}{A} = \frac{Q/60}{\pi(d/12)^2/4} = 0.4085\frac{Q}{d^2} \, ft^3/s \quad \textbf{(III-2)}$$

The opportunity for error is much greater using mixed units. It is much better to follow the advice given above and to convert all units before beginning a computation.

Velocity head

Velocity head represents the kinetic energy of the flowing fluid at a cross section. It is converted to feet or meters of head units by hydraulicians. It

CHAPTER THREE | **Pumps — Theory and Equations**

is easier for other disciplines to use energy per unit mass units. To make the conversion, it is probably good to remember the kinetic energy is numerically equal to the distance a body, starting with zero velocity, must fall in order to reach flowing velocity.

$$h_v = \alpha \frac{U^2}{2g_c} \quad \text{(III-3)}$$

In pump work, the velocity, U, is the average velocity across a pipe section in feet or meters per second. The coefficient, α, is close to one, and is usually taken as such without too much error. The dimensional constant, g_c, is 32.17 ft-lb_m/lb_f-s^2 in customary U.S. units. If the units are checked using the U.S. system, it is seen that head has the dimensions of energy per unit mass, ft-lb_f/lb_m. The units of this term are numerically, but not dimensionally, equal to feet of fluid.

With SI units the dimensional coefficient is not needed; it is equal to one without units. The coefficient, α, is again usually taken as one. The dimensions of the head term are velocity squared or meters squared per second squared. Substitution of Newton's law shows these units to be equivalent to newton-meters per kilogram. In other words, the units are energy per unit mass. These units are not numerically equivalent to meters. In order to obtain meters, divide by the acceleration of gravity, 9.805 m/s^2 (see Chapter I, Section 10).

Viscosity effects

The resistance to motion of liquids, the viscosity, tends to decrease with rising temperature. Viscous liquids increase the power requirements because of the conversion of mechanical energy to internal energy in the flowing system. Viscous liquids reduce the differential head (and differential pressure) of a given pump due to internal irreversibilities. Effectively, very viscous fluids reduce pump capacity, head and efficiency.

Work performed in pumping fluids

In order for liquid to flow, work must be expended. A pump converts energy at its shaft to fluid energy. This fluid energy is used to:
1. raise liquid to a higher elevation;

2. overcome a high pressure in a vessel;

3. overcome viscous resistance in conduits;

4. increase the velocity of the fluid;

5. or, all of the above.

Pumps — Theory and Equations | CHAPTER THREE

Pump efficiency

Pump efficiency is usually expressed as the ratio of power out (fluid power) to the power in (shaft power). Power is the rate of doing work. Work is force moving through a distance. The power out is the difference between the fluid power leaving the pump and that entering. The power in is the shaft power to the pump.

Efficiency is of importance economically because reduced efficiency means expenditures of work energy are leaked to the surroundings as heat energy instead of being used to perform the intended work. Efficiency is important in other ways. A pump with low efficiency is a hot pump. There may be unexpected bearing failures. Increased maintenance may result.

The pump manufacturer is not normally the motor manufacturer. Pump efficiency is not motor efficiency. When the pump manufacture specifies the efficiency of his pump, he does not know necessarily what motor will serve as the driver, or even if it will be an electrical motor or a steam turbine, for instance.

A general, arbitrary classification of pumps follows.

1 A. Radial pumps

Even though the single stage radial pump looks simple, it is not. It is a carefully engineered device.

The first distinction to consider is the impeller. It can be open, meaning the pump casing completes the channel walls, or closed or shrouded, meaning the fluid passing through the impeller is completely contained by the impeller. The shrouded impeller is more efficient, but has smaller passages. The process will dictate the choice, for instance if the fluid contains solids, an open impeller will be less prone to problems because of the extra clearance.

The second distinction is whether the pump is a single suction or double suction pump. Normally, the impeller receives liquid along its axis, at the eye, from one side. It accelerates this liquid radially while turning it through ninety degrees from the axial to the radial direction. The reactive forces involved are obviously unbalanced and thrust bearings are necessary. On larger capacity pumps, sometimes it is advantageous to have what are effectively back-to-back impellers on the same shaft. Equal vol-

CHAPTER THREE | **Pumps — Theory and Equations**

umes of liquid are fed axially from both sides. The thrust is better balanced. Discharge is to a common casing.

Casings

The casing of a pump is usually classified as circular type, volute type or diffuser (guide vanes) type.

The circular type of casing simply encloses the impeller with some clearance. There is a single discharge point. It is the least efficient of the three types due to the shock and turbulence of fluid leaving the impeller. It is seldom used.

The volute type has an increasing casing radius in the direction of the discharge point, much like a sea snail. The shape and the increment of the radius per degree are chosen to minimize shock losses.

The diffuser type improves on the volute concept by adding stationary guide vanes surrounding the tips of the impeller. These guide vanes smoothly direct flow to the discharge. The diffuser type pump is the most efficient. Diffusers are used in multistage, high head pumps to direct flow to the suction of the next stage, as well as to the discharge.

1 B. Axial pumps

Axial pumps have a lifting or propeller action. In the strictest sense, they are not centrifugal pumps. They move liquid axially, not centrifugally. Figure III-2 is an example of a large axial pump used as a closed loop recirculation pump in an evaporator.

This type of pump typically has a very high flow rate with a fairly low differential pressure across the pump. Flow can be in excess of 450 cubic meters per hour (2,000 gpm) and the differential head can be 15 meters or about 50 feet (with water, about 20 psi).

In evaporators and calandrias the pump casing is often comprised of a large piping elbow. Flow control is frequently by design of the overall hydraulic system. The pump simply runs out along its curve until the system resistance exactly balances the pump head.

1 C. Mixed flow pumps

Mixed flow pumps combine both radial and axial elements. They are usually vertical, multi-stage pumps and in this case are known as turbine pumps. However, the term "turbine pump" is also used for a pump that uses the mechanical impulse principle combined with centrifugal force to impart momentum to the fluid.

The vertical mixed flow pump is capable of about 100 feet or 30 meters per stage (43 psi with water). By its nature, it is a self-priming pump. The liquid at the suction is intended to be above the eye of the bottom impeller. The pump is frequently chosen in order to avoid bottom connections to a vessel. It is also used in sumps.

The regenerative turbine pump is a low capacity, high head pump with a very steep, almost straight head-capacity curve. It has a very sharp peak on the efficiency curve. Because of the turbulence generated during the impulse, it can only be used on clean fluids. Erosion can present problems.

1 D. Multistage pumps

Multistage pumps overcome the speed (and, therefore, pressure) limitations of a single stage by combining more than one stage, usually on the same shaft. They are used as deep well pumps, as high-pressure water supply pumps, as boiler feed water pumps, as firewater pumps and as charge pumps to high-pressure refinery processes.

The discharge pressures can be as high as 40 MPa (6,000 psi). The more stages a pump has, the more delicate and costly the pump is and the more limited its turndown capabilities. The pump engineer should always be consulted for his knowledge about the specific characteristics of such pumps.

2. Positive displacement pumps

A positive displacement pump either has a reciprocating action or a rotary action. The simplex reciprocating pump allows liquid to flow into a cylinder during one stroke. Flow is through one set of check valves. It expels the liquid through another set of check valves during the opposite stroke. The rotary pump traps a quantity of liquid in a cavity at the pump suction and squeezes the same quantity out at the discharge. Positive displacement pumps include diaphragm pumps and screw pumps. Screw pumps are rotary positive displacement pumps whose discharge can be extremely smooth.

A reciprocating pump has an inherent pulsation associated with the passage of fixed quantities of liquid in a periodic fashion. The rotary pump can have quite a smooth output, depending on its design.

CHAPTER THREE | **Pumps — Theory and Equations**

One characteristic that is common to positive displacement pumps is the fact they cannot have their discharge throttled. If flow to the user must be controlled, it must be done by speed control or by recirculating a quantity of fluid around the pump. Throttling would result in excessive backpressure on the pump that could stall the driver, rupture the piping or casing, or cause a safety valve to lift.

Pump characteristics always find their utility. In pipeline pigging operations, the positive displacement pump will force the pig through a momentary obstruction whereas a centrifugal pump might not generate sufficient excess pressure to overcome the obstruction. The pipe has to be protected against overpressure by a safety valve when a positive displacement pump is used.

2 A. Piston and cylinder pumps

The first pumps were piston and cylinder pumps. They were made of wood and they frequently had square section cylinders and pistons. Bronze, cast iron and steel eventually replaced the wood. It was natural to connect a reciprocating steam engine to the reciprocating pump for such applications as mine drainage. In the early days, pumps were used in the suction mode, so it did not take long to learn one could only lift water about 10 meters or 30 feet before suction was lost. However, it was also learned that in a discharge mode, as long as the cylinders were kept primed and the discharge piping and pump casing were strong enough, water could be raised to much greater elevations.

Loss of prime

It is worthwhile to discuss "loss of prime" as a preliminary to detailed discussions on net positive suction head. In the suction mode, the (approximate) 10-meter or 33-foot limitation applied only as long as the check valves were in good shape and there was no leakage of air into the suction piping. If any of the above occurred, the pump operating in the suction mode was observed not to be able to raise the liquid to the usual level.

The word "prime" refers to the presence of liquid in the eye of the impeller. Without liquid, the pump could be trying to operate on much less dense gas or vapor. The presence of gas would be due to a leak into the suction. The vapor could be generated by the low pressure associated with the displacement of the piston.

Pumps — Theory and Equations | CHAPTER THREE

When pumping technology developed to the point where there was little or no leakage into the suction, it was found there was still a limitation on the maximum suction lift. This limitation was found to be associated with the vapor pressure of the liquid. The vapor pressure is essentially due to the escaping tendency of molecules from a liquid. The vapor pressure is a function of temperature and of the fluid.

Normally, the surrounding pressure is greater than the vapor pressure, thus limiting the rate at which molecules can escape. When the surrounding pressure becomes less than the vapor pressure, the escaping rate of the molecules of liquid increases in order to maintain the vapor pressure associated with a particular temperature. Heat flow from the environment supplies the energy necessary for vaporization to occur.

In the older, slower, simple piston pump operating in a suction mode, loss of prime signifies inability to pump, and not much else. In a centrifugal pump, loss of prime can have devastating consequences. In particular, problems arise as the pressure approaches the vapor pressure of the liquid at the inlet to the pump. Subsequent acceleration of the liquid can cause the pressure to drop below the vapor pressure and cavities (vapor bubbles) to form in the liquid. Subsequent deceleration can cause the pressure to increase and the cavities to implode, suddenly. It is this sudden implosion that generates the enormous point pressures that are the cause of metal erosion in a pump. The phenomenon is usually called cavitation.

Specific control requirements of piston and cylinder pumps

We will only briefly discuss the control requirements relating to fluid flow problems. As always, it is wise to divide control requirements into those associated with machinery protection and those associated with flow measurement and control. We wish to concentrate on the latter, but one cannot ignore the former. They must at least be addressed so we are aware of them.

Overpressure

Most reciprocating pumps are capable of discharge pressures sufficiently high to do damage if not controlled. Above 15 psig (103 kPa), most boiler and pressure vessel codes call for mandatory use of a pressure safety valve. The safety valve should be set at the design pressure of the weakest member of the protected system. It should have a capacity that allows it to function without exceeding this pressure by a stated safety margin. In the case of a positive displacement pump, this capacity is normally the

CHAPTER THREE | **Pumps — Theory and Equations**

rated capacity of the pump, unless there is another relieving passage in the system.

Pulsation

All reciprocating pumps have pressure and flow pulsations in the discharge and suction lines. Pressure (and flow) dampeners are usually associated with such pumps. These dampeners are usually volume tanks. Frequently the tanks have bladders filled with or surrounded by gas. The function of the tank is to act as a capacity tank during the high pressure half cycle and to relieve the accumulated volume through the system resistance during the low pressure half cycle. Figure III-7 shows the typical discharge characteristic of a positive displacement pump. The suction characteristic will be similar with smaller amplitude

The purpose of most pulsation dampeners is to prevent mechanical damage to the system. Usually residual pulsations are present that can cause difficulties for measurement and control devices. Dampening is applied usually for mechanical protection. This smoothing of pulsations is frequently not sufficient for measurement and control purposes.

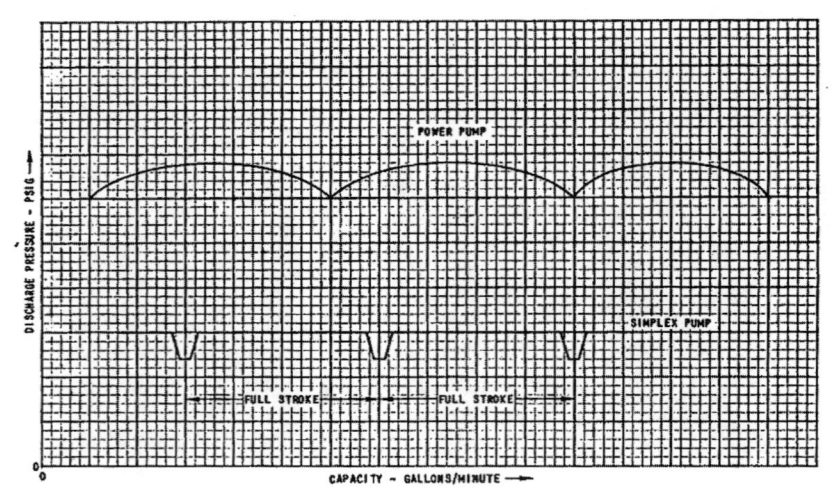

Figure III-7. Typical characteristic curves for reciprocating type pumps

Flow control

Flow control must be by diverting some of the volumetric flow from a constant volume pump or by controlling the speed of the driver. Usually, the decisions are made on the basis of process, mechanical and economic

124   Flow of Industrial Fluids—Theory and Equations

considerations. If the driver is a fixed speed one, flow diversion is usually the only solution. If the driver is a variable speed one, diversion may still be necessary. This occurs when the combined turndown of the pump and driver is limited or if the pressure requirements of the system prevent the use of speed control.

2 B. Rotary pumps

Rotary pumps vary from simple peristaltic pumps, associated with laboratory dosing operations, through gear pumps to lobe pumps and screw pumps. Their pressure and flow characteristics vary from mild pulsations to no pulsations. They still, however, are positive displacement pumps and as such are subject to the same overpressure and flow control requirements as mentioned above. Only the pulsation-damping requirement may be relaxed in some cases.

Frequently, rotary pumps have internal controls that divert some flow to suction when the backpressure becomes excessive. This sometimes leads to the assumption that rotary pumps can have their discharge throttled. In effect, a combination of throttling and diversion is being used. It is well to remember when any pump operates in a recirculation mode, energy is being added continuously to a (partially) closed system. Inefficiencies will cause the system to heat up. If the excess heat is not removed adequately by natural or forced cooling, problems will result.

2 C. Diaphragm pumps, metering pumps

Diaphragm pumps and metering pumps are positive displacement pumps. They are classified together only because they use a diaphragm and check valves. They may be large, air-driven pumps used extensively as portable pumps in chemical plants for a large variety of services. They may be highly precise metering pumps for dosing small quantities of fluid into a larger process.

One problem to bear in mind is the maximum pressure the pump is capable of delivering when the main system is shut in. It may be necessary to provide for local safety valves if this pressure is too high. The small, innocent looking metering pump is often quite capable of rupturing equipment and even piping.

Miscellaneous pumping devices

There are many specialized devices that have been used for pumping. Some are not too efficient, but are used for practical reasons. We will

CHAPTER THREE | Pumps — Theory and Equations

briefly describe some of them in order to complete the picture and to supply some food for thought. The theory of such devices is frequently enlightening.

3 A. Gas lifts

A gas lift (Figure III-4) works on the principle that, if a column of water exists, and, if gas of sufficient pressure can be introduced into the bottom of the column, then that gas will reduce the overall density of the column of liquid by virtue of mixing with it. If the column is separated from the surrounding fluid by a wall except at the bottom, the heavier density surrounding fluid will force the lighter density contained fluid up the column. The liquid and gas mixture within the column will rise to a higher elevation than the more dense liquid. The length between the normal level of the liquid and the bottom of the column is called submergence. The length between the normal surface of the liquid and the point of discharge from the column is called lift.

The use of gas lifts (air lifts) was very common in mine drainage applications before the advent of efficient centrifugal pumps. Gas lifts are still used in the oil extraction industry.

3 B. Jet pumps

Jet pumps (Figure III-5) are a class of pumping device that makes use of momentum transfer between two compatible fluids. In this chapter we are only considering liquids, but the principle also applies to gases.

Jet pumps, like steam engines, evolved in a highly empirical manner. Externally they have the appearance of a simple piping tee. Internally, they consist of four parts. The first part is a nozzle that accelerates the motive fluid to the entrance to the mixing section. The second part is an inlet chamber that brings the driven fluid more or less smoothly to the entrance to a mixing section. The third part is a mixing section where the two fluids are mixed and are decelerated and accelerated by momentum transfer. The fourth part is a diffusing section that decelerates the mixed fluid further and increases the static pressure.

Jet pumps are given different names depending on their application, on their user and on their manufacturer. They are called injectors, ejectors, eductors and elevators and when they are "pumping" gases they are called gas compressors or exhausters. In principle, they are the same. In this section, we shall refer to them as jet pumps.

Pumps — Theory and Equations | CHAPTER THREE

Jet pumps are essentially static devices. All parts are of fixed dimensions, therefore it can be expected that they be designed around one operating point and that they will be less efficient when operated away from the design point. The addition of an adjustable needle in the inlet nozzle gives some degree of control of motive flow and efficiency in some cases.

If a sufficiently high-pressure motive fluid can be introduced to the inlet chamber of the jet pump, the fluid can be made to accelerate through a nozzle and thus gain kinetic energy and momentum at the expense of the pressure energy. The static pressure drops as the fluid is accelerated. The design of the jet pump is such that the suction inlet is connected physically to the low pressure point of the accelerated fluid. Therefore, the normal pressure of the fluid at the suction inlet forces it into the accelerated stream. The mixed stream is now decelerated in the mixing section where it gains pressure. It is further decelerated in an expanding nozzle where it gains even more pressure. This pressure can be used to move the mixed fluid to another location.

Efficiency considerations

At first glance, it may seem a jet pump is a wonderfully simple, inexpensive and, therefore, efficient device to use. It is true the device is simple and inexpensive; whether or not it is efficient depends upon the alternative applications.

The first point to be made is the jet pump is not too efficient thermodynamically. It is inherently a mixing device and mixing is one of the main causes of thermodynamic inefficiency. In addition, turbulence and shock cannot be overcome as well as with centrifugal pumps, so the efficiency is much less.

The second point is the motive fluid has to be pumped in order for it to gain the pressure required at its flow rate. There are losses associated with this pump and with the piping system to the device. The combination of inefficiencies results in a low overall efficiency.

Utility considerations

If the jet pump is so inefficient, why is it used at all? The answer to this rhetorical question lies in the simplicity of the device and in the possible availability of cheap (excess) motive power.
- The device is simple. It can be designed to accept solids, limited in size, mixed with the pumped liquid at its suction inlet. It can therefore pump slurries.

CHAPTER THREE | **Pumps — Theory and Equations**

- It is small and can be made of exotic materials at little extra expense. It can, therefore, handle corrosive and erosive fluids with less capital cost than can a normal pump.
- If motive power is available that cannot be economically used elsewhere, there is no need to pay for a more expensive pump.

Control considerations

The jet pump is a relatively static device. The motive fluid can be throttled only over a limited range before losing its ability to transfer momentum within the fixed configuration of the pump. The fluid being pumped can be throttled as a means of controlling its flow. What suction throttling does is to create a resistance to flow into the low-pressure area of the vena contracta of the motive fluid's jet. Throttling at this point can be done with relative impunity.

Applications when pumping liquids

Jet pumps can pump liquids using other liquids as the motive stream, or they can use steam as the motive stream. This section will discuss only the use of liquids as a motive force.

When pumping liquids, jet pumps are used either in a suction mode or in a discharge mode. In the suction mode the vapor pressure and the suction friction losses limit the height to which the liquid can be elevated. With water, six meters (20 feet) is the approximate limit. The additional discharge elevation is normally limited to about one half a meter when the jet pump is operating in the suction mode. In the discharge mode, the jet pump is located within one half a meter of the liquid to be elevated and it is capable of raising the liquid a further approximately nine meters (30 feet). One manufacturer gives an elevation of two meters per bar gauge (100 kPa, 14.5 psig) of motive pressure.

In spite of its low efficiency when compared to a centrifugal pump, a jet pump is frequently used in place of a centrifugal pump. This is particularly the case when the pumping operation is infrequent such as when a sump must be emptied periodically or when a centrifugal pump must be primed by pulling a vacuum on its casing. The jet pump is also used when abrasive solids would otherwise damage a more expensive pump and when corrosive materials must be diluted or neutralised while they are being pumped. Sometimes it is used simply because a stream of high-pressure liquid is available whose energy would otherwise be wasted.

Commercially available jet pumps are usually off-the-shelf items sold in a choice of fixed sizes. Each size is designed around a fixed operating point. The user picks the size closest to his application and accepts the additional inefficiency due to the pump operating away from the design point.

One manufacturer gives performance specifications as two sets of charts: one for the suction mode and one for the discharge mode (same jet pump). In the discharge mode the suction elevation is limited to one half a meter. The charts give curves for discharge elevations from two to nine meters. Suction and motive flow rates can be read corresponding to motive pressures. Corrections are applied for different densities and for suction elevations greater than one half meter. In the suction mode, the discharge elevation is limited to one half meter. The charts give curves for suction elevations from one to six meters. Again, corrections must be applied for non-standard conditions.

The use of such performance charts is similar to the use of performance curves for a centrifugal pump. The major difference is the fact the motive stream is added to the suction stream. It effectively goes along for the ride. In Appendix AIII, we will use the above mentioned curves in order to compare jet pump efficiencies with centrifugal pump efficiencies. While we seem to be emphasising efficiency, it must be remembered jet pumps are frequently chosen in spite of their inefficiency, because they fit other requirements better than do centrifugal pumps.

3 C. Liquid metal pumps

The motor principle is that a conductor carrying an electric current at right angles to a magnetic field will have a force imposed upon it. The force will be perpendicular to both the magnetic field and the electric current. In a normal electrical motor, the windings carrying the current are distinct. In a liquid metal pump, there are no windings. The liquid metal is the conductor. The current is made to pass at right angles to the conduit by suitable electrodes placed opposite one another on the conduit. The magnetic field is established at right angles to both the electrodes and the conduit. The direction that is mutually perpendicular to both the magnetic field and the imposed current is in the axis of the conduit. This is the direction of the force and of fluid motion.

Electromagnetic pumps have been developed with differential heads of 300 psi (2 MPa) and flow rates up to 10,000 gpm (2.3 m^3/h).

CHAPTER THREE | **Pumps — Theory and Equations**

These pumps are highly specialized and are generally limited to heat transfer applications in the nuclear industry. We will discuss them briefly in Appendix AIII.

III-6: INHERENT AND INSTALLED CHARACTERISTICS OF PUMPS

Pumps, like control valves, have inherent characteristics and installed characteristics. It is well to remember the distinction. We will give a rather detailed technical description in Appendix AIII.

Inherent characteristics

An inherent characteristic is one that is proper to the pump when all external influences have been eliminated. It is what the pump manufacturer uses to describe his pump. This is done so a maximum of information can be transmitted in the most simplified form. The inherent characteristics are usually shown on the pump curve, and most of us are reasonably familiar with that of the centrifugal pump. The basic curve is that of developed head in feet or meters of fluid versus the volumetric flow rate. It is up to the user to convert this basic data to a more useful form.

If the reader gets into the habit of thinking of the developed head as energy per unit mass instead of feet or meters of fluid, he or she will avoid many potential pitfalls. The typical centrifugal pump curve gives the energy per unit mass that the pump transfers to the fluid plotted against the volumetric flow rate. The pump curve will give the appropriate units in feet or meters and in gallons (U.S.) per minute or cubic meters per hour.

Since the developed head is the energy per unit mass that the pump delivers to the fluid, this energy can manifest itself as additional kinetic energy, additional pressure (static) energy or additional potential energy (lift). Depending on the physical characteristics of the pump, elevation differences and velocity differences at the suction and discharge, the developed head read from the pump curve might or may not correspond directly with the measured differential pressure across the pump. The measured differential pressure must be converted to the same units as are used on the pump curve. The measured quantity must be corrected for the changes due to elevation differences and to velocity differences at the inlet and outlet of the pump.

Pumps — Theory and Equations | CHAPTER THREE

Installed characteristics

An installed characteristic depends upon the system in which the particular pump functions. The most common installed characteristic curve described in the literature is that of the centrifugal pump. Again, remember centrifugal pumps have a wide variety of installed characteristic curves. Positive displacement pumps have completely different ones and positive displacement pumps are subject to the constraint that they cannot be directly throttled.

Constant speed centrifugal pumps

We will start with the installed characteristic curve for a constant speed centrifugal pump. The standard figure consists of two curves. The first is the inherent pump curve. The second is the system curve.

Usually the system is a simple one consisting of a fixed configuration, but the resistance also can be a constant pressure with very little variable resistance or even a mixture of the two. In analyzing a fluid system, it is often found the system may have different configurations at different times when subsystems are valved into and out of the main system. In this section we will only consider the simple fixed configuration. The other systems will be discussed in Appendix AIII.

The system curve is usually superimposed upon the inherent curve. The inherent curve represents energy per unit mass flowing added by the pump. The system curve represents energy per unit mass flowing required to lift the fluid to a particular elevation, to overcome pressure differences, and to overcome viscous resistance due to the nature of the fluid being pumped. The requirement of overcoming viscous resistance represents useful work transformed to internal energy.

The inherent curve is dependent on pump design. It usually curves downward with increasing flow, but it may have a fairly sharp peak before doing so. For the moment we will consider a monotonically decreasing curve. This type of curve represents the fact that energy transferred to the fluid per unit mass flowing decreases with increasing flow through the pump. The maximum fluid energy per unit mass transferred to the fluid is at a flow close to shut off, when there is little flow (and little cooling effect). As the flow rate is increased, the energy transferred drops off because of internal inefficiencies within the pump. The pump is designed to maximize its efficiency at a fixed flow rate.

CHAPTER THREE | **Pumps — Theory and Equations**

The system curve usually starts off above zero on the abscissa. The ordinate represents the minimum amount of energy per unit mass that must be transferred to the fluid to lift it against gravity and the system static pressure differential without flow. As flow increases, the system curve also curves upward at approximately a square law rate. This is due to the energy requirements to overcome irreversibilities generated by flow and to create the kinetic energy being added to the static energy requirements. Note we have not drawn a distinction between pumps that operate under positive suction head and those which must supply suction lift. Once the pump is operating within its limits, there is no difference between the two modes of operation. The energy transferred to overcome pressure, to create elevation and kinetic energy is recoverable. The energy transferred to overcome irreversibilities is not.

The two curves (inherent and system) will meet at a common point. At this point, the energy transferred to the fluid by the pump per unit mass exactly equals the energy necessary per unit mass to produce flow and to raise the fluid to some elevation against some pressure within the particular system. The pumping system described by this curve will operate at this point and no other. If the pump is turned on, it will automatically settle at the flow rate and total dynamic head associated with this point.

The difference between the inherent pump curve and the system curve at a lower flow rate than that of the common point represents the energy per unit mass that must be absorbed by a valve or turbine if flow is to be controlled at this lower flow rate. Higher flow rates than the one associated with the point of convergence of the two curves are impossible – you cannot get there from here.

It is conceptually important to remember that a control valve converts useful mechanical energy into internal (heat) energy, although it is not often apparent. The difference between the two curves at points with flow rates lower than that of the common point represents the energy converted to internal energy in order to maintain the given, lower, flow rate. Throttling is not free, although when all costs and efficiencies associated with other methods of control are factored into the equation, throttling is still the most common method of controlling flow.

If an economic means could be found to recover the mechanical energy converted to internal energy by a control valve, it would be used. Liquid turbines, for instance, are used in some high-pressure let down situations. They are called power recovery turbines.

Variable speed centrifugal pumps

The curves for a typical variable speed pump consist of essentially parallel plots for different rotational speeds. At first glance, it is tempting to think speed control is an extremely good way of adjusting the head versus capacity requirements of a system and of saving energy.

However, there are a few points that cause many variable speed pumps to be operated in a fixed speed mode:
- From a control point of view, variable speed drives usually have much greater inertia than control valves, so their response to a control signal is much more sluggish. It is possible to use a variable speed drive to slowly optimize the response of a faster acting control valve, and this is sometimes done. Steam turbine drivers are usually much more responsive than drivers that use gear reducers, but steam must be available. Variable frequency electrical drives may not have the problem associated with inertia, but their cost must be considered.
- From an efficiency point of view, as has been previously stated, centrifugal pumps are designed about a fixed operating point. Their efficiency drops off as the fixed speed curve is traced in either direction away from the design point. Examination of variable speed curves reveals that at lower speeds, the design maximum efficiency is not even reached.
- In addition to the pump efficiency, the driver efficiency must be taken into account. The match between the pump and its driver may not be perfect and, even if quite adequate at one point, the efficiency drops off quite radically at other points. Remember the overall efficiency is the product of the pump efficiency and the driver efficiency. Both efficiencies have curves, so the product must be plotted versus flow to obtain the true picture.

Distinction between centrifugal pumps operating with suction lift and those under a positive suction pressure

Regardless of the type of installation, the pressure at the inlet to a pump, *or within the pump*, must not dip below the vapor pressure of the fluid. If this occurs, two distinct possibilities exist.

The first one is the pump simply loses prime. It ceases to pump liquid and simply churns whatever liquid is in the pump casing against a closed check valve. The second possibility is flow continues and pressure is recovered within the pump. The vapor bubbles (cavities) created when pressure dips below the vapor pressure implode violently, creating point

CHAPTER THREE | **Pumps — Theory and Equations**

pressures that can exceed 10,000 psig (69 MPa). This second possibility is known as cavitation. Most pumps do not last too long if cavitation is allowed to occur within them.

Since the pump manufacturer knows the inherent characteristics of his pump, he establishes the minimum pressure that must be available at the inlet to the pump to prevent problems with cavitation. This pressure is given in terms of head. It is called the net positive suction head required, NPSH(R). The system designer computes a suction head available within a given piping configuration. The formula will be given in Appendix AIII. This head is designated, NPSH(A). The NPSH(A) must always exceed the NPSH(R) *at all points along the pump curve*.

Some pumps require more head than others do at their suction. It should be evident a pump operating with suction lift should be designed to have a very low NPSH(R). Once operating within its design parameters, the pump does not "know" what mode it is operating in – lift or positive suction head. From an operating point of view, establishing the NPSH(R) for a particular pump and making sure the NPSH(A) always exceeds the NPSH(R) is paramount.

Pumps operating in the suction lift mode have another problem – that of prime. To get the pump to start pumping, liquid has to be present within the pump casing and suction piping. Priming the pump can be done manually, by opening valves to add liquid and vent gases, prior to starting or it can be done by control logic. It can also be performed by pulling a momentary vacuum on the pump casing. Priming has to take place each time the pump is stopped more than momentarily. In some cases, foot valves have to be present to allow the piping to be filled. Vent valves are necessary to allow purging gases in the suction piping and pump casing.

Positive displacement pumps

The relationship that exists between differential pressure across a centrifugal pump and volumetric flow does not exist with positive displacement pumps. By its nature, the positive displacement pump, or volumetric pump, moves a fixed volume of fluid with each cycle or revolution. Increasing backpressure does cause slip to occur. Slip is leakage past cylinders and rotary members that reduces the forward flow on increasing backpressure. It is an incidental phenomenon; it cannot be used for control purposes.

The constant speed positive-displacement pump discharges a fluid that is essentially incompressible at an average rate that is fixed. The average rate is pulsating or it can be quite smooth, depending on the type of pump. For a positive displacement pump, the head is strictly a function of the system backpressure. Head influences capacity only through slip.

III-7: CONTROLLING FLOW THROUGH PUMPS

When we talk of controlling flow through pumps, we tend to concentrate on the flow control devices. We tend to neglect the aspects of control that are part of the design decision-making process. In this section, we will discuss pumps as they interact with the system in which they are installed.

Fixed speed centrifugal pumps

When operating at a fixed speed, the volumetric flow through a centrifugal pump is a function only of the total dynamic head imposed on the pump by the system. This mechanical energy is exactly equal to that transferred to the fluid by the pump. The system resistance is made of dynamic elements (such as control valves and varying pressures in vessels) and static elements (such as pipe and equipment friction drops).

Self-regulation

As has been pointed out by Les Driskell[xvii] (*Control Valve Sizing and Selection,* ISA), the ideal method of control, when feasible, is self-regulation. He also points out self-regulation does not come without thought. Furthermore, he emphasises the importance of studying the possible interactions of the entire system before choosing a control scheme.

Driskell gives an example of self-regulation of a centrifugal pump. The system is that of Figure III-8 which is taken from the above work. It is worth analyzing, as it fixes the fundamental principles. We will put some real numbers on the analysis in Appendix AIII.

CHAPTER THREE | **Pumps — Theory and Equations**

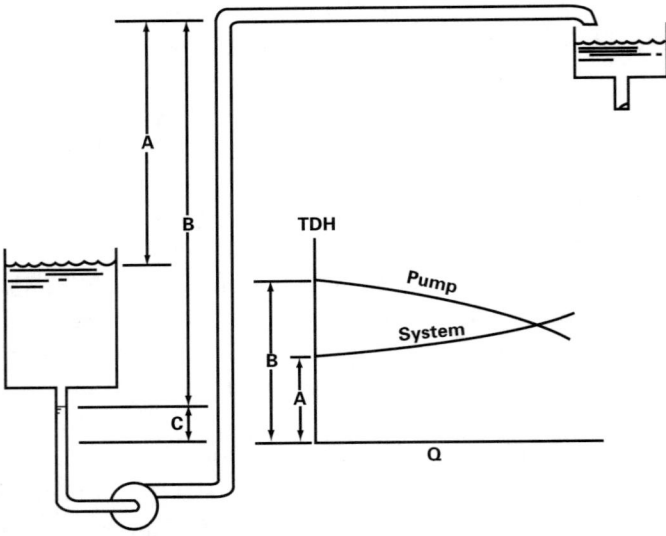

Figure III-8. Self-controlled pump

Source: Control Valve Sizing and Selection, L.R. Driskell, ISA

If the pump has a smoothly decreasing head-versus-flow curve it can be made to operate over a wide range of flow rates without external controls. The head imposed on the pump is due to the differences between the suction and receiver pressures and elevations and the differences between the inlet and outlet kinetic energies and losses due to irreversibilities associated with flow. The suction elevation varies with inflow to the vessel; the discharge elevation is fixed. In the example, the maximum flow rate is governed by the minimum difference between the two elevations. The minimum flow rate is governed by the maximum difference.

We do not have to worry too much about the maximum flow rate. It normally, will be chosen so it is on the pump curve (this assumption may be worth checking, however). We may have to take precautions to avoid the loss of NPSH. These precautions can be as simple as having a low-level switch trip the pump (with automatic restart on a higher level). An automatic recirculation valve would achieve the same protection without stopping the pump.

The flow rate in this case will vary smoothly between the maximum and the minimum depending only on the imposed head. This type of control system will settle out at a flow rate that is in balance with the flow into the sump. It needs no controls other than those that protect against the suction loss mentioned above.

136   Flow of Industrial Fluids—Theory and Equations

Pumps — Theory and Equations | CHAPTER THREE

Discharge throttling

A common method of flow control is discharge throttling by placing a control valve in the pump discharge line. A valve in the suction will cause the pump inlet pressure to decrease and may cause cavitation. The control valve in the discharge line adds to the system losses so the total dynamic head imposed on the pump may be varied. Note the irreversibilities created by the valve must be real losses not apparent ones. A high recovery valve such as a butterfly valve or an angle valve will have lower real losses than the ones that appear to correspond to the measured differential pressure immediately across the valve. A high recovery valve would have to be closed to a greater extent than would a low recovery one.

The energy per unit mass converted to internal energy by the valve can be depicted by a straight vertical line between the inherent pump curve and the system curve at a given flow rate. More exactly, the energy per unit mass of flowing fluid imposed on each side of the pump by the system pressures and friction losses govern the flow rate. Reducing the differential head increases the flow rate; increasing the differential head reduces the flow rate. The relationship is not linear.

There is another way of looking at the pump inherent characteristic versus the system characteristic relationship. It is to realize the throttled control valve simply adds to the permanent losses of the system – to the irreversibilities that were estimated with the valve wide open.

By throttling the valve, we have created a new system with greater losses. Since the shutoff head remains the same, and since the system curve still follows a square law relative to flow, one can simply rotate the system curve counter clockwise so it meets the pump's inherent curve at the corresponding flow rate.

Unless there is a mismatch between the pump and the system, the maximum flow rate is rarely a problem with typical centrifugal pumps. Pipe fluid velocities in suction and discharge piping have already been discussed. These are controlled by the design of the system. When all control and manual valves are wide open, the pump simply settles down to the point where the system curve meets the pump curve – where the energy delivered by the pump is completely converted to static, kinetic, potential, internal energy and to system irreversibilities.

CHAPTER THREE | **Pumps — Theory and Equations**

Some centrifugal pumps require special attention. For instance, high head, high-speed pumps frequently have a maximum in their inherent curve. Such a pump has two different flow rates that correspond to the same total dynamic head. Since it is the total dynamic head that governs the rate, this pump can oscillate between the two rates, much as a centrifugal compressor is subject to surging.

Such pumps are usually very expensive and very delicate. Surging can ruin the driver's gear train and damage the assembly. Therefore, a minimum flow recirculation, frequently through a cooler, is often required. The pump vendor may already have a standard minimum-flow recirculation valve designed into his pump. If no such system exists then one must be designed. The pump vendor will have established what the minimum flow should be. An independent safety shutdown trip may also be required.

Centrifugal pumps in parallel

Just because two pumps are piped in parallel does not mean they are intended to operate in parallel. They may be spare pumps.

If two pumps are operated in parallel, their respective head curves must be matched. The curves must be parallel and close to one another. Not only must the two curves be matched but also the suction and discharge piping to the common points must be almost identical hydraulically. All of this to make sure the differential head across each pump is as close as possible. If the pumps and the hydraulic configurations are not matched, it is quite conceivable a high discharge head of one pump may cause the other pump to back up along its curve until it is shut off. It will overheat.

It must not be thought two pumps in parallel give twice the flow as one pump. If an identical pump is added in parallel to an operating pump, it initially will increase the flow. The increased flow will cause an increased backpressure proportional to the square of the increased flow rate. The increased backpressure is translated into increased total dynamic head that causes both pumps to back up on their curves. Each pump, individually, will produce a lower flow rate than the original pump did. The sum of the flow rates of the two pumps will not by twice the original one; it will depend upon the shape of the head versus capacity curves.

Centrifugal pumps in series

When two centrifugal pumps are connected in series, one is frequently considered a booster pump. For instance the pump on a fire truck takes

Pumps — Theory and Equations | CHAPTER THREE

its suction from the fire hydrant fed by the city water pumps. The purpose of the city water pumps is to circulate water to all hydrants at a reasonable pressure. The purpose of the fire booster pump is to increase the pressure so the pressure energy can be converted to kinetic energy. A high velocity jet of water results.

In pipeline work, pump stations exist along the pipeline in order to limit the maximum pressure to which the line will be exposed. Each pump adds the head that was lost in overcoming irreversibilities and head changes between one station and the next. If fewer pumps were used, the discharge pressure of each pump would increase and the pipe rating would have to increase.

In plant work, a booster pump is sometimes added to a larger system that contains many users, some at lower pressure, some at higher pressure. The main problem with pumps in series is to make sure they are controlled so their NPSH(R) requirements are always met. Each pump must have controls that prevent it from starting before the requirement for a minimum suction pressure is satisfied and that trip the pump when the minimum suction pressure is in danger of not being satisfied. Automatic recirculation can be used, but cooling has to be considered. The normal starting sequence is from upstream to downstream.

Variable speed centrifugal pumps

Operating a centrifugal pump at variable speeds affords the opportunity to change flow relatively slowly between static conditions. We have already discussed the problems of sluggishness and the fact this inertia frequently causes a variable speed drive to be used for trimming purposes only. The fact the variable speed drive gives the opportunity to match the pump discharge pressure to the system pressure requirement is often of interest.

The control characteristics of a variable speed pump are essentially those of its driver. If the driver is a directly coupled, low inertia device (a variable frequency motor, for instance), what was said above regarding sluggishness may be ignored. Control of flow can be by control of pump speed. If the driver has considerable mechanical inertia, or if the driver has minimum turndown problems due to overheating at low rotational speeds, then the characteristics of the driver may overwhelm those of the pump. In any case, the total picture must be considered. The characteristics of each element must be known.

CHAPTER THREE | **Pumps — Theory and Equations**

Start-up of centrifugal pumps

Most single stage, constant speed centrifugal pumps are started remotely under automatic control. Limit switches are used to ensure automatic valves are open at the appropriate time. Level switches or pressure switches are used to ensure adequate suction head. The pump is frequently started against a closed discharge valve that is opened once the pump is running. This is done to prevent the pump impeller from being turned in the wrong direction because of a leaking check valve. It is also the unloaded state for a centrifugal pump – minimum horsepower.

The more complicated the pump, the more expensive it is and the more sophisticated the controls. The machinery expert may insist on operator presence at start-up, even though the control systems expert may think he can devise a perfectly safe automatic start-up sequence. The machinery expert will usually win.

Positive displacement pumps

There are really only two means of controlling flow when positive displacement is involved. Some of the constant flow can be partially diverted back to suction or to a receiver or the speed of the driver can be changed. If the driver will not stall before the pipe or the casing rating is exceeded, a pressure safety valve is necessary.

An interesting phenomenon that occurs with piston and cylinder pumps is one that is due to the acceleration of the piston and that of the fluid in the suction line. The piston has a maximum acceleration at bottom dead center and top dead center and zero acceleration as crosses the mid point of its travel. The pulsating flow in the suction piping is not necessarily synchronized with the movement of the piston. When the inlet valve opens, the fluid must be accelerated to catch up with the receding piston. The pressure required to achieve this acceleration is called "acceleration head". If the fluid is insufficiently accelerated, there will be a momentary vapor space behind the piston that will be eliminated suddenly somewhere in the fill stroke – usually about two thirds of the travel. A knock occurs and this knock can damage the pump. The acceleration head must be supplied by suction pressure over and above that required by a centrifugal pump.

Start-up of all types of pumps

We should always distinguish between start-up conditions and operating conditions. This applies to all types of pumps, centrifugal or positive displacement.

Starting a positive displacement pump against a high backpressure usually requires the pump be started in an unloaded mode – recirculation valve open, discharge valve closed. Once the pump is operating smoothly in the recirculation mode, the discharge valve is opened and the recirculation valve gradually closed. This allows the build up of pressure to overcome the backpressure and to open any check valves. Frequently, positive displacement pumps are started locally so the operator can witness any abnormal conditions.

III-8: HYDRAULIC TURBINES

Hydraulic turbines are used in process work where an incoming line carries liquid at much higher pressures than is required by the local user. The pressure may be required for transportation purposes by other users, for instance. Letting down the liquid through a control valve means wasting useful mechanical energy. It also might mean a special letdown valve in order to avoid cavitation and noise.

If the flow rate and pressure are great enough, a hydraulic turbine can convert otherwise wasted energy into useful energy. It does so while letting down the pressure to that of the final user. A suitable user of the mechanical power would be an air compressor or a pump in continuous service or an electrical generator.

Hydraulic turbines are sometimes described as pumps that run in reverse. This is conceptually descriptive, but not exactly true physically. They do reverse the action of a pump. They take energy available in a fluid stream and convert it into mechanical energy at an output shaft. What is connected to the shaft depends on the ingenuity of the mechanical engineer.

Physically, the turbine is designed to accept high-pressure fluid smoothly and to discharge it at some pressure that will be useful to the user. The vector equations used to describe turbines are the same ones used to describe centrifugal pumps. The Bernoulli[i] equation also applies in an identical fashion.

Hydraulic power recovery turbines are at least as reliable as pumps. Their maintenance costs may be less.

Matching the turbine to the system

It must be remembered the turbine imposes its own requirements on the process. The efficient operating range, for example, may be very narrow.

CHAPTER THREE | **Pumps — Theory and Equations**

In fact, if the turbine is connected as the driver of an electrical generator that is connected to a grid, it can be a power absorber rather than a power producer. It can then operate as a pump during process upsets. This is not what was intended by the designer.

In refinery service, it is not unusual to see a conventional driver connected through a one-way torque coupling to the ultimate power consumer. The conventional driver can supply power to the user until the hydraulic turbine has developed enough torque to satisfy the user's requirements. The driver is then shut down.

There are no major fluid flow problems associated with hydraulic turbines different from those associated with pumps. Certainly, the computations are no different from those associated with flow caused by pumps. The problems are those of matching the turbine to the system. Questions that must be asked are:
- Does the narrow range of the turbine fit the larger range of the process? How do we control flow to maintain it in this range?
- Is a high-pressure letdown valve necessary to replace the turbine during upset conditions? If so, are noise and cavitation a problem?
- Is an independent conventional driver necessary during process upsets? If so, how is it integrated into the system?
- How is everything started up in an integrated fashion? What happens when we lose one element?

The technical problems associated with turbines from the point of view of fluid flow measurement and control are of less severity, but similar, to those associated with pumps. Net positive suction head is not normally a problem. Turndown limitations may result in motoring (the turbine becomes a pump). Generally, the operating limits must be clearly defined, and the conditions imposed upon the turbine by the fluid must be controlled so the machine always remains within these limits. Once the operating limits of the specific turbine are understood, the problem usually resolves itself to a "what if" analysis.

III-9: WORKED EXAMPLES

Example of Customary U.S. units: Total discharge head

Sixty-degree Fahrenheit water is being pumped. A pressure gauge on the discharge flange of a centrifugal pump reads 40 psig. The center line of the gauge is 2 feet above the center line of the pump. The barometric pressure at the pump location (usually an average number obtained from the local airport) is 100 kPa. The discharge line is 2 inch, Sch. 40. The flow rate is 100 U.S. gpm. What is the total discharge head in foot-pounds force per pound mass and in feet? It is worth remembering the definition of total discharge head as being the sum of the absolute centerline discharge pressure and the kinetic energy both in the same units.

The first problem arises because of the use of weight density, γ, in place of mass density, ρ, in the hydraulic version of the Bernoulli[1] equation. Weight density has units of pounds force per unit volume. Mass density has units of pounds mass per unit volume. The densities are numerically identical, but the units are different. Dividing pressure in pounds force per square foot by weight density in pounds force per cubic foot leads directly to units of feet. Dividing the same units of pressure by mass density leads to units of foot-pounds force per pound mass, or energy per unit mass.

The energy units are not quite so simple as the hydraulic ones, but they are more meaningful. In the first example we will use both units, after that we will use only energy per unit mass.

Discharge

Weight density	62.37 lb_f/ft^3
Mass density	62.37 lb_m/ft^3
Gauge pressure	40 psig, 5,760 psfg
Feet of fluid	92.352 ft
Energy/mass	92.352 ft-lb_f/lb_m

Atmospheric pressure

Barometer	100 kPa
U.S. units	14.5 psia, 2,088 psf
Feet of fluid	33.487 ft
Energy/mass	33.487 ft-lb_f/lb_m

CHAPTER THREE | Pumps — Theory and Equations

Correction to absolute units
(discharge gauge plus atmospheric conditions)
    Pressure   54.5 psia, 7,848 psf
    Feet   125.830 ft
    Energy   125.830 ft-lb$_f$/lb$_m$

Correction to pump center line. If the gauge center line is above the pump center line, the correction is positive. If it is lower, the correction is negative. See explanation above.
    Distance   (+) 2.0 ft
    Energy/mass   (+) 2.0 ft x g/gc = 2.0 ft-lb$_f$/lb$_m$

Velocity head ($\alpha \sim 1$)
    Flow rate   100 gpm,   0.2228 cfs
    Flow area   0.0233 ft^2
    Velocity   0.2228/0.0233 = 9.56 fps
    Kinetic energy   U^2/2g$_c$ = 9.56^2/(2 x 32.17) = 1.42 ft-lb$_f$/lb$_m$
    Velocity head   U^2/2g = 9.56^2/(2 x 32.17) = 1.42 ft

Total discharge head, h$_d$

	feet	energy/mass (ft-lb$_f$/lb$_m$)
Gauge reading	92.352	92.352
Barometer	33.487	33.487
Correction	(+) 2.0	(+) 2.0
Velocity energy	1.42	1.42
Total	129.259	129.259

It is to be noticed the correction to the pump center line (potential energy) and velocity energy each are less than two percent of the total discharge head in this case. The percentage is even less with higher pressure pumps and, as will be shown later, is partially subtracted in subsequent computations. For this reason, these two factors are often neglected. It is wise not to neglect them so as not to fall into conceptual traps.

Notice also the sum of the gauge pressure and the atmospheric pressure is an absolute pressure. A source of error is the omission of the barometric correction (because in most cases involving incompressible fluids, the barometric pressure cancels). When dealing with compressible fluids, the correction is mandatory. Therefore, it will always be included in this book.

Pumps — Theory and Equations | CHAPTER THREE

Example of metric units: Total discharge head

Four hundred liters per minute of 25°C water are pumped at a flange discharge gauge pressure of 250 kPa. What is the total discharge head in meters and in joules per kilogram or newton-meters per kilogram? The gauge center line is two meters below the pump center line. The local barometer reading is 100 kPa. The pipe is 2 inch, Sch. 40 (not unusual, even in metric areas).

Discharge

Mass density	997 kg/m^3	
Gauge pressure	250 kPa (kN/m^2)	
Meters of fluid	P/(ρg) (1,000 x 250)/(997 x 9.805) = 25.574 m	
Energy/mass	P/ρ (1,000 x 250)/997 = 250.752 joules/kg (N-m/kg)	

Atmospheric pressure

Barometer	100 kPa	
Meters of fluid	(1,000 x 100)/(997 x 9.805) = 10.229 m	
Energy/mass	(1,000 x 100)/997 = 100.301 J/kg (N-m/kg)	

Correction to absolute units

Pressure	350 kPa
Meters	35.803 m
Energy	351.053 J/kg (N-m/kg)

Correction to pump (or pipe) center line

Distance	(-) 2.0 m
Energy/mass	(-) 2.0 m x g/g_c = 2.0 x 9.805/1 = 19.610 J/kg (N-m/kg)

Velocity head ($\alpha \sim 1$)

Flow rate	400 l/m, 0.00667 m^3/s
Flow area	0.00216 m^2
Velocity	0.00667/0.00216 = 3.088 m/s
Kinetic energy	$U^2/2g_c$ = 3.088^2/(2 x 1) = 4.678 (m/s)^2 (N-m/kg) (J/kg)
Velocity head	$U^2/2g$ = 3.088^2/(2 x 9.805) = 0.486 m

CHAPTER THREE | **Pumps — Theory and Equations**

Total discharge head, h_d

	meters	energy/mass (N-m/kg, J/kg)
Gauge reading	25.574	250.752
Barometer	10.229	101.301
Correction	(-) 2.0	(-) 19.610
Velocity energy	0.486	4.678
Total	34.289	337.121

The difference between these two examples is that, in the first, the discharge pressure gauge was above the datum line (the center line of the pump) and the head or energy per unit mass had to be added. In the second, the gauge was below the center line of the pump and the head or energy had to be subtracted.

The second difference is that, in the U.S. customary system, the head and energy per unit mass are numerically identical. The problem comes with using these units in subsequent equations. This is the reason this book uses energy per unit mass units. In the metric system, the head unit, the meter, differs from the energy per mass unit, the newton-meter per kilogram or the joule per kilogram by the factor 9.805, the acceleration due to gravity, whose units are meters per second squared.

Example of customary U.S. units: Total suction head

We will build on the previous examples. Assume the inlet piping is 4 inch Sch. 40. The measured pressure at the pump inlet, one foot above the center line, is 6 psig. What is the total suction head?

Suction

Weight density	62.37 lb_f/ft^3
Mass density	62.37 lb_m/ft^3
Gauge pressure	6 psig, 864 psf
Feet of fluid	13.853 ft
Energy/mass	13.853 ft-lb_f/lb_m

Atmospheric pressure

Barometer	100 kPa
U.S. units	14.5 psia, 2,088 psf
Feet of fluid	33.478 ft
Energy/mass	33.478 ft-lb_f/lb_m

Pumps — Theory and Equations | CHAPTER THREE

Correction to pump center line
Distance	(+) 1.0 ft
Energy/mass	(+) 1.0 ft x g/g_c = 1.0 ft-lb_f/lb_m

Velocity head ($\alpha \sim 1$)
Flow rate	100 gpm,	0.2228 cfs
Flow area	0.0884 ft^2	
Velocity	0.2228/0.0884 = 2.520 fps	
Kinetic energy	$U^2/2g_c$ = 2.52^2/(2 x 32.17) = 0.0987 ft-lb_f/lb_m	
Velocity head	$U^2/2g$ = 2.52^2/(2 x 32.17) = 0.0987 ft	

Total suction head, h_s

	feet	energy/mass (ft-lb_f/lb_m)
Gauge reading	13.853	13.853
Barometer	33.487	33.487
Correction	(+) 1.0	(+) 1.0
Velocity energy	0.0987	0.0987
Total	**48.439**	**48.439**

Example of metric units: Total suction head

Assume the inlet piping is 4 inch Sch. 40 (Again, the reader is reminded it is not unusual to see American piping in metric areas). The measured pressure at the pump inlet gauge, one meter above the center line, is 40 kPa. What is the total suction head?

Suction
Mass density	997 kg/m^3
Gauge pressure	40 kPa (kN/m^2)
Meters of fluid	$P/(\rho g)$ (1,000 x 40)/(997 x 9.805) = 4.092 m
Energy/mass	P/ρ (1,000 x 40)/997 = 40.120 joules/kg (N-m/kg)

Atmospheric pressure
Barometer	100 kPa
Meters of fluid	(1,000 x 100)/(997 x 9.805) = 10.229 m
Energy/mass	(1,000 x 100)/997 = 100.301 J/kg (N-m/kg)

Correction to pump (or pipe) center line
Distance	(+) 1.0 m
Energy/mass	(+) 1.0 m x g/g_c = 1.0 x 9.805/1 = 9.805 J/kg (N-m/kg)

CHAPTER THREE | Pumps — Theory and Equations

Velocity head

Flow rate	400 l/m,	0.00667 m³/s
Flow area	0.00821 m²	
Velocity	0.00667/0.00821 = 0.812 m/s	
Kinetic energy	$U^2/2g_c = 0.812^2/(2 \times 1) = 0.330$ (m/s)²	
	(N-m/kg) (J/kg)	
Velocity head	$U^2/2g = 0.812^2/(2 \times 9.805) = 0.0337$ m	

Total suction head, h_s

	meters	energy/mass (N-m/kg, J/kg)
Gauge reading	4.092	40.120
Barometer	10.229	101.301
Correction	(+) 0.3	(+) 2.941
Velocity energy	0.0337	0.330
Total	14.655	144.692

III-10: CHAPTER SUMMARY

The purpose of this chapter was to give a narrative description of centrifugal pumps, positive displacement pumps, jet pumps and turbines and to describe how they function when integrated into hydraulic systems. The functions and the installed and inherent characteristics of pumps were discussed in fair detail. Centrifugal pumps were given much more attention than the other types because of their importance in the process industries. An expansion of the technical details will be found in Appendix AIII.

- Pumps were described in terms of their primary function – the transfer of mechanical energy to a liquid. Transfer of energy is sometimes lost sight of in our discussions of head and pressure.
- Six different means of energy transfer were described with the emphasis given to the most common one in the process industries – centrifugal force.
- A brief history of pumps was given to show they indeed are a very mature technology. "Mature" means well documented and trustworthy.
- A commonly accepted, although somewhat arbitrary, classification of pumps was given in order to facilitate further description.
- The characteristics of various pumps were described. The emphasis was placed on what is important to those who wish to understand those characteristics that are important to operations and maintenance personnel.
- It was shown that centrifugal pumps have turndown limitations and

Pumps — Theory and Equations | CHAPTER THREE

they are sometimes subject to surging problems. It was stressed that positive displacement pumps cannot be throttled. Flow control must be by varying the speed of the driver or by diverting some of the flow to the pump suction or elsewhere.

- Miscellaneous types of pumps such as gas lift pumps, jet pumps and electromagnetic pumps were presented in order to complete the picture.
- Inherent characteristics were compared to installed characteristics of centrifugal pumps. The problem of suction lift and net positive suction head was touched upon.
- Control of flow through pumps was introduced. Emphasis was on the most common process pump, the centrifugal pump. It was pointed out centrifugal pumps operating in parallel must have carefully matched head curves and one pump can actually force the other to the shut off point if they are not well matched. Pumps in series, booster pumps, were also discussed.
- It was shown many variable speed drives operate in a fixed speed mode because of sluggishness of the driver and because of the fact impellers are designed to be most efficient at one flow rate.
- Hydraulic turbines used for power recovery were presented briefly.
- The chapter closed with a few worked examples.

CHAPTER FOUR IV

Compressible Fluid Flow

IV-1: SCOPE OF CHAPTER — COMPREHENDING COMPRESSIBLE FLOW

Chapter IV will build upon the previous discussions of incompressible flow in an attempt to render the phenomena of compressible flow comprehensible. It will discuss and explain the various conceptual models used to simplify the analysis of compressible flow problems. It will attempt to show the utility and limitations of these models, including the various equations-of-state. The equations developed for incompressible fluid flow will be shown to be applicable with modifications to many plant situations. The modifications and the range of application will be discussed.

Choked flow

The concept of choked flow will be introduced and explained. In particular, the important contribution of Peter Paige[xxiv] to the understanding of the choking phenomenon will be presented. The first step in any compressible flow computation should be to investigate the possibility of choked flow. If this is not done, the subsequent computation may be seriously in error in the unsafe direction. The concept of choked flow will be introduced in this chapter and specific computation methods will be given in Appendix AIV.

Mixed phase flow

Compressible flow includes mixed phase flow. Mixed phase flow is not easily measured and is controlled only indirectly. It does have to be understood, however, and accommodated. Control systems professionals encounter mixed phase flow when they size pressure relief devices and the associated downstream piping. They also come across it when sizing control valves in which flashing flow is occurring. Extensive work has been done on mixed phase flow resulting in some rather complicated theories. Fortunately, Paige[xxiv], Richter[xxv] and others have been able to develop working equations that make many mixed phase problems more tractable. Their work will be discussed in fair detail.

CHAPTER FOUR | **Compressible Fluid Flow**

Mach number

A new parameter, the Mach number, will be introduced. The Mach number is important in computations for flow velocities in excess of about 0.3 Mach and for noise computations. Mach 1 is the speed of sound at the conditions of temperature and pressure of the fluid. We will point out the use of the Mach number concept is limited within the terms of reference of this book. In order to introduce the application of the Mach number in solving higher velocity flow problems, the ideal nozzle will be presented briefly. We will then develop three basic models for compressible flow. The adiabatic model is the one most applicable to the ordinary industrial flows covered in this book.

More complicated mathematics

In trying to explain the concepts of compressible flow, it is not possible to avoid some rather complicated mathematics. However, to use the equations developed for compressible flow, it is not necessary to fully understand the derivations. It is necessary to have a grasp of the limitations involved with certain assumptions so as not to fall into the trap of using cookbook methods where they do not apply. The mathematical content in this chapter tends to be greater than in other chapters. We will, however, try to explain the logic behind the mathematics.

IV-2: DIFFERENCES BETWEEN COMPRESSIBLE AND INCOMPRESSIBLE FLOW

Many equations of compressible flow are modifications of those of incompressible flow. Therefore, it is important to have a grasp of the differences between the two types of flow in order to be able to judge the fields of application of the equations.

Density change

An incompressible fluid flowing through a conduit has its density changed only slightly, often negligibly, due to temperature changes. The temperature change can be due to fluid flow irreversibilities or due to heat exchange. The first cause results in a small temperature rise that often may be ignored. The second cause, heat exchange, can result in viscosity changes that alter the Reynolds[v] number and the friction factor. Even in the second case, problems can be solved with a fair degree of accuracy by using average friction factors over segments of conduit.

Density variations become very important in compressible flow. The Reynolds number remains important, but at high velocities, the Mach

Compressible Fluid Flow | CHAPTER FOUR

number becomes important. Acceleration effects absorb available energy and can result in the choking phenomenon both for gases and flashing liquids.

Acceleration in uniform conduits

In steady state flow of incompressible fluids in uniform conduits, there is essentially no acceleration of the fluid. In steady state flow of compressible fluids in uniform conduits, there is acceleration of the fluid. Although the velocity at a point is constant, with compressible flow the velocities increase in the direction of flow. As the molecules of the compressible fluid separate due to the acceleration, both the temperature and the density of the fluid decrease. These effects are not found with flowing liquids.

An unaccelerated fluid in a conduit has static energy and internal energy associated with temperature. Mechanical energy and thermal energy are loose terms for two forms of energy. Mechanical energy is recognized as pressure-volume energy, as kinetic energy or as energy by virtue of elevation. Thermal energy is energy associated with temperature change. It becomes internal energy and is often dissipated due to a temperature difference. Only mechanical energy is useful in producing flow (if we ignore convective effects from density changes). An unavoidable consequence of flow is the conversion of some mechanical energy to internal energy that is no longer useful in terms of flow.

Acceleration requires energy, and this energy is obtained from the available static energy of the fluid. The result of continued acceleration is more and more energy is required from the available static energy and less and less is available for the conversion process – to overcome the irreversibilities, the losses, the "fluid friction".

Major difference between compressible and incompressible flow

A major difference between compressible and incompressible flow is, with incompressible flow, the pressure at one extreme of the system is always known. The problem resolves itself to computing forward or backward from the known pressure. With compressible flow, it frequently happens that neither the upstream nor the downstream pressure is known. That is to say, the source pressure and the sink pressure may be known, but neither the pressure inside the inlet of a conduit nor inside the discharge are known. A solution to this problem will be given in this chapter.

CHAPTER FOUR | Compressible Fluid Flow

IV-3: USING MODELS

Models are useful conceptual tools for simplifying analysis and as an aid for performing computations. Ideally, in the industrial situation, the use of a model will produce a reasonable worst-case estimate of the real situation. There are many kinds of models. Some large-scale computer models make use of many simpler smaller-scale models.

Equations-of-state

Equations-of-state are the primary models relating pressure, temperature, volume and composition of compressible fluids. A grasp must be had of the limitations of each equation-of-state. The ideal gas model is the easiest to use, but it is the most inaccurate. It is fair to say that the more accurate the equation-of-state, the more complicated it becomes. Also, the more complicated, the greater the number of parameters needed for its solution. Sometimes, all parameters are not available. A little knowledge allows one to make a judgement as to when it is necessary to abandon a simpler model for a more complicated one. We will compare the ideal gas model, the original Redlich-Kwong[xxviii] model, and the virial model in this chapter.

Adiabatic and isothermal models

There are two general models used for compressible flow in closed conduits – the adiabatic model and the isothermal model. These are simplifications that allow further computations. They are not equations-of-state. For plant work involving relatively short distances with relatively large changes in pressure in insulated conduit, the adiabatic model is useful. For pipeline work involving relatively long distances and relatively small changes in pressure, or for long, uninsulated, small diameter pipes, the isothermal model is useful.

The adiabatic model is one in which the heat transfer to or from the fluid across the conduit wall is considered negligible. In the isothermal model, sufficient heat transfer takes place to keep the temperature of the flowing fluid constant.

Unsatisfactory models

There are times when neither the adiabatic nor the isothermal model is satisfactory. When flow is through heat exchangers, or when extremely cold fluids are subject to heat transfer due to warmer ambient temperatures, changes in density and even vaporization can make the use of the simpler models completely incorrect. In this case, a model involving heat transfer is clearly needed.

Compressible Fluid Flow | CHAPTER FOUR

Polytropic model

In turbomachinery work, the polytropic model is frequently used. This model is not useful for industrial piping networks because the heat flow across the conduit walls is not sufficiently predictable. We will not be concerned with this model in our discussions.

Main models

The main models with which to be concerned in the industrial setting are the equations-of-state, the adiabatic model, the isothermal model and the model for heat transfer. Whenever a model is being used in place of exact computations, it should be examined critically to see it really applies to the situation at hand.

The adiabatic model assumes there is no energy flow to or from the fluid under a temperature difference. This may be a reasonable assumption in many cases when the conduit is short or is well insulated and, when the fluid is hotter than the environment; it may even give a factor of safety. However, when very cold or cryogenic fluids are flowing (flare headers, for instance), there is a flow of energy from the environment to the fluid. The enthalpy (internal energy plus Pv energy) of the fluid increases, as does its specific volume. The friction factor is greater. Choking may occur. The adiabatic model, in this case, can cause errors in conduit sizing in a direction that is not conservative

Given the fact it is almost impossible to predict overall heat flow in a piping system with any accuracy and given the fact the adiabatic model is conservative in most cases (except the one mentioned in the previous paragraph), it is a good model for most plant situations. However, it is always wise to check the assumptions. If the temperature difference between the fluid and the outside environment is not too great or if the conduit is well insulated, the adiabatic model is usually a good one. If a gas or vapor is deliberately being heated or cooled, the adiabatic equations do not apply; at least, not at the point of the heat exchange. Note the word "adiabatic" simply means the absence of heat energy flow. This situation may be obtained by deliberately heat tracing a conduit or a vessel and controlling the temperature of the jacket so its temperature is the same as the fluid.

In the average plant, most flows are kept below about 0.3 Mach to avoid excessive industrial noise and potential damage due to vibration. This is the flow range where modified compressible flow equations (adiabatic) are most useful.

CHAPTER FOUR | Compressible Fluid Flow

The isothermal model is a tempting one to use because isothermal equations are simpler than the adiabatic ones. It is clearly incorrect to use this model for most plant computations involving large changes in pressure and temperature. The fluid densities predicted by this model may be grossly in error almost everywhere but at the starting point of the computations.

Using computer simulations

It is not possible to perform more than trivial compressible flow computations without the use of a computer. Usually, there are too many variables that change simultaneously to allow hand computations. The simplifying assumptions involving ideal gases do not always hold. Some mixtures (but not all) of varying composition require some complicated equations-of-state. A good computer simulation program can reduce an otherwise untractable problem to manageable proportions.

It is the author's strong conviction that lack of knowledge of basic principles on the part of some users causes computer simulations to be run in a cookbook fashion. Unchecked, incorrect assumptions can lead to major errors – even when computers are used. A good grasp of the basic equations of compressible flow gives a comfort level that otherwise does not exist. The careful reader should obtain this understanding of the basic equations from this book and should achieve the ensuing comfort level.

IV-4: TREATING MIXTURES

For didactic purposes, it is convenient to deal with pure (single component) fluids. In real industrial life, such fluids are rare. Even natural gas, which we tend to think of as a pure substance somewhat related to methane, varies in its properties depending on its source. It varies so much that gases from different sources frequently are blended in order to obtain a more uniform mixture.

In refineries, in particular, the gases can be mixtures of many different components: H_2, CO, CO_2, O_2, N_2, H_2S, CH_4, C_2H_6, C_3H_8, C_4H_{10}, and many, many more. In order to measure the flows, one must know the mixture properties as a function of the component properties. Even air is a mixture and one tends to forget the trace components such as argon and carbon dioxide.

The properties of gases are functions of their temperature, pressure and composition. These properties influence the flow characteristics of the

Compressible Fluid Flow | CHAPTER FOUR

mixture and the response of measuring and controlling devices largely through the overall viscosity and overall density.

Composition of a gas or vapor must be known or be estimated before meaningful computations can be made. The process flow diagram and its associated material balance are the normal sources of information on composition.

The equation-of-state used to describe a pure component must be modified to describe a mixture. Some equations-of-state describe pure substances quite well and mixtures very poorly. The original Redlich-Kwong[xxviii] equation works with reasonable accuracy for both pure substances and for mixtures.

Mixing rules

A set of "mixing rules" must be decided upon. These rules give the correlation between the single component properties and the mixture properties. In many instances, these rules are laid down by an organization – a company, a gas association, etc. In other cases, they are chosen based on the need for accuracy or based on balancing accuracy against complexity. We will give some general guidelines later.

IV-5: EQUATIONS OF COMPRESSIBLE FLOW OF AN IDEAL GAS

It is frequently convenient, when dealing with complicated relationships, to start from an idealization. Once the basic equations are worked out, it is possible to compare the ideal situation with the real one and to make allowances for the differences. This is why we start with the ideal gas.

An ideal gas is simply one that follows the simple relationship $Pv = RT$, where P is the absolute pressure, v is the volume per unit mass, R is the particular gas constant on a mass basis (the universal constant, R_o, molar basis, divided by the molecular weight) and T is the absolute temperature. Most gases follow the ideal gas relationship (often loosely called a law) at sufficiently low pressures and high temperatures. Corrections are necessary for higher pressures and lower temperatures.

Some gases such as carbon dioxide and the vapors of organic acids deviate strongly from the above ideal gas relationship – even at low pressures and high temperatures. With these gases and vapors, the ideal gas equation is inaccurate. Nevertheless, it is still a very useful relationship, especially as a first approximation – as a base for further computations.

CHAPTER FOUR | **Compressible Fluid Flow**

Usual assumptions for the equations of ideal gas flow

We will state the usual assumptions used in an ideal gas analysis in closed conduit, as follows:

1. steady-state, one dimensional flow;
2. ideal gas as defined above with a constant heat capacity, c_p;
3. no shaft work (we are dealing with flow through conduits, not pumps, compressors or turbines);
4. gravitational affects are negligible;
5. velocity gradients within a cross section are negligible (the coefficient alpha in the Bernoulli[i] equation equals approximately one).

Basic relationships to be developed

Some relationships are more important than others are. The ones we are interested in are:

1. the equation of continuity;
2. the steady state, total energy balance;
3. the total energy balance with irreversibilities due to shear;
4. the equation for the velocity of sound in a fluid;
5. the ideal gas equation-of-state.

We will now develop the above relationships with some explanations.

Continuity

The continuity relation states that, under the assumption of steady state flow without accumulation of matter, the mass flow rate is constant. If the mass flow rate is measured at several points it will be found to be the same. However, the mass flow rate can be factored into velocity, conduit cross section area and fluid density. Average velocity, U, changes in compressible flow. The area of the conduit, A, may or may not change. Density definitely changes. The product of all three terms remains constant in steady-state mass flow. We can take logarithms in order to separate the terms and differentiate in order to get rid of the constant terms and to produce a differential equation as follows:

Compressible Fluid Flow | CHAPTER FOUR

$$\dot{m} = UA\rho = const \tag{IV-1}$$

$$\ln \dot{m} = \ln U + \ln A + \ln \rho$$

$$d\ln \dot{m} = d\ln U + d\ln A + d\ln \rho = 0$$

$$\frac{dU}{U} + \frac{dA}{A} + \frac{d\rho}{\rho} = 0$$

This differential equation (IV-1) allows exploring the relationships among average velocity across a section, area and density. The only assumption was that of a constant mass flow rate. The differential equation also allows the development of integrated equations that are simpler to grasp.

Total energy balance, steady state

The first law of thermodynamics is a total energy balance. It is not so directly useful as the Bernoulli[i] equation for fluid flow work. However, it is necessary to have an understanding of the first law. The first law of thermodynamics expressed in terms of open (flowing) systems on a rate basis is,

$$\dot{Q} + \frac{\dot{W}_f}{J} = \dot{m}\left[\frac{U_2^2 - U_1^2}{2g_c J} + (h_2 - h_1) + \frac{g(Z_2 - Z_1)}{2g_c J}\right] \tag{IV-2}$$

In SI work, the reader is again reminded the dimensional constant, g_c, equals one and has no dimensions. In addition, the "mechanical equivalent of heat", J, is 778.16 ft-lb$_f$/Btu in customary U.S. units and is one and is dimensionless in SI units. The above equation is therefore valid for both systems. The two enthalpy terms, h_1 and h_2, are not hydraulic heads. All of the terms in the brackets have the same units of energy per unit mass.

The heat energy and fluid work terms, Q and W_f, are those amounts of energy that cross the control volume during the period that the amount of mass transits the same control volume. A dot above the three terms, Q, W_f and m, puts them on a rate basis, heat energy flow, work flow and mass flow per unit time. If equation IV-2 is divided through by the mass flow rate, the heat energy and shaft work terms are on a unit mass flowing basis. Lower case letters would then designate them. Note fluid work is often incorrectly called shaft work. We have used the term fluid work to emphasize that mechanical work to or from the shaft differs from mechanical work to or from the fluid because of inefficiencies within the pump or turbine mechanism or because of irreversibilities in the fluid. We can impose constraints on the above equation such as limiting the

CHAPTER FOUR | Compressible Fluid Flow

analysis to one in which there is no change in elevation, $Z_2 = Z_1$. This constraint usually does not introduce a major inaccuracy when the fluid is not too dense and the elevation change is not too great. (This assumption should always be checked with real numbers). We have already stated that another constraint can be that there is no fluid work, W_f, done on or extracted from the fluid, so we can drop the term containing W_f.

The above constraints allow us to simplify the first law and to put it into a differential equation as follows:

$$\frac{\dot{Q}}{\dot{m}} = q = \left[\frac{U_2^2 - U_1^2}{2g_c J} + h_2 - h_1 \right] \quad \text{(IV--3)}$$

$$\frac{\delta \dot{Q}}{\dot{m}} = \delta q = \frac{dU^2}{2g_c J} + dh$$

The Greek delta on the heat energy transferred, Q or q, is to remind us thermal energy is not an exact differential, in other words it must be integrated over a particular path. The value of the integral does not depend only on the limits of integration. The last equation states that differential heat energy transfer per unit mass flowing is equal to the algebraic sum of differential kinetic energy and the differential enthalpy.

Again, we have to remember the constraints or limitations we have accepted. If necessary, we can always go back and restore variables that have been neglected. The above equation relates changes in heat transfer across the wall of a conduit per unit of mass flowing to changes in the fluid's kinetic energy and enthalpy. In the case of adiabatic flow, we can impose a further constraint by making the heat energy flow term equal to zero. We will do this later in the analysis of adiabatic flows.

Mechanical energy balance modified to include 'losses'

The Bernoulli[i] equation is often referred to as a mechanical energy balance. However, when the loss term, h_L is included, the equation also contains a term describing mechanical energy converted to internal energy or heat flow, or both.

$$\frac{P_1}{\rho_1} + \frac{\alpha_1 U_1^2}{2g_c} + \frac{gZ_1}{g_c} + w_f = \frac{P_2}{\rho_2} + \frac{\alpha_2 U_2^2}{2g_c} + \frac{gZ_2}{g_c} + h_f \quad \text{(IV--4)}$$

The above form of the equation is valid for liquids and for gases whose density change is known. The alpha correction to each kinetic energy term adjusts this term to correct for velocity profiles across a section of conduit. The velocity in the kinetic energy term is the average velocity across the section. All terms are on a per unit mass basis. The first three terms on each side of Equation IV-4 are functions of the particular sections. The fluid work term represents work supplied to the fluid anywhere between points 1 and 2. Similarly, the "loss" term is the accumulation of mechanical energy converted to internal energy, enthalpy, heat flow and anything else, between sections 1 and 2 — it is not "lost", in spite of the name.

'Skin friction'

Equation IV-4 can be made more applicable to turbulent, compressible flow in a conduit by considering alpha to be equal to one, no compressor in the segment under consideration, constant density over a short length and all the losses to be due to "skin friction". The term "skin friction" is a synonym for irreversibilities in a conduit, and it arises from the fact there is a more rapid change in velocity profile close to a solid wall and a less rapid one remote from the wall. Most of the shear resistance to flow arises nearest a wall. However, in a conduit, shear exists across the entire cross section. It is simply more intense near the wall. The subscript, fs, sometimes used on the loss term emphasises this fact.

$$\frac{P_2 - P_1}{\rho} + \frac{g(Z_2 - Z_1)}{g_c} + \frac{U_2^2 - U_1^2}{2g_c} + h_{fs} = 0 \quad \text{(IV–5)}$$

Putting Equation IV-5 into differential form and considering elevation changes to be negligible gives

$$\frac{dP}{\rho} + d\left(\frac{U^2}{2g_c}\right) + dh_{fs} = 0 \quad \text{(IV–6)}$$

Equation IV-6 describes changes over a very short (differential) length of horizontal conduit.

The differential form of the Darcy[vii] equation, Equation II-3, for horizontal conduits of circular section is repeated as Equation IV-7.

$$dh_{fs} = f_M \frac{U^2}{2g_c} \frac{dL}{D} \quad \text{(IV–7)}$$

In Equation IV-7, the average velocity, friction factor and diameter are considered constant over the differential length of conduit.

For non-circular conduits in turbulent flow, $D = 4r_H$. The equivalent diameter equals four times the hydraulic radius. The hydraulic radius is defined

CHAPTER FOUR | Compressible Fluid Flow

as the conduit cross sectional area divided by the wetted perimeter.

Combining equations IV-6 and IV-7 and using the concept of the hydraulic radius gives Equation IV-8.

$$\frac{dP}{\rho} + d\left(\frac{U^2}{2g_c}\right) + f_M \frac{U^2}{2g_c} \frac{dL}{4r_H} = 0 \qquad \text{(IV–8)}$$

This last equation is a differential equation suitable for developing relationships for compressible flow in a straight, horizontal conduit of any section. Because this is a differential equation, the assumption of incompressibility can be removed. If the conduit is circular, four times the hydraulic radius may be replaced by the diameter.

Sonic velocity

Most industrial flows are deliberately kept to Mach numbers below about 0.3 Mach in order to avoid excessive noise and vibration in piping. Therefore, it is rare the Mach number concept is needed in the context of ordinary industrial flow in conduits. The Mach number is important when dealing with turbo-machinery. We will not dwell on the subject beyond the time necessary to grasp the concept. We will also give simpler methods of handling related problems in a later section.

The velocity of sound in the compressible fluid becomes important when computing the parameter called the Mach number. The velocity of sound in a fluid was first given by Laplace[xxvi] as a correction to Newton's original hypothesis. Newton had assumed the transmission of sound in fluids followed an isothermal process. Laplace showed that, *in order to make Newton's theory match experimental data*, the process had to be considered to be adiabatic, not isothermal.

The resulting acoustic velocity is given by the differential equation, Equation IV-9.

$$a = \sqrt{g_c \left(\frac{\partial P}{\partial \rho}\right)_s} \qquad \text{(IV–9)}$$

The velocity of sound is equal to the square root of the product of the dimensional constant and the partial derivative of pressure with density at constant entropy. The Mach number is defined as the ratio of the actual fluid velocity to the above sonic velocity. Mach one is reached when a fluid travels with a velocity equal to the velocity of sound in the fluid. It is

Compressible Fluid Flow | CHAPTER FOUR

usually more familiar as the velocity of an object, such as an aeroplane, travelling in a fluid, such as air.

It is to be noted the choked flow velocity and sonic velocity are not always the same. More on this a little later.

Ideal gas equations

Ideal gas equations simplify analysis. For many industrial situations they are not too far from reality. However, this last assumption must be re-examined each time it is made. We will discuss this problem more thoroughly in Section IV-6: Ideal and Non-Ideal Gases – Comparison of Some Equations-of-State

An ideal gas is one that obeys the following equation-of-state:

$$Pv = RT \qquad \text{(IV–10)}$$

$$P = \frac{R_o}{(MW)} \rho T$$

The first gas constant, R, is the specific gas constant for a particular gas. Its units are energy per unit mass per degree. The numerical value of the gas constant depends upon the units used in the formula. The second gas constant, R_o, is the so-called universal gas constant. The "universal" gas constant is on a molar basis. It must be divided by the molecular weight to obtain the specific gas constant on a mass basis. In customary U.S. units, R_o equals 1546 foot-pounds-force per pound-mole-degree Rankin. In SI units, it equals 8,314.3 joules per kg-mole-degree Kelvin.

Should a gas constant using a Btu as the energy unit be chosen, the "mechanical equivalent of heat", J, must be a factor

By taking logs and differentiating, we can separate the individual terms and get rid of the constant terms in IV-10.

$$\ln P = \ln \frac{R_o}{(MW)} + \ln \rho + \ln T \qquad \text{(IV–11)}$$

$$d \ln P = d \ln \rho + d \ln T$$

$$\frac{dP}{P} = \frac{d\rho}{\rho} + \frac{dT}{T}$$

CHAPTER FOUR | Compressible Fluid Flow

An assumption frequently made for the sake of simplifying the analysis is the specific heat at constant pressure is independent of temperature over the range of the analysis. When this assumption is made for an ideal gas, the specific enthalpy becomes proportional to temperature only. We can obtain the following differential equation:

$$h = h_o + c_p(T - T_o) \qquad \text{(IV–12)}$$
$$dh = c_p dT$$

That the constant heat capacity assumption must be carefully made can be verified by looking at graphs of heat capacity versus temperature. Assumptions must not be made arbitrarily. Fortunately, the ratio of specific heats, c_p/c_v or γ, varies with temperature less than does either of the two specific heats. This is the justification for the ratio of specific heats to be taken as constant in what follows.

The word 'isentropic'

The term isentropic means constant entropy. Practically, it means the process (the change being considered) is close to ideal, without noticeable irreversibilities, and is adiabatic. There is no heat energy flow. The only way to change the value of the quantity known as entropy in an ideal process is by allowing energy to cross the boundary under a temperature gradient. This flow of energy under a temperature gradient is called heat flow. The energy involved is often loosely called "heat".

Ideal gases with constant γ, adiabatic, reversible processes

The reader should consult any text on basic thermodynamics if he or she wishes a more thorough understanding of some of the derivations that follow. For an ideal isentropic (adiabatic, reversible) process involving an ideal gas with a constant ratio of specific heats, the following relationships are valid:

$$\gamma \equiv \frac{c_p}{c_v} = \frac{c_p}{c_p - R_o/(MW)J} \qquad \text{(IV–13)}$$

$$Pv^\gamma = const$$

$$P\rho^{-\gamma} = const$$

$$\frac{T}{P^{1-1/\gamma}} = const$$

$$TP^{-(1-1/\gamma)} = const$$

Compressible Fluid Flow | CHAPTER FOUR

The last three constants are not identical and the adiabatic, reversible constraint must be followed. A gas flowing isentropically has to follow an adiabatic process because of the definition of entropy as δq/T. The molar mass is introduced so as to put the gas constant on a mass basis and J, "the mechanical equivalent of heat", is introduced because the heat capacities, in North America, generally have Btu's as their energy unit.

We can use Laplace's[xxvi] equation, Equation IV-9, to find the sonic velocity of an ideal gas under the isentropic constraint as follows,

$$P\rho^{-\gamma} = const = \frac{P}{\rho^{\gamma}} \tag{IV-14}$$

$$d\ln P - d\ln \rho^{\gamma} = d\ln const = 0$$

$$\frac{dP}{P} - \gamma \frac{d\rho}{\rho} = 0$$

$$\left(\frac{\partial P}{\partial \rho}\right)_s = \gamma \frac{P}{\rho}$$

The isentropic pressure dependence on density is seen to equal the product of the ratio of the specific heats and the ratio of the absolute pressure to density for an ideal gas. This equivalence can be plugged into Laplace's equation, Equation IV-9.

$$a = \sqrt{g_c \left(\frac{\partial P}{\partial \rho}\right)_s} = \sqrt{g_c \gamma \frac{P}{\rho}} = \sqrt{g_c \gamma T \frac{R_o}{(MW)}} \tag{IV-15}$$

Equation IV-15 can be used to obtain the velocity of sound in air taken as an ideal gas at 60°F (520 R) with the ratio of specific heats equal to 1.4, the molecular mass equal to 29, the universal gas constant equal to 1546 ft-lb$_f$/lb$_{mol}$-R and the dimensional coefficient equal to 32.17 lb$_m$-ft/lb$_f$-s^2. The velocity of sound in air under these conditions is 1117 ft/s or 340.5 m/s.

The acoustic velocity is seen to be a function only of the absolute temperature for a given ideal gas with constant ratio of heat capacities. The Mach number for an ideal gas can be written as follows (the variable, a, is the acoustic velocity),

$$N_{Ma} = \frac{U}{a} \tag{IV-16}$$

$$N_{Ma}^2 = \frac{U^2}{a^2} = \frac{U^2}{g_c \gamma T R_o/(MW)} = \frac{\rho U^2}{g_c \gamma P}$$

Flow of Industrial Fluids—Theory and Equations

CHAPTER FOUR | **Compressible Fluid Flow**

Equation IV-16 might look fairly simple, but average velocity, U, temperature, T, pressure, P, and density, rho, must all be measured, or estimated at flowing conditions at the point of interest.

IV-6: IDEAL AND NON-IDEAL GASES – COMPARISON OF SOME EQUATIONS-OF-STATE

Ideal gas

The ideal gas model assumes gases are made up of spherical molecules that do not interact with one another except during elastic repulsions. At low pressures and high temperatures this model is not too bad an assumption for most gases. However, molecules do interact with one another to a degree dependent on their size, shape and physical or chemical affinity. At pressures and temperatures approaching the critical point (the peak of the vapor dome), the ideal gas equation is so inaccurate as to be unusable.

Corresponding states

Corresponding states theory holds that gases, when reduced to fractions of their critical values, all behave in the same manner. The theory gives an approximation of true behavior. A dimensionless correction often is applied to the ideal gas law, $Pv = RT$, in the form of the compressibility factor, Z. This gives $Pv = ZRT$. The correction factor, Z, can be estimated from corresponding states considerations. Its use can change an accuracy of plus or minus 20% to plus or minus 5%, as a broad example involving highly non-ideal gases.

Cubic equations-of-state

The Dutch physicist, van der Waals[xxvii], circa 1870, was one of the first to try to improve the ideal gas model with a model that involved two parameters. One parameter allowed for the volume occupied by the molecules and the other allowed for molecular interaction – simultaneous repulsion and attraction. The van der Waals model stimulated several improvements of which the original Redlich-Kwong[xxviii] equation, circa 1950, is probably still the most famous. The original R-K equation is particularly useful for mixtures of gases. It is still only a two-parameter equation, but it gives surprising accuracy for many gas mixtures. The van der Waals equation now been largely replaced by other two and three parameter models.

Compressible Fluid Flow — CHAPTER FOUR

We will give the original R-K equation here, since it is the equation with which we will work most frequently in this book. The gas "constant" and the volume are molar quantities. Mass quantities could be used with a mass gas constant.

$$P = \frac{RT}{\underline{v}-b} - \frac{a}{\underline{v}(\underline{v}+b)T^{0.5}}$$

(IV–17)

The constants, a and b, are functions only of the critical pressures and temperatures of a substance. In addition, fairly straightforward mixing rules allow the constants to be altered to accommodate mixtures.

$$a = 0.42748 \underline{R}^2 T_c^{2.5} / P_c$$

(IV–17A)

$$b = 0.086640 \underline{R} T_c / P_c$$

$$a_{mixture}^{0.5} = \sum_i y_i a_i^{0.5}$$

$$b_{mixture} = \sum_i y_i b_i$$

The first set of coefficients, a and b, refers to pure components. The second set refers to mixtures. The summations involving the pure components allow computation of the corresponding coefficients for the mixtures. Mixtures are more common than pure components in the industrial situation. Even dry air is a mixture generally given as nitrogen (0.7809), oxygen (0.2095), argon (0.0093) and carbon dioxide and trace elements (0.0003). Air is usually treated as a pure component that obeys the ideal gas equation. At cryogenic temperatures, this is no longer acceptable. It has to be treated as a mixture.

In order to improve the accuracy of the R-K equation, a third parameter was introduced. This parameter was supposed to give a measure of the non-spherical nature of many molecules. This is the so-called acentric factor. Soave modified the R-K equation using this parameter. The result was the R-K-S equation. The Peng-Robinson equation is very similar in structure to the R-K-S equation. Reid, Prausnitz and Poling[xxix] is a good source of the most widely used equations-of-state.

Additional parameters called interaction parameters have been introduced to describe the interactions between different binary pairs of molecules.

CHAPTER FOUR | **Compressible Fluid Flow**

Low density gas molecules are assumed to interact between only two molecules at a time – hence, binary interactions. All of these parameters improve accuracy at the expense of computational effort. They make the use of the computer mandatory.

The virial equation

The above mentioned equations-of-state are basically empirical. An equation-of-state with a firm theoretical background is the virial equation. However, in spite of being preferred by many because of its theoretical basis, it is not possible to use it over wide ranges of pressure. Nevertheless, Prausnitz[xxx] developed a solid database for its use in normal plant situations – less than several atmospheres and at temperatures between 200°K and 600°K.

These equations and others are described in any good text on thermodynamics. Here we will draw comparisons among the more common equations.

Comparisons among common equations-of-state

Mixed gases consist of molecules of different sizes and shapes moving at velocities inversely proportional to their molecular mass. The pressure on the walls of a container is proportional to the number of repulsions of molecules per unit time. The density of a gas is dependent on the number of molecules contained in a given volume and on the molecular masses. The temperature of a gas is related to its internal energy. An equation-of-state is a relationship among pressure, temperature and molar (or mass) volume. Mixtures of gases complicate the problem by introducing molecules that have different sizes and different affinities for one another.

To demonstrate the considerations that go into the choice of an equation-of-state, Table IV-1 has been prepared. This table is based on a common gas, but not necessarily a well behaved one – carbon dioxide. The table takes experimental data published in Perry[x] and gives the error in pressure by the application of three equations. The equations are the ideal gas equation, the virial equation truncated after the second term and the original R-K equation.

The table covers a range of common process pressures and temperatures, 300°K to 600°K and one bar to 60 bar (a maximum pressure of 20 bar is more common). We have limited the computations to three temperatures, 300°K, 450°K and 600°K. These three temperatures should be sufficient to show trends.

Table IV-1. Comparison of equations-of-state-for CO_2 with experimental data

Actual Data P, bar, versus v, m3/kg	Temperature		
	300°K	450°K	600°K
1.0	0.564	0.849	1.133
5.0	0.111	0.169	0.226
10.0	0.054	0.084	0.113
20.0	0.025	0.042	0.056
30.0	0.016	0.027	0.037
40.0	0.011	0.020	0.028
50.0	0.008	0.016	0.022
60.0	0.006	0.013	0.019

Ideal Gas Computations

300 K		450 K		600 K	
P, bar	ERR%	P, bar	ERR%	P, bar	ERR%
1.005	0.50	1.001	0.08	1.000	0.01
5.124	2.48	5.027	0.54	5.006	0.13
10.515	5.15	10.108	1.08	10.022	0.22
22.225	11.12	20.435	2.18	20.097	0.48
35.644	18.81	31.026	3.42	30.226	0.75
51.521	28.80	41.877	4.69	40.337	0.84
70.842	41.68	52.801	5.60	50.601	1.20
97.713	62.85	63.917	6.53	60.613	1.02

Virial Equation Computations

300 K		450 K		600 K	
P, bar	ERR%	P, bar	ERR%	P, bar	ERR%
1.000	0.01	1.000	-0.02	1.000	-0.00
4.995	-0.10	5.000	0.00	5.002	0.03
9.970	-0.30	9.999	-0.01	10.002	0.02
19.792	-1.04	19.987	-0.06	20.018	0.09
29.386	-2.05	29.993	-0.02	30.048	0.16
38.447	-3.88	39.996	-0.01	40.020	0.05
46.123	-7.75	49.812	-0.38	50.102	0.20
50.686	-15.52	59.536	-0.77	59.900	-0.17

Redlich-Kwong Computations

300 K		450 K		600 K	
P, bar	ERR%	P, bar	ERR%	P, bar	ERR%
1.000	0.02	0.999	-0.06	1.000	-0.03
4.999	-0.02	4.993	-0.15	4.995	-0.10
9.993	-0.07	9.970	-0.30	9.976	-0.24
19.945	-0.28	19.879	-0.61	19.917	-0.41
29.918	-0.27	29.766	-0.78	29.829	-0.57
39.897	-0.26	39.626	-0.93	39.646	-0.89
49.624	-0.75	49.293	-1.41	49.539	-0.92
59.307	-1.15	58.879	-1.87	59.125	-1.46

NOTE: Actual data from Perry - Computation by author

Source: R. Perry and D. Green, Perry's Chemical Engineers' Handbook, 1984, reprinted by permission McGraw-Hill Companies

CHAPTER FOUR | **Compressible Fluid Flow**

It can be seen that carbon dioxide gas approaches the ideal state at sufficiently high temperatures. All gases do this. By "approaching the ideal state" is meant the gas behavior can be described by the ideal gas equation-of-state. It can also be seen that even when temperatures are low, carbon dioxide gas approaches the ideal state when the pressure is sufficiently low. This is also true for all gases.

The "critical point" is a critical parameter. It is the highest point on the liquid-vapor dome. It is the point at which it is impossible to distinguish between a liquid and its vapor. The closer the state is to the critical point, the more difficult it is to obtain good accuracy with an equation-of-state.

Ideal gas equation

The ideal gas equation-of-state is the simplest and potentially the most inaccurate one. Simply by including the compressibility factor, Z, the accuracy of this equation can be improved several fold. The compressibility factor is obtained by computing the reduced pressures and temperatures – the ratios of the actual pressures and temperature to the critical ones. These ratios are then applied to the compressibility charts found in Perry[x] or in any thermodynamics textbook in order to obtain the compressibility. Note the compressibility factor is not constant; it varies with temperature and pressure.

Example IV-1: Corrected ideal gas equation

Carbon dioxide at 300°K (26.85°C) and 20 bar has a specific volume of 0.0255 cubic meters per kg. When this specific volume is used in the ideal gas equation without the compressibility correction, it has been shown that the computed pressure is plus 11.12% in error. Recompute the pressure using the compressibility factor, Z. The critical pressure and temperature of carbon dioxide are 73.8 b and 304.1°K respectively.

$P_r = P/P_c = 20/73.8 = 0.271$
$T_r = T/T_c = 300/304.1 = 0.987$

From the generalized compressibility charts of Nelson and Obert, the compressibility, Z, can be read as approximately 0.89. From Table 3-166 in Perry, it can be interpolated as 0.8871. The corrected pressure is then,

$P = ZRT/v = 0.8871(0.08314)300/(0.0255)(44.01) = 19.717$ bar

Error $= 100(19.717 - 20.0)/20.0 = -1.417\%$

The error in the pressure has been reduced from plus 11.12%, using the ideal gas equation, to minus 1.42% using the compressibility correction. This type of accuracy is adequate for many problems such as sizing control valves. It permits quick results of acceptable accuracy. It is not adequate for flow measurement, when custody transfer is involved, for instance. Also, when the pressure becomes very high or the temperature becomes low, greater inaccuracy may be expected.

The virial equation

The virial equation-of-state is an equation in reciprocal molar volume. The equation is given as

$$Z = 1 + \frac{B}{\underline{v}} + \frac{C}{\underline{v}^2} + \frac{D}{\underline{v}^3} + \ldots$$

$$Z = \frac{P\underline{v}}{R\underline{T}}$$

The virial coefficients, B (second), C (third), etc., are functions only of temperature and composition. A great deal of effort has been expended on developing the coefficients, but only the second virial coefficient has extensive data available for it. By truncating the virial equation after the second term, a very simple equation of surprising accuracy results. Reasonable industrial accuracy is limited to densities less than about half the critical density (Prausnitz[xxx]).

The virial equation is theoretically based and its coefficients are simply related for mixtures. For instance, the virial equation truncated after the second term has a mixture coefficient given by

$$B_{mixture}(T, y_i, \ldots y_m) = \sum_{i=1}^{m} \sum_{i=1}^{m} y_i y_j B_{ij}$$

Note inherent in the summation of only binary pairs is the assumption that only binary interactions are possible with low density gases.

The coefficient B_{ii} is a pure component coefficient. The coefficient B_{ij} is called the cross coefficient and B_{ij} is equal to B_{ji}. The cross coefficient is computed using mixing rules (See Perry[x] 3-268 for an extensive and easily followed discussion).

CHAPTER FOUR | Compressible Fluid Flow

Prausnitz[xxx] gives a rule of thumb for the pressure and temperature limitations of the truncated virial equation for use in typical process situations.

$$P \leq \frac{T}{2} \frac{\sum_{i=1}^{m} y_i P_{ci}}{\sum_{i=1}^{m} y_i T_{ci}}$$

If we apply the above rule of thumb to carbon dioxide at 300 K, we find the pressure should be limited to, P ≤ (300/2)73.8/304.1 = 36.4 bar. This is the point in Table IV-1 where the absolute error exceeds about two percent. This pressure corresponds to approximately 528 psia so we see the first column of the table gives a false impression of the virial equation at higher pressures when not limited by the above rule of thumb.

Frequently encountered concepts

There follow two concepts frequently encountered and sometimes useful when making approximations to real cases.

The asterisk condition

Equations sometimes can be simplified by referring them to the same normalized condition, the sonic velocity. The asterisk condition is defined as being one in which all properties represent those associated with the speed of sound in the medium.

$$N_{Ma}^* = 1, P^*, T^*, \rho^*, h^* \tag{IV-18}$$

The stagnation condition, adiabatic processes

Another frequently used reference condition is the stagnation condition. The condition is defined as the group of properties that results when fluid traveling at high speed is brought to rest adiabatically with no developed work. The action may be hypothetical. The first law equation can be written for this condition as,

$$\frac{\dot{Q}}{\dot{m}} = q = h_2 - h_1 + \frac{U_2^2}{2g_c J} - \frac{U_1^2}{2g_c J} \tag{IV-18A}$$

Subscript 2 represents the stagnation condition when the fluid is brought to rest. Subscript 1 is the upstream condition. If the process is adiabatic, Q and q equal zero. When the fluid is brought to rest, U_2 equals zero. The conversion factor, J, is the "mechanical equivalent of heat".

Compressible Fluid Flow | CHAPTER FOUR

For the adiabatic case, the description of the stagnation process is

$$0 = h_2 - h_1 - \frac{U_1^2}{2g_c J} \tag{IV–19}$$

Subscript 2 is usually changed to an S (or ST) to represent the stagnation condition. For a gas that follows the ideal gas equation, enthalpy is a function only of temperature and, if the specific heat is reasonably constant over the temperature range, the equation is rearranged as follows,

$$h - h_s = -\frac{U^2}{2g_c J} = c_p (T - T_s) \tag{IV–20}$$

$$T - T_s = -\frac{U^2}{2g_c J c_p}$$

$$T_s = T + \frac{U^2}{2g_c J c_p}$$

$$h_s = h + \frac{U^2}{2g_c J}$$

The specific heat at constant pressure does not appear in the last equation.

For an adiabatic process, T and h (flowing conditions) will vary, but T_s and h_s (stagnation conditions) will be constant.

IV-7: MODEL PROCESSES FOR COMPRESSIBLE FLOW

The purpose of most models is to isolate the essentials for study and to eliminate factors that have minimum impact on the results of the analysis. McCabe and Smith[xxxi] give three basic models for compressible flow. These models are shown in Figure IV-1.

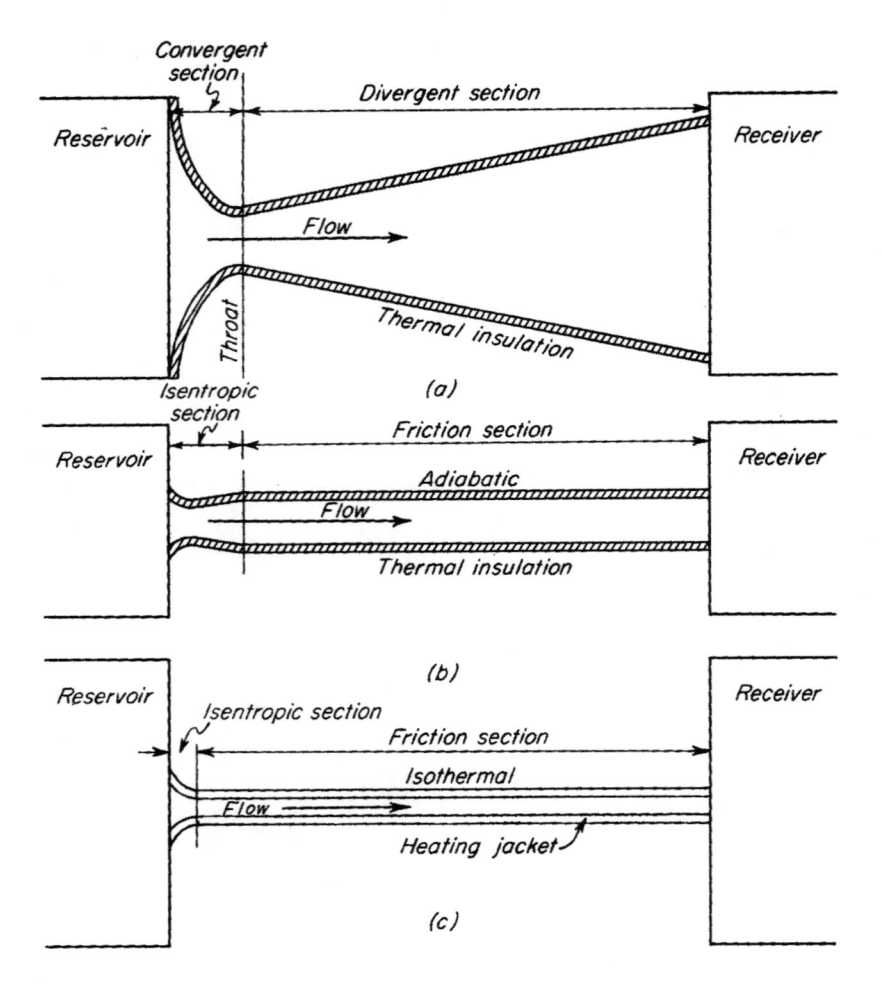

Figure IV-1. Three models of compressible flow: (a) Isentropic expansion in convergent-divergent nozzle. (b) Adiabatic frictional flow. (c) Isothermal frictional flow.

Frequently, models are combined. For instance the three basic models shown in Figure IV-1 may be combined with the ideal gas model. This latter model may be further restricted to a gas that has a constant heat capacity independent of temperature. All of these restrictions must be borne in mind when using equations developed from the use of models. The accuracy of the results depends upon the closeness to reality of the assumptions.

Each model has a different utility. The first model, an isentropic, adiabatic expansion, helps us understand flow in turbomachinery and gives us the equations for use with near sonic, sonic and supersonic flow. The second one, adiabatic flow with irreversibilities, is useful as the normal envelope for plant flows. It gives us the limits within which the compressible flows of most plants will be found. The third model, isothermal flow with irreversibilities, is useful for pipeline work and for work involving long, small diameter, uninsulated conduits.

The models of Figure IV-1 will now be described.

Isentropic, adiabatic expansion

An adiabatic expansion is one that takes place without energy transfer due to a difference in temperature between the fluid and the environment. Isentropic means that, in addition to the adiabatic constraint, there are no irreversibilities – no conversion of mechanical energy to internal energy, no losses. The process is considered ideal. If there are no losses and no heat energy transfer, the entropy property remains constant. The real-world approximation is a well-insulated, well-designed, smooth nozzle.

The principal causes of entropy change are heat transfer, turbulence due to wall asperities or to sudden expansion. The first cause is already mostly eliminated by the use of a well-insulated conduit. The remaining causes can be virtually eliminated by the use of a carefully shaped nozzle.

The ideal nozzle carefully contracts the flowing stream in the convergent segment so as to accelerate it to the throat. Then, it even more carefully expands the stream in the divergent segment to either recover the original pressure or to accelerate the flow even more in the case of supersonic flow.

The essentials of an ideal nozzle are:
- the shape;
- an upstream reservoir, the source, and a downstream receiver, the sink, where conditions are essentially stagnant;

CHAPTER FOUR | Compressible Fluid Flow

- a smoothly turned convergent segment;
- a throat, the narrowest portion of the nozzle, which is the transition between the convergent and the divergent segments;
- a divergent segment with a very small angle of divergence (less than about seven degrees);
- good insulation.

From the previous discussion of stagnation temperature, it can be seen that T_s is constant throughout the ideal, isolated nozzle at steady state. The measured temperature, T, and the velocity will vary along the nozzle.

Adiabatic irreversible flow

The essentials of the model for adiabatic flow with friction (conversion of mechanical to thermal energy) are:
- the same two reservoirs connected by a conduit of constant cross section;
- a smoothly turned, isentropic inlet to eliminate disturbances which are not due to conduit irreversibilities;
- good insulation.

This is the model that will concern us most of the time, but one should not assume it is the only model.

The stagnation temperature, T_s, is still constant in this model at steady state. The flowing temperature, T, will vary.

Isothermal irreversible flow

This model results in equations simpler to use than the adiabatic frictional flow model. It is therefore frequently used by default. Assumptions should always be checked. The essentials are:
- the same two reservoirs connected by a conduit of constant cross section;
- a smoothly turned, isentropic inlet to eliminate disturbances which are not due to conduit friction;
- a constant temperature jacket with good heat transfer.

The stagnation temperature is not constant in this model at steady state. The flowing temperature is constant.

Compressible Fluid Flow | CHAPTER FOUR

From the equation,

$$T_s = T + \frac{U^2}{2g_c J c_p} \tag{IV-21}$$

It can be seen that, with a constant flowing temperature, T, the stagnation temperature must vary with changes in the velocity squared. The velocity being measured is at the same point that the temperature is measured.

Flow of gases through ideal nozzles, qualitative discussion

Flow through an ideal nozzle is the starting point for understanding flow through venturis, orifice plates, restriction orifices, chokes and safety valves.

The convergent segment of an ideal nozzle is usually short because separation of the flowing fluid from the conduit boundary cannot occur. Flow separation is a major cause of irreversibilities because of the subsequent destruction of eddies that may occur remote from their place of generation. The divergent segment is long because the angle must be small to avoid separation of the flowing fluid from the boundary.

In the convergent segment the velocity increases, the pressure, density and temperature all decrease. In the divergent segment, the velocity may decrease or increase. At the throat, the velocity can reach a maximum of sonic velocity. Mach numbers greater than unity are not observed in a convergent segment or at the throat.

When the velocity reaches choked velocity in the throat, the mass flow rate is observed to remain constant in spite of decreasing downstream pressures. The only way to increase the mass flow rate once sonic velocity is obtained is to increase the upstream, reservoir pressure. Decreasing the receiver pressure no longer has any affect. This is the choked-flow condition.

This phenomenon is shown in Figure IV-2 which is given by McCabe and Smith[xxxi]. When the ratio of the receiver pressure to the reservoir pressure is equal to one, there is no mass flow. As the receiver pressure is decreased so that the ratio decreases, the mass flow rate increases as shown in the figure. A point is reached, however, around a ratio of 0.5, depending on the gas, when the mass flow rate becomes constant, independent of the downstream pressure. The flow is said to be choked.

CHAPTER FOUR | **Compressible Fluid Flow**

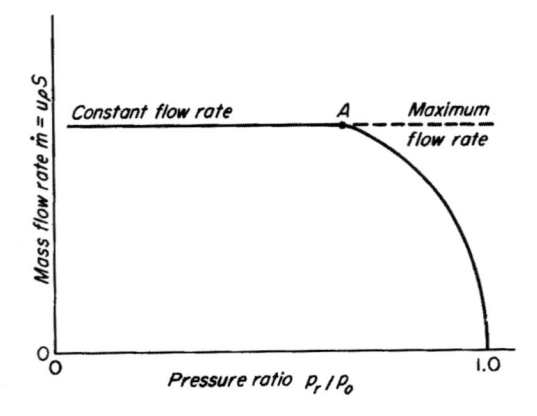

Figure IV-2. Mass flow rate through nozzle

The nozzle flow regimes are depicted in Figure IV-3. The meaning of the symbols must be understood. The abscissa is the length along the ideal nozzle. The left ordinate is the ratio, P/P_0, of the pressure at a point (not the receiver pressure) to the reservoir pressure. The right ordinate is the receiver to reservoir pressure ratio. It simply fixes the end points of the various curves.

The reservoir pressure remains fixed at P_0. The variable pressures are those down the length of the nozzle and in the receiver.

The following phenomena occur:

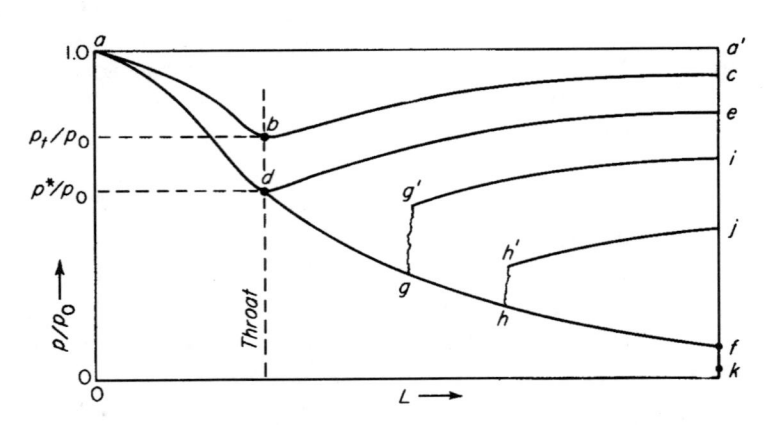

Figure IV-3. Variation of pressure ratio with distance from nozzle inlet

Compressible Fluid Flow | CHAPTER FOUR

- **Line a – a′.** This line describes the situation where P_r equals P_o. The pressure, P, is constant along the entire length and is equal to both the reservoir and the receiver pressures. No flow occurs.
- **Line abc.** When the receiver pressure is slightly lowered, flow occurs. The pressure, P, varies with distance along the nozzle, L. The throat pressure is the lowest pressure in the nozzle (vena contracta pressure). The line abc is representative of many lines that could be drawn so that they terminate between points a′ and e. They all have similar characteristics: a throat where the pressure is lowest and the velocity highest and some pressure recovery as the velocity subsequently decreases. Mass flow increases as the line dips toward point d, at the vena contracta.
- **Line ade.** This line represents the situation of choked flow with sonic velocity at the throat but with pressure recovery and subsonic velocities in the divergent segment.
- **Line adf.** If the receiver pressure is dropped to the point f, the mass flow rate, represented by the constant mass flow rate line of Figure IV-2, remains choked at its maximum. However, the velocity in the divergent segment increases. It becomes supersonic. The sound "barrier" has been crossed.
  - Once sonic velocity is reached at the throat, depending on the receiver pressure being at point e or point f, flow in the divergent segment of the nozzle can be either subsonic or supersonic. The mass flow rate is the same in both cases because of the choking phenomenon, even though the velocity changes.
  - This line is an isentropic one in a well-designed nozzle. There is no heat energy flow and there are no irreversibilities. Supersonic velocity only occurs along the line dghf; the mass flow remains choked.
  - If the receiver pressure is reduced below point f, for instance to point k, no increase in velocity occurs but a sudden pressure drop occurs after the discharge of the nozzle into the receiver.
- **Lines adgg′i and adhh′j.** If the downstream receiver pressure is increased from point f to points j or i, the flow will be supersonic from point d to points h or g. It will then suddenly drop to subsonic velocities with an associated pressure increase to points h′ or g′. The subsequent downstream curves represent normal subsonic pressure recovery.

Further discussion of isentropic, adiabatic nozzles may be found in Appendix AIV.

CHAPTER FOUR | Compressible Fluid Flow

Adiabatic flow in conduits with irreversibilities, qualitative discussion

When heat transfer through a conduit wall is negligible, either due to good insulation or to control of the external temperatures so they equal the internal temperatures, the process can be considered adiabatic — there is not heat energy flow to or from the system. What follows is subject to the adiabatic constraint.

In experimental situations, in order to minimize entrance effects, it is usual to have an isentropic segment at the beginning of a conduit. This isentropic segment is the converging segment of a nozzle. An isentropic nozzle is necessarily adiabatic.

Experimentally, it is possible to obtain supersonic velocities within a straight conduit by attaching the conduit to the discharge of an ideal nozzle. In this case, the velocity will decrease with distance along the conduit, but it will not cross the so-called "sound barrier" in the subsonic direction. If the conduit is lengthened and the same upstream pressure is applied, the mass flow rate will decrease, but the velocity will remain supersonic in the conduit and constant at the exit from it. The phenomenon of a maximum limit to mass flow in spite of lower downstream pressure is known as "choking".

Similarly, if the flow is subsonic and an attempt is made to increase the velocity by lowering the pressure at the end of the long, constant section conduit, sonic velocity can be obtained at the end of the conduit. However, supersonic velocities cannot be obtained no matter how much the downstream pressure is lowered. Here again, we are dealing with the phenomenon of choking.

The sonic barrier can be crossed only in a carefully designed isentropic nozzle as described in the previous section.

The governing differential equation for flow of a compressible fluid with irreversibilities is given in IV-22.

$$\frac{dP}{\rho} + d\left(\frac{U^2}{2g_c}\right) + f_M \left(\frac{U^2}{2g_c}\right)\frac{dL}{4r_H} = 0 \quad \text{(IV–22)}$$

This equation makes the usual simplification that elevation changes are negligible when gases or vapors are concerned and there is no compressor or turbine in the length of conduit. *It is only when the equation is integrated*

Compressible Fluid Flow | CHAPTER FOUR

that the constraints of an adiabatic or an isothermal process become important in relating property changes.

For circular conduits, the diameter is equal to four times the hydraulic radius, so D replaces 4_{r_H} in the last term on the left.

In gas flow under an adiabatic constraint, the temperature changes with expansion and turbulence. The viscosity varies with temperature, so the Reynolds[v] number and the friction factor are not constant. However, the changes are small, and it is often possible to use average values or, more commonly, to step up or down the conduit adjusting the values at each step.

Friction factors in supersonic flow are not well documented, but it is thought they are about half those of subsonic flow for the same Reynolds number. We need not be concerned with them in this book because we limit consideration to normal plant flow rates — which are deliberately held to less than 0.3 Mach to avoid too much noise and vibration.

Equation IV-22 will be developed further in Appendix AIV.

Equations involving the Mach number

Even with the ideal gas assumption, the equations governing the flow of gases with irreversibilities under adiabatic conditions at velocities greater than about 0.3 Mach are not simple. Therefore, their development will be moved to Appendix AIV. Here, we will jump to the final, integrated result.

$$\frac{(L_2 - L_1)f_M}{4r_H} = \frac{1}{\gamma}\left[\frac{1}{N_{Ma1}^2} - \frac{1}{N_{Ma2}^2} - \frac{\gamma+1}{2}\ln\frac{N_{Ma2}^2}{N_{Ma1}^2}\frac{1-\frac{\gamma-1}{2}N_{Ma1}^2}{1-\frac{\gamma-1}{2}N_{Ma2}^2}\right] \quad \text{(IV-23)}$$

Equation IV-23 applies to an ideal gas flowing adiabatically with a reasonably constant ratio of specific heats. It gives the relationship among the length of a straight, constant section conduit, the Moody[viii] friction factor, the ratio of specific heats and the Mach numbers at two points in the conduit. Average values of the friction factor may be used or step computations may be made using incremental values.

The velocity of the fluid cannot cross the sound barrier, in either direction, in straight conduit. If flow is supersonic, it remains supersonic and can only decrease to Mach 1. If it is subsonic, it remains subsonic and can

CHAPTER FOUR | Compressible Fluid Flow

only increase to Mach 1. At Mach 1, the mass flow rate will choke at its maximum value.

IV-8: CHOKED FLOW AND THE MACH NUMBER

The phenomenon of choked flow must be understood, especially by those who size PSV's and relief headers. It is an observed fact of nature that when a gas or vapor is allowed to flow through a conduit or restriction from some fixed upstream pressure to a lower downstream pressure, the mass flow rate increases with decreasing downstream pressure.

This increase in mass flow rate has a limit. There will come a point, when the downstream pressure is about half the upstream pressure, when the mass flow rate ceases to increase with decreasing downstream pressure. The mass flow rate will remain fixed no matter how much lower the downstream pressure becomes. The flow is said to be "choked". The choked flow condition is a common occurrence in PSV's, vent headers and control valves. The practical importance of the choked flow phenomenon is that, for a given mass flow rate, the pressure within the exit of the conduit at the choke point will be higher than immediately downstream. All upstream pressures consequently will be higher than those computed without considering the phenomenon.

Note the choked mass flow rate can be made to increase by increasing the upstream pressure, but not by decreasing the downstream pressure. It should not be assumed the choked flow rate is the same under all conditions.

In the usual explanation of the choked flow phenomenon it is stated that small pressure disturbances in gas and vapor travel with the speed of sound in the particular fluid. Once the molecules are themselves flowing at the speed of sound, it is not possible for a pressure disturbance, associated with the lowering of the downstream pressure, to be propagated upstream. It is sometimes stated that the upstream molecules cannot "know" the downstream pressure has been lowered. This explanation smacks of animism and does not explain the same phenomenon in mixed phase fluids.

A more logically satisfying explanation can be found from analyzing an ideal nozzle. It can be shown that a velocity equal to sound in a gas can only be reached within the throat area of an ideal nozzle, when the change in area along the length of the nozzle, dA, equals zero. Upstream, the velocity is subsonic and dA is negative. Downstream, it can be subsonic or supersonic and dA is positive.

Compressible Fluid Flow — CHAPTER FOUR

In a conduit of constant cross section, the change in area along the length, dA, is zero. Since increased velocity is obtained by decreased pressure, and since pressure increases in the upstream direction, the velocity must decrease upstream. The maximum attainable velocity must be obtained at the discharge of the conduit. If the maximum velocity attainable in the throat of an ideal nozzle is Mach 1, it cannot be greater than this in a conduit or in any orifice. If a converging-diverging nozzle is connected to the entrance to a conduit, supersonic velocity can exist within straight conduit.

Flashing liquids

The choking phenomenon also occurs with flashing liquids. In this case the fluid is a mixture of liquid and gas. Sonic velocity in liquid is much greater than that in gas. With mixed phase flashing flow it is not possible to compute a sonic velocity. Indeed, it is not necessary to compute one to establish the choking properties.

Peter Paige[xxiv] (*Chemical Engineering*, Aug 14, 1967) gave a more rational explanation of the choking phenomenon than that generally given involving sonic velocity. He simply reasoned from observation of the pressure profiles of flashing fluids in conduits and the fact that the length to achieve a fixed pressure drop decreased downstream, the incremental energy available (from pressure drop) to accelerate the fluid was converted to kinetic energy with nothing left to overcome additional irreversibilities. The flowing system is in balance at this point. The flow rate ceases to increase, no matter how much the downstream pressure is decreased.

Examination of the differential form of the Bernoulli[i] equation with irreversibilities, Equation IV-22, for horizontal flow shows there are three differential energy terms in balance. These terms are those representing pressure (static) energy, kinetic energy and irreversibilities. As the terminal pressure is reduced, more pressure energy is converted to kinetic energy and the length over which this conversion takes place is reduced. This can be readily seen by examining the pressure-distance history of flow, proceeding towards the end of a conduit. At the point where the length increment necessary to balance the equation is zero, all the pressure energy is converted to kinetic energy and none is left to overcome any incremental losses associated with distance down the conduit. The energy relationships represented by the Bernoulli equation are now in balance. This is the choked condition. The choked-flow phenomenon applies to all compressible fluids, and it includes mixtures of compressible and incompressible ones. We will develop the relationships in Appendix AIV.

CHAPTER FOUR | Compressible Fluid Flow

IV-9: EQUATIONS FOR ADIABATIC FLOW WITH IRREVERSIBILITIES NOT INVOLVING THE MACH NUMBER – THE PETER PAIGE[XXIV] EQUATION

The importance of the Peter Paige equation is it may be used for adiabatic flow of gases, vapors and flashing mixtures. The equation is particularly useful for understanding choked flow for gases and flashing mixtures. It will be given now but its development and a more detailed discussion will be left to Appendix AIV.

Starting with the differential form of the Bernoulli[i] equation, Equation IV-22, for horizontal flow, Paige re-arranged the equation to solve for incremental length.

$$\Delta L = L_2 - L_1 = \frac{2g_c}{G^2} \frac{D}{f_M} \left(-\int_1^2 \rho dP\right) - \frac{D}{f_M} \ln\left(\frac{\rho_1}{\rho_2}\right)^2 \quad \text{(IV–24)}$$

The quantity G is the mass flux or mass flow per second per unit area and is equal to ρU.

For small, fixed increments of pressure, the following approximation can be made.

$$\Delta L = L_2 - L_1 = \frac{2g_c}{G^2} \frac{D}{f_M} \rho_{avg}(P_1 - P_2) - \frac{D}{f_M} \ln\left(\frac{\rho_1}{\rho_2}\right)^2 \quad \text{(IV–25)}$$

This last equation is sometimes expressed in the equivalent form,

$$\Delta L = \frac{2D}{f_M}\left[\frac{g_c}{G^2} \rho_{avg}(P_1 - P_2) - \ln\left(\frac{\rho_1}{\rho_2}\right)\right] \quad \text{(IV–26)}$$

Compressible Fluid Flow | CHAPTER FOUR

This approximation becomes more and more exact as the pressure difference examined becomes smaller. It is therefore extremely useful for computer simulations. In these simulations, the equation is solved for the incremental length associated with a small pressure drop and the corresponding change in density. The lengths are then added until the actual length of interest is reached. The Moody[viii] friction factor may be computed for each incremental length or an average value may be used.

Paige[xxiv] considered the driving force for flow to be the first term in the brackets on the right of Equation IV-26. He considered the associated acceleration term to be the second one on the right. The length over which this transformation takes place, due to the balance between these competing terms, is the term on the left.

It is interesting to examine how these terms are related as we proceed down a conduit. In an experiment performed on flashing water, Benjamin and Miller[xxxii] examined fixed differences of 1 psi and the corresponding lengths. They found the length necessary to create this difference in pressure upstream was 20.6 feet. Downstream, at the exit from the conduit, it was only 0.57 feet.

These data allow us to reason that, for fixed pressure drops proceeding down the conduit, the average fluid density falls and the first term on the right in the bracket of Equation IV-26 becomes smaller. The second term on the right becomes smaller also, but at a slower rate. The difference becomes smaller and tends to zero. The length necessary to balance the equation decreases proceeding down the conduit.

At the conduit exit, all the available pressure energy is converted to kinetic energy and none is left to overcome the irreversibilities at the exit, so the mass flow rate cannot increase with decreasing downstream pressure.

Explanation and use of the Peter Paige equation

Equation IV-26 gives the increment in length of a constant diameter conduit with an inside diameter of D that will cause a pressure drop from the upstream pressure, P_1, to a downstream pressure, P_2. The mass velocity, G, and the upstream and downstream densities, ρ_1 and ρ_2, must be known and a friction factor must be computed for each step. The equation is particularly useful for computer simulations, but it can be used with a hand calculator.

CHAPTER FOUR | Compressible Fluid Flow

The equation is often used for relief system computations involving high pressure liquid being relieved from a process vessel. The liquid may flash on depressurization creating a two-phase mixture within the conduit. The conduit normally vents to a knockout drum where the liquid is separated from the vapor for recovery. The vapor frequently is vented via a flare.

To use the equation practically, fluid properties must be known as a function of absolute pressure and certain assumptions must be made (and checked). The easiest assumption to make is that involving the use of the adiabatic model, if the conduit is insulated or reasonably short. The next assumption is the mixture is homogeneous and there is no slip (different velocities) between the phases. This is a reasonable assumption for most relief situations where conduit is close to horizontal on the mixed phase side. This assumption allows simplification of the computation of the friction factor.

The equation helps solve the peculiar problem of computing a conduit cross section just large enough to allow evacuation of a given rate of fluid through a known length of conduit when neither the upstream nor the downstream effective pressures are known. The pressures in the process vessel, the source, and in the knockout drum, the sink, will be known, but their difference is not necessarily the driving differential pressure.

If the flow is great enough to be choked, the pressure within the conduit at the discharge end will be higher than that in the sink. If the flowing fluid exits through a relief valve, the pressure on the downstream side of the valve will be lower than that in the process vessel.

When a relief valve is used, an additional consideration is that the backpressure must not be great enough to prevent it functioning adequately.

Choking can occur at any point in the system where there is an enlargement. The obvious one is at the discharge into the sink. Less obvious ones are at junctions of smaller conduits to larger headers. Easily overlooked is the relief valve orifice. The fluid properties at the choked (critical) condition are governed by the mass flux, G, and the upstream enthalpy. The fluid properties immediately downstream of the choke point are governed by the backpressure imposed by irreversibilities within the conduit to the discharge or to the next choke point.

Compressible Fluid Flow | CHAPTER FOUR

Requirements for solution of the Peter Paige equation

To start the solution to the Peter Paige[xxiv] equation, it is necessary to know the specific enthalpy of the fluid at the source. If this fluid can be approximated by an ideal gas, the enthalpy can be computed directly from the temperature and the specific heat. If the fluid is liquid that will pass through its bubble point, flash calculations are necessary to establish the relationship that exists between absolute pressure and temperature and mixture density. For non-ideal gases the enthalpy at the sources may be estimated using the departure function and the R-K equation.

Computations involving flashing liquids are simpler than those involving gases because the equilibrium flash strictly relates temperature and pressure. From flash calculations, the liquid and vapor densities are found as a function of pressure. The mixture temperature is also found as a function of pressure. From the chosen mass flow rate and the data from the flash calculations, a Reynolds[v] number and friction factor can be computed for each increment. The Peter Paige equation may then be applied.

It is to be noted the method is very tedious if applied with a hand calculator. The curve fitting capabilities of a computer make the solution easier.

The equation will give negative increments if the integration is carried beyond the choke point. This a physical impossibility, but not a mathematical one. This inversion of sign gives a very good flag to indicate critical (choking) pressure. The choked pressure is the starting point for backward (against the flow) computations.

Normally, the computation is carried out backwards from the sink. At a chosen flow rate in a given size conduit, the equation will give positive length increments if the flow is not choked. In this case the sink pressure is the valid one to use and there is a loss coefficient (a K value) to be applied at the exit.

If the flow is choked, the length increment will be negative and the pressure must be increased until the length becomes positive (zero is the cross over point). The associated pressure is the choked flow pressure or critical pressure that must be used in subsequent computations. In this case an exit loss coefficient is not needed. The effective conduit length is less than that for the non-choked flow situation.

CHAPTER FOUR | Compressible Fluid Flow

Once the valid pressure is established at the exit from the conduit, the equation is applied stepwise up the conduit and the increments are summed until the computed length equals the actual length. If the calculated upstream pressure is greater than the allowable maximum pressure, the chosen conduit diameter was too small. If the computed upstream pressure is too low, a smaller conduit diameter may be tried.

If the computed pressure is lower than the allowable pressure, and a restriction such as a pressure safety valve exists, the mass flow will also be choked across the PSV. This is the normal relief situation. The backpressure must be checked to see it is not too great so the chosen PSV will function adequately.

If no restriction exists (a wide-open gate valve or a full port ball valve is used), the mass flux will be greater than the assumed flux. In this case, the assumed flux can be incremented and the computations can be done again until the computed upstream pressure equals the actual pressure.

Peter Paige[xxix] developed another equation to calculate the choked flow pressure directly. This equation is of theoretical interest, so its development is given in Appendix AIV. However, as Richter[xxv] pointed out, by incrementing the trial choked pressure, P_c, until the difference between the computed lengths switches from negative to positive, the critical pressure can be found without the use of this second equation. Mulley also solved the same problem using the Redlich-Kwong[xxviii] equation-of-state so he could use an analytic method instead of an incremental method for computer simulations. This equation is given in Appendix AV.

Friction factor computations in mixed flow

Extensive work has been done on the problem of finding friction factors for use with various two-phase regimes (Duckler, A.E. et al, "Frictional pressure drop in two-phase flow", AIChE J., 10 44-51 (1964)[xxxiii]. The friction factor depends on the type of flow regime (See DeGance and Atherton[xxxiv], *Chemical Engineering*, March 23, 1970[xxxiv]). Fortunately, for most relief header and valve sizing problems, we can use simplifying assumptions.

IV-10: EQUATIONS FOR ISOTHERMAL FLOW WITH IRREVERSIBILITIES

The isothermal model of flow with friction is only approximated in long, relatively small diameter, uninsulated conduits that contain fluid flowing at low Mach numbers. The equations are simpler than those for adiabatic flow. This fact contributes to their popularity, but not to their accuracy.

Compressible Fluid Flow | CHAPTER FOUR

Comparison of isothermal and adiabatic acoustic velocities

The adiabatic acoustic velocity has already been given for an ideal gas as,

$$a = \sqrt{\frac{g_c \gamma P}{\rho}} \qquad \text{(IV–27)}$$

If the flow is isothermal, the maximum attainable velocity is given by IV-28.

$$a' = \sqrt{\frac{g_c P}{\rho}} \qquad \text{(IV–28)}$$

Newton first derived this formula on the assumption that sonic velocity was due to isothermal changes. Even Newton could be wrong.

By taking the ratio of the two equations, it can be seen that,

$$a = a'\sqrt{\gamma} \qquad \text{(IV–29)}$$

For air, the ratio of specific heats, γ, is approximately 1.4. The square root of this ratio is approximately 1.2. Laplace[xxvi] had to correct Newton's formula for the velocity of sound in air by this much, 20%, to match experimental fact. The isothermal homolog of the Mach number is $N_i = U/a'$.

Mass flux (mass velocity)

If the temperature change of the fluid over the length of a constant cross section conduit is small, we can use the mass velocity in the Bernoulli[i] equation as follows. First we transform the Bernoulli equation by multiplying through by ρ^2.

$$\frac{dP}{\rho} + \frac{UdU}{g_c} + \frac{U^2 f_M dL}{2g_c(4r_H)} = 0 \qquad \text{(IV–30)}$$

$$\rho dP + \frac{\rho^2 UdU}{g_c} + \frac{\rho^2 U^2 f_M dL}{2g_c(4r_H)} = 0$$

Then we introduce the concept of mass velocity (mass flow per unit area or mass flux). The mass velocity along a constant cross section conduit is a constant. The following development only applies to constant cross section conduits.

Flow of Industrial Fluids—Theory and Equations

CHAPTER FOUR | Compressible Fluid Flow

$$G \equiv \rho U \tag{IV-31}$$

$$U = G\rho^{-1}$$

$$dU = -G\rho^{-2}d\rho$$

$$UdU = -G^2\rho^{-3}d\rho$$

We will again restrict our model to that of an ideal gas.

$$\rho = \frac{(MW)P}{R_o T} \tag{IV-32}$$

The density is mass per unit volume. We can now substitute mass velocity and the ideal gas relationship into the modified Bernoulli[i] relationship to give,

$$\frac{(MW)PdP}{R_o T} - \frac{G^2 d\rho}{g_c \rho} + \frac{G^2 f_M dL}{2g_c(4r_H)} = 0 \tag{IV-33}$$

Isothermal mass flux

If the temperature and the friction factor are reasonably constant, the above equation can be integrated to give the second of the set IV-34.

$$\frac{(MW)}{R_o T}\int_1^2 PdP - \frac{G^2}{g_c}\int_1^2 \frac{d\rho}{\rho} + \frac{G^2 f_M}{2g_c(4r_H)}\int_1^2 dL = 0 \tag{IV-34}$$

$$\frac{(MW)}{2R_o T}\left(P_2^2 - P_1^2\right) - \frac{G^2}{g_c}\ln\frac{\rho_2}{\rho_1} + \frac{G^2 f_M}{2g_c(4r_H)}(L_2 - L_1) = 0$$

The density ratio in the second term may be replaced by the pressure ratio when the ideal gas law is followed and the term $4r_H$ may be replaced by D in circular conduits.

$$\frac{(MW)}{2R_o T}\left(P_2^2 - P_1^2\right) - \frac{G^2}{g_c}\ln\frac{P_2}{P_1} + \frac{G^2 f_M}{2g_c D}(L_2 - L_1) = 0 \tag{IV-34A}$$

Equation IV-34 *for isothermal flow* is frequently used even for adiabatic flows at low Mach numbers (below about 0.3). In fact, the equation is often cited without identifying the fact that it was derived from isothermal considerations only. It is wise to check assumptions. The difference between the adiabatic equation and the isothermal equation is in the treatment of gas densities and friction factor.

Compressible Fluid Flow | CHAPTER FOUR

Energy transfer in isothermal flow

As a gas expands, its temperature decreases. Flow of a compressible gas is a process that involves expansion. Therefore, to maintain isothermal conditions, energy must flow from the environment to the gas.

Energy transfer per unit mass between two points on a conduit is equal to the difference in stagnation enthalpy between these same two points when the flow is isothermal. The stagnation enthalpy is the sum of the enthalpy at flowing conditions and the kinetic energy change on bringing the fluid to rest adiabatically. A value for stagnation enthalpy can be computed even for processes that are not adiabatic since the concept is hypothetical.

The stagnation temperature is defined as the sum of the temperature measured at the flowing condition and the change in kinetic energy divided by the specific heat at constant pressure on bringing the fluid to rest adiabatically.

By substituting these relationships into the first law equation for an open system we can establish a formula for the heat transfer per unit mass. Note that in an isothermal system the flowing temperature is constant, but not the stagnation temperature. The constraint is that of steady state with no fluid work and no elevation change.

Equation IV–35 gives various forms of the first law, open system energy balance under the constraints mentioned in the previous paragraph.

$$\frac{\dot{Q}}{\dot{m}} = q = h_2 - h_1 + \frac{U_2^2}{2g_c J} - \frac{U_1^2}{2g_c J}$$

$$\frac{\dot{Q}}{\dot{m}} = q = h_{s2} - h_{s1} = (T_{s2} - T_{s1})c_p$$

$$T_s \equiv T + \frac{U^2}{2g_c J c_p}$$

$$T_1 = T_2$$

$$\frac{\dot{Q}}{\dot{m}} = q = (T_{s2} - T_{s1})c_p = \frac{U_2^2}{2g_c J} - \frac{U_1^2}{2g_c J}$$

(IV–35)

CHAPTER FOUR | Compressible Fluid Flow

We have proven that, in isothermal flow, energy transfer to the fluid per unit mass flowing is equal to the change in kinetic energy between two points.

By substituting the mass velocity equation for the isothermal, steady state kinetic energy change of Equation IV-35, we can obtain a relationship involving the mass velocity and the density. The mass velocity (or mass flux) of a compressible fluid is constant in a constant section conduit,

$$\frac{\dot{Q}}{\dot{m}} = q = \frac{U_2^2}{2g_c J} - \frac{U_1^2}{2g_c J}$$ (IV-36)

$$G = U\rho, \quad U^2 = \frac{G^2}{\rho^2}$$

$$\frac{\dot{Q}}{\dot{m}} = q = \frac{G^2}{2g_c J}\left(\frac{1}{\rho_2^2} - \frac{1}{\rho_1^2}\right)$$

This is the heat transfer per unit mass necessary to maintain the isothermal state. The only variable is the density.

IV-11: CHAPTER SUMMARY

Chapter IV has introduced the basic concepts used in compressible flow computations. The fact that the concepts of compressible flow were based upon those of incompressible flow was emphasised. The important differences between the two types of flow were shown. The use of models was discussed and caveats were laid down as to when to question a particular model.

The utility of the ideal gas model was outlined both as a conceptual tool, for discussion purposes, and as a tool in real life situations. Using this model, five basic groups of equations were presented: continuity, the total energy balance, the mechanical energy balance, sonic velocity, and the ideal gas equations.

The simplest non-ideal gas equation was developed using the compressibility, Z. The virial equation was discussed briefly. The very useful Redlich-Kwong[xxviii] equation was given. The other equations-of-state will be given in Appendix AIV.

Compressible Fluid Flow | CHAPTER FOUR

Model processes for compressible flow were discussed in the following order: an isentropic, adiabatic expansion, adiabatic frictional flow, and isothermal frictional flow. Their limitations were pointed out.

The Peter Paige[xxiv] equation for adiabatic flow was introduced and its great utility for choked flow computations was remarked upon.

Isothermal flow with irreversibilities was discussed, and it was pointed out these equations were limited to long, uninsulated pipelines.

CHAPTER FIVE V

Compressible Fluid Flow – Complex Systems

V-1: SCOPE OF CHAPTER — COMPUTATIONS FOR COMPLICATED COMPRESSIBLE FLOW SYSTEMS

Chapter V will lay the groundwork to permit computing pressure drops and flows of some very complicated compressible flow systems. The scope will be again confined to the problems of the industrial plant.

We will choose a typical piping system and a typical problem common to most chemical plants so the reader may then make the necessary extrapolations to his own systems. The system is a safety relief vent header. The problem is how to get there from here or, to put it more technically: how is flow computed when neither the upstream nor downstream pressures, nor the fluid compositions are immediately known? The system and the problem were chosen because they are representative of many similar systems and problems involving fluid flow in industrial situations. These situations often demand solutions that go beyond experience and available data. One must then use engineering judgment, testing and safety factors in arriving at a solution to the problem.

Some of the main concepts have already been discussed in Chapter IV and in Appendix AIV. We will put these concepts into the context of a complicated system in this chapter and will demonstrate solutions to the associated problems in the accompanying appendix.

The education engineers and technicians receive usually prepares them for problems involving single pipes and single, pure fluids. Real life presents problems of networks of pipes of different dimensions and fluid mixtures of different cats and dogs. The relief vent header is a typical problem consisting of a complicated piping network in which, because of fluid mixing, the fluid composition can change throughout the system of piping segments.

CHAPTER FIVE | **Compressible Fluid Flow — Complex Systems**

The methods discussed in this chapter and associated appendix evolved over time. The search began with an isothermal model. Subsequently, an adiabatic gas or vapor model was developed. Later, an adiabatic mixed flow model (flashing fluid) evolved. We will concentrate on the description of the adiabatic gas or vapor model.

The author would like to acknowledge a collaborator many years ago when the ideas presented in this chapter and the associated appendix were being developed. His name is Ishwar Davé. We both worked at Dravo Chem Plants.

V-2: DESCRIBING THE PIPING NETWORK

The overall piping configuration of a vent header system generally is fixed by the plant layout. The major design variable for process engineers and control systems engineers is the diameter of each individual pipe segment. This variable must be chosen as a function of safety and economy. An optimum choice must be made of design pressures of vessels, relief device types, and pipe diameters. These elements must be integrated into a system.

A computer program requires that the physical network be described by a simple coding procedure. Any combination of possible flows can then be imposed on the system to study the effects of changed piping configurations, changed diameters, or different combinations of flows. Many design disciplines benefit from the output data from such a program. In addition to information of interest to process engineers and control systems engineers, the program can give velocities, densities and pressures at every fitting. This information can be used in studies of piping supports and pipe stress.

Two typical isometric figures, Figure V-1, and Figure V-2, are intended to demonstrate the methodology.

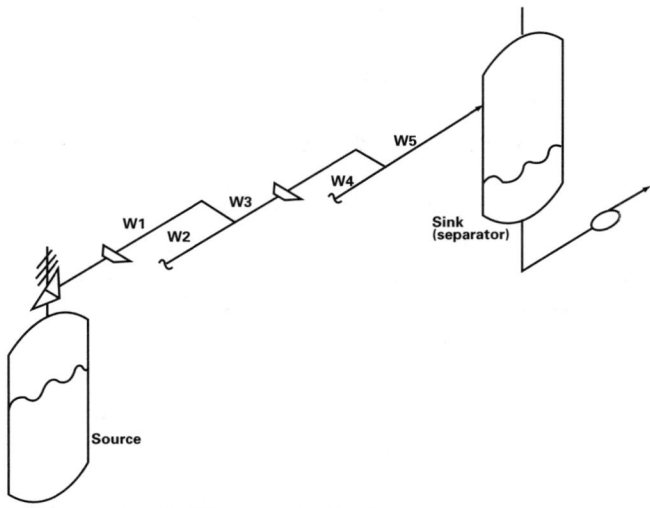

Compressible Fluid Flow – Complex Systems | CHAPTER FIVE

Figure V-1. First example of isometric sketch

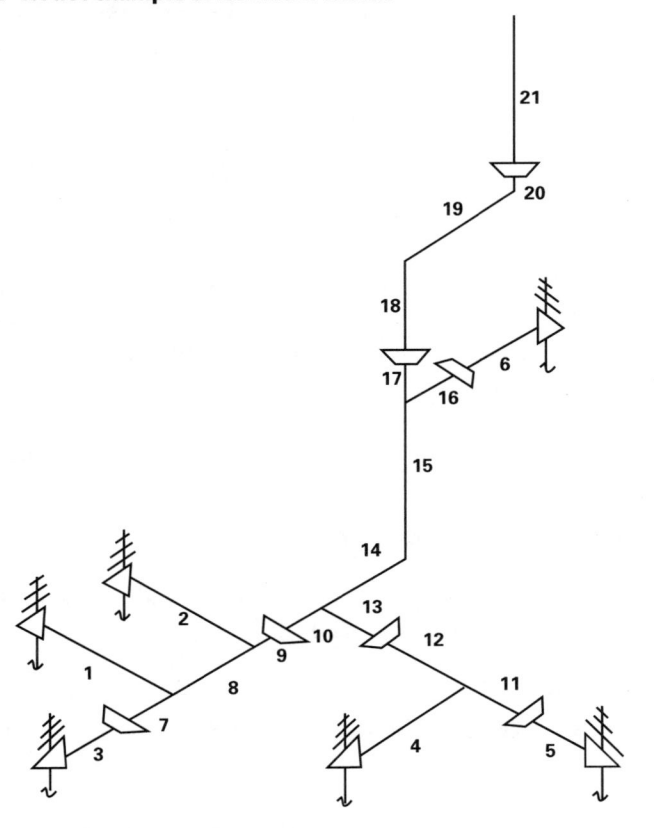

Figure V-2. Second example of isometric sketch

CHAPTER FIVE | **Compressible Fluid Flow – Complex Systems**

The two sketches, V-1 and V-2, are meant to show complex systems can be reduced to the essentials and, before attempting simulations, the entire physical system has to be described. A sketch, no matter how crude, allows description and peer review. An "as-built" sketch permits a personal verification during a "walk-down" of an actual network.

All relief devices in a system should be shown on a sketch. The network is described by numbering each individual segment of same diameter conduit from number one, starting at the sources of relief flow, to the highest number reached at the single segment that finally exhausts to a sink. The sink is usually the point of lowest consistent pressure. It can be the atmosphere or a knockout pot. A requirement in our scheme is that each intermediate segment must have its feeders numbered before it receives a number. A segment is described as a conduit of constant section not joined by any other conduit together with all bends included as equivalent lengths, and including the terminal fitting. Expanders on safety valves are treated as segments of zero length consisting of a "K" value only. Note that, although the numbering begins at the sources, the numbers refer to the segments. Bends may be included in a segment as a "K" factor or may be described separately, if one requires data on forces operating on them.

The methods of Benedict et al[xxxv], which can replace the K factor methods, are discussed in Appendix AV.

Once the piping network has been described on an isometric sketch, it can be coded into a computer. The input is simply in order of the numbered segments. The required data for each segment are length, internal diameter, pipe roughness, the sum of all "K" values for bends and fittings except the terminal fitting, the code for the terminal fitting and, finally, the identifying numbers of up to four feeder segments.

V-3: DESCRIBING THE FLOW REGIME

It is not necessary to size all vent headers for simultaneous relief of all safety devices. It is recommended to divide the system into groups of devices that may relieve together under similar conditions such as fire exposure (by fire zone), cooling water failure, power failure, etc., and then to investigate the groups for the worst case at each source.

If it is necessary to provide additional data for piping support or stress analysis, segments can be described as being terminated by bends and fittings where the data are needed.

Compressible Fluid Flow – Complex Systems | CHAPTER FIVE

It is necessary to decide on a model of fluid flow: isothermal or adiabatic vapor or gas, or flashing flow. The most general model will involve multi-component mixtures. The model chosen for demonstration will be the multicomponent, adiabatic vapor model.

V-4: PLAN OF ATTACK

Each simulation project must have its plan of attack. Our plan is to:
1. establish the mass flow rate for each safety device based on the worst case failure of each vessel;
2. sum the mass flow rates of each successive downstream conduit;
3. convert mass flow rates to mass velocities (mass flux, G);
4. convert mass flow to mole fractions and total moles;
5. establish the molar heat capacity of each mixture by using the ideal gas heat capacity coefficients for each component and the departure functions;
6. establish the pseudocritical properties for each mixture;
7. establish the constants of the Redlich-Kwong[xxviii] equation-of-state for each segment;
8. compute the enthalpy at each source and sum for the subsequent enthalpies in each segment;
9. identify the choke points based on the mass velocities (mass fluxes) of each segment (there may be more than one choke point);
10. estimate the downstream temperature and specific volume of the fluid in each segment from pressure and stagnation enthalpy using Newton-Raphson iteration (This involves computation in the backwards direction from the flow starting from sink pressure plus the pressure difference across the exit or from the choked pressure.);
11. compute the upstream pressure, temperature and specific volume of the fluid in each segment using the Peter Paige[xxiv] equation as one of the simultaneous equations;
12. pass the computed data to the upstream pipe segment or use choked downstream properties and repeat the computations upstream until the safety device discharge is reached;
13. compare the computed pressures downstream of the relief devices with the required maximum discharge pressures and make any necessary changes to segment sizes;
14. make changes and iterate to a satisfactory solution.

CHAPTER FIVE | Compressible Fluid Flow – Complex Systems

Fortunately, most of the computations and the data manipulation can be programmed to be executed automatically.

V-5: MANIFOLD FLOW

Manifold flow is defined as the splitting of a single stream into two or more streams or the combining of two or more streams into a single one. Safety relief vent headers contain many fittings that fall under this general designation. Safety vent headers deal mostly with combining flows.

The Bernoulli[i] equation still applies in manifold flow. If a velocity reduction across a fitting produces a negative change in kinetic energy greater than the "friction (mechanical energy) losses", the downstream static pressure will increase. This is a fairly common, though often totally unexpected result.

The losses of mechanically useful energy are often expressed in terms of "velocity heads", meaning a multiplying factor times the kinetic energy at a specific location. The location must be associated with a "K" value and the user must know whether it is a downstream or an upstream location.

Caveat — Negative K values

Data obtained from experimental work on manifold flow consistently give some negative values for the coefficient, "K". Some analysts have attributed this fact to error in the use of the average velocity, U. The author believes this phenomenon occurs because one smaller stream has been entrained by another and has gained specific energy from it. Computations on both streams will reveal an overall reduction in mechanical energy.

Note it is also possible that those who established the data may have neglected pressure recovery. Erroneous data may result when measurements are taken too close to the fitting concerned, before the velocity profile has had a chance to re-establish itself. This is why projected values should be used. The reader is referred to the distinction already made between ΔP and $\Delta(\Delta P)$ and to the projection method of overcoming the error.

Terminology

A common terminology is needed to describe flow in manifolds. The one used here evolved over some time to be as concise as possible. There are two types of manifold flow: combining, which is the usual case of vent headers, and dividing. Physically, these manifold flows may be represented by the tees of Figure V-3. Usually, the run of a contiguous tee is constant diameter. The branch may be of different diameter than the run. The volumetric flow will vary depending on conditions, so a ratio, Q, of minor flow to major flow will be set up.

Combining Flows

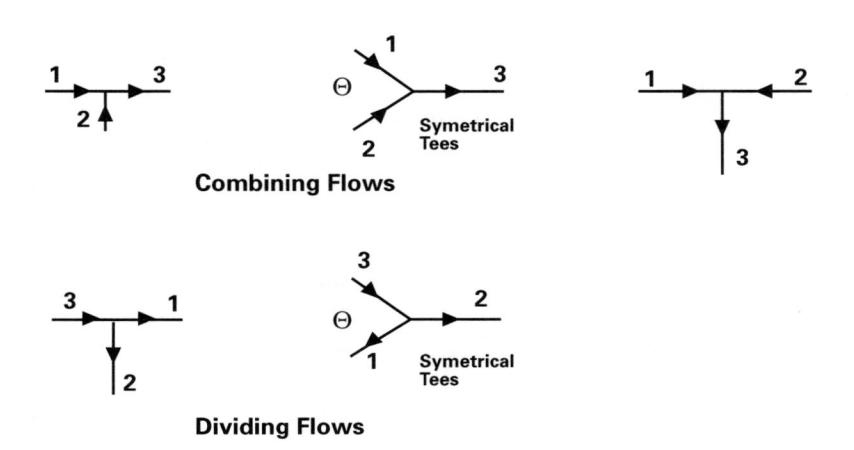

Dividing Flows

Figure V-3. Manifold flows

Caveat — Definition of K factors

The conventional "K" factors or "loss" coefficients are usually defined as the change in specific mechanical energy across a path through a fitting divided by the specific kinetic energy of the flow before it was divided or after it was combined. Other definitions exist; therefore, when using K factors, the reader must know which set of equations to apply. Care should be taken with Miller's[xix] variation which defines the loss coefficient as the change in "total" pressure divided by the velocity pressure at a specific location. The velocity pressure is the product of the kinetic energy and the density.

Benedict et al[xxxv] correlated total pressure ratios with inlet and outlet pressure ratios to arrive at a straight-line relationship for sudden expansions and sudden contractions. The relationship between entropy increase and total pressure ratio across an adiabatic conduit with irreversibilities can also be shown to be linear. The author has reason to believe this line of investigation will be profitable in the future. We will enlarge upon this concept in Appendix AV.

CHAPTER FIVE | Compressible Fluid Flow – Complex Systems

Combining or dividing tees should have their connections numbered so the major flow is always in connection number 3. This way, correlations can be performed on experimental data and the data can be reduced. Charts can be drawn of the K factors versus the fractional volumetric flow. Figure V-4 is typical of Miller's[xix] K factors. Since the denominator in the definition is always the major flow, the abscissa varies from zero to one while the ordinate can vary from about minus one to about plus 1.5.

Figure V-4 is representative of many sketches that could be drawn from the relationships between the K factor coefficients (Miller's, in this case) and the split in flow in a combining tee. Note the volumetric flow ratio is used. This is probably because the relationship must represent many pipe sizes, and venturis or orifice plates, at the time of this writing, are the most common flow measuring devices to be used in small and large diameters. These are essentially volumetric devices. Note also there are two curves: one to correlate losses in mechanical energy through the run and the other losses through the branch. A further point worthy of notice is the sketch has the notation A2 = A3, meaning the run has a constant section.

Curves such as that depicted in Figure V-4 are useful because they can represent many sizes and configurations of fittings. However, the constraints must be specified. Such curves can be used in curve fit programs in simulations either directly or via a polynomial or other equation.

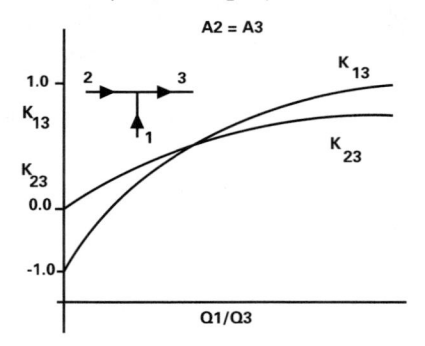

Figure V-4. Miller's K factors

'Loss' coefficients

Before automatically using loss coefficients, it is necessary to check that choked flow does not exist. Choked flow is more likely in the reduced branch of a tee rather than in the run. Should choked flow exist in a branch, the loss coefficient for that branch is not used because the coefficient has no influence on the choked pressure.

Compressible Fluid Flow – Complex Systems | CHAPTER FIVE

D.S. Miller[xvi] offers insight into the differences between loss coefficients used in incompressible flow and those used in compressible flow. He points out the vena contracta can actually "migrate" in a fitting, and the area of the contracted flow can change under changing relative flow rates. He states that correcting some incompressible loss coefficients to apply to compressible flow may not be very satisfactory. However, until experimental data are available, it is necessary to adapt incompressible coefficients to a form suitable for compressible flow.

Incompressible fluid loss coefficients

Assure yourself all the terms are understood. The subscripts are especially important. The original "losses" were simply the differences in static, kinetic and potential energies (difference in mechanical energy) across a device. A loss coefficient was a dimensionless term equal to the ratio of the difference in mechanical energy across a path through a fitting to the kinetic energy at a specific location, upstream or downstream, of the fitting.

The original loss coefficients were ratios of mechanical energy to kinetic energy. Mechanical energy is measured in terms of pressures, densities, velocities and elevations. The association with pressure, in particular, leads to misunderstanding since some pressure "losses" may be recovered, but mechanical energy "losses" are not.

Miller's definition of a loss coefficient for incompressible flow is given in V-1.

$$K_i = \frac{P_{t1} - P_{t2}}{\rho U_i^2 / 2 g_c} \qquad \textbf{(V-1)}$$

This equation may be more understandable if it realized density is constant in an incompressible fluid and is the divisor of the numerator. The equation is equivalent to the difference in the sums of static (pressure) energy and kinetic energy divided by kinetic energy. It is identical to the conventional K factor equation for incompressible fluids. The subscript, t, means total; the subscript, i, means the coefficient refers to the section where velocity is measured. The total pressure is the sum of the static pressure and the velocity pressure that is derived from energy considerations and by neglecting the elevation change across the fitting, as follows:

$$\frac{P_t}{\rho} = \frac{P}{\rho} + \frac{U^2}{2 g_c} \qquad \textbf{(V-2)}$$

$$P_t = P + \frac{\rho U^2}{2 g_c}$$

CHAPTER FIVE | Compressible Fluid Flow — Complex Systems

Compressible fluid loss coefficients

For compressible flow, the density is no longer constant across the fitting and Miller[xix] changes the definition of a loss coefficient to accommodate this fact.

(V-3)

$$\frac{P_{t1}}{\rho_1} = \frac{P_1}{\rho_1} + \frac{U_1^2}{2g_c}$$

$$P_{t1} = P_1 + \frac{\rho_1 U_1^2}{2g_c}$$

$$P_{t1} - P_1 = \frac{\rho_1 U_1^2}{2g_c}$$

$$P_{t2} - P_2 = \frac{\rho_2 U_2^2}{2g_c}$$

From the Bernoulli[i] equation with losses, the original loss coefficient is

(V-4)

$$\frac{P_1}{\rho_1} + \frac{U_1^2}{2g_c} = \frac{P_2}{\rho_2} + \frac{U_2^2}{2g_c} + h_L$$

$$h_L = \frac{P_1}{\rho_1} + \frac{U_1^2}{2g_c} - \left[\frac{P_2}{\rho_2} + \frac{U_2^2}{2g_c}\right]$$

$$K_1 = \frac{h_L}{\left(\frac{U_1^2}{2g_c}\right)}, \quad K_2 = \frac{h_L}{\left(\frac{U_2^2}{2g_c}\right)}$$

The two loss coefficients are subscripted 1 and 2 to indicate which velocity should be used. Equation set V-4 gives the traditional definitions of the loss coefficients.

Miller writes his definition in terms of "total" and actual pressures as follows.

(V-5)

$$K_1 = \frac{P_{t1} - P_{t2}}{P_{t1} - P_1} = \frac{1 - P_{t2}/P_{t1}}{1 - P_1/P_{t1}}$$

$$K_2 = \frac{P_{t1} - P_{t2}}{P_{t2} - P_2} = \frac{P_{t1}/P_{t2} - 1}{1 - P_2/P_{t2}}$$

For compressible fluid flow, Miller's coefficients differ from the traditional ones by the ratio of the downstream to upstream densities applied to the second term of the normal mechanical energy loss equation. If the density is constant, they are identical.

Compressible Fluid Flow – Complex Systems | CHAPTER FIVE

The two expressions for the "loss" coefficients in the set V-5 give a means of establishing them experimentally from the measurable quantities: pressure, velocity and density. Elevation change is neglected. It is inconsequential for compressible fluids in most vent headers.

Note each loss coefficient is established by experiment, by measuring, or estimating from measurements, pressure, density and velocity at a point in a specific type of fitting. The general results are published in the form of graphs (Figure V-4) or formulae. The loss coefficients should be correlated from specified conditions (adiabatic) and these conditions should be stated.

The traditional correlation allows the coefficient to be used as a multiplying factor of kinetic energy at a section to obtain the mechanical energy converted to thermal energy across the fitting. This "lost" energy can then be plugged into the Bernoulli[i] equation and, with the appropriate downstream quantities, the upstream quantities may be computed. Miller's[xix] correlation for K_1 allows the coefficient to be multiplied by the kinetic energy to produce the difference between upstream mechanical energy and the product of downstream mechanical energy and density ratio.

The two definitions in the set V-5 do not give identical results and the coefficients must be used with the appropriate equations.

Combined loss coefficients

When fittings are close together, even experimental observations can be in error due to pressure recovery and the fact data is taken without considering velocity profile. When sizing vent headers, one does not normally have the luxury of exact data and one is normally under time constraints, so safety factors are used. For instance, the assumption that all tees are right-angled is usually safe. A distance between fittings of greater than thirty pipe diameters can sometimes be specified by the piping designers.

Benedict et al correlations

Benedict et al[xxxv] use gradients in place of K factors, or loss coefficients. We hope to show, in subsequent work, this is the most accurate means of estimation of "losses".

CHAPTER FIVE | **Compressible Fluid Flow – Complex Systems**

V-6: DATA COLLECTION AND VERIFICATION

The heading of this section refers to data collection and verification. The verification part must be emphasised. The author remembers vividly his shock on "walking down" a complex piping system and finding it had not been installed according to his carefully conceived isometric drawing. The field personnel had not seen the logic in the design, nor had they deigned to inform the home office of the changes they had wrought.

Revisions are made to data and to specifications for safety devices. A verification must be made of all "as-built" data to see the data match the design's intent.

Piping isometric map

The map of the system is the piping isometric. It identifies every device, every segment, each diameter, each fitting and each bend in the system from source to sink. It permits the organization of input data to the simulation.

Data sheets of all input sources

Under this heading, we include rupture disk and PSV information and drawing and design data on vessels. Someone must check that what was bought fits what it was designed to fit. The flowing quantities and properties at the accumulated pressure are required for each source.

Piping specifications, process and engineering flow diagrams

The piping specifications are a valuable source of data on pipe internal diameters and pipe types. The information contained is necessary for pressure drop computations.

The process flow diagram will contain design data and the design material balance. This document is a guide that must be interpreted in the light of relieving conditions.

Engineering flow diagrams are variously called EFD's, MFD's, P&ID's, or they may have other names. Their purpose is to conveniently present an engineering design. They normally show every piece of equipment in a system. They show lines and line numbers and give information that indicates where one can find other information. In case you are tempted to think they can replace the piping isometric drawing, the answer is no; they are too cluttered. Their purpose is different from that of the isometric drawing.

Compressible Fluid Flow – Complex Systems | CHAPTER FIVE

Personal 'walk down'

Last but not least is a personal walk down of the as-built piping system to verify it conforms to drawings and to make sure the isometric drawing conforms to the physical system. Note most of the simulation work may have been done prior to this activity, but the check has to be made. The simulation can be revised and rerun if necessary.

We will detail further information on complex compressible flows in Appendix AV.

V-7: CHAPTER SUMMARY

This chapter has tried to outline the most important aspects of simulating complex systems. The method chosen was to describe a common industrial situation involving a complex piping network and complex mixtures of components. The system was a common vent header.

The importance of the following was underlined:
- describing the piping network by isometric sketches;
- describing the flow regimes by grouping devices that may relieve simultaneously;
- having an organized plan of attack;
- establishing the choke points;
- understanding manifold flow and when to make use of worst-case data;
- understanding the "K" factors used and the importance of picking the correct kinetic energy term. Alternatively, using the Benedict et al[xxxv] approach;
- the use of engineering judgement in influencing design to avoid unknowns;
- the significance of the various drawings and data sheets;
- verification of data personally by a final walk down of the system.

APPENDIX AI

Equations Of Incompressible Fluid Flow And Their Derivations

AI-1: PURPOSE — PROVIDING CHAPTER I DETAILS

This appendix describes some of the information given in Chapter I in much greater detail. It gives information regarding SI and customary U.S. units and fluid properties. Pressure is defined, as is the concept of hydrostatic equilibrium. Complete derivations are given for the more important equations. Of particular interest is the explanation of fluid irreversibilities on a rational, thermodynamic basis. Some of the information given may be thought to be trivial, but the author has found it useful to be reminded about trivia from time to time.

The main derivations are for:
- the steady-state force-momentum equation. Many of the other derivations are based on this relationship, and much insight is gained from it;
- the Darcy[vii] equation. The Darcy equation gives the permanent head or mechanical energy loss, the irreversibility, in terms of the length to diameter ratio of the pipe, the kinetic energy of the fluid and the friction factor. The friction factor is simply the coefficient, the ratio of the irreversibility to the product of the L/D ratio and the kinetic energy;
- the Bernoulli[i] equation. The Bernoulli equation relates all of the mechanical energy terms involved with a flowing fluid. The modern version includes a term for the permanent head loss, the irreversibility. This term can be obtained from the Darcy equation. It also includes a term for an external energy source or energy consumer such as a pump or a turbine. It is probably the most widely used equation in fluid mechanics;

APPENDIX AI Equations of Incompressible Fluid Flow

- the thermodynamically derived Bernoulli[i] equation. We will show the Bernoulli equation can be derived from first and second law considerations. This derivation gives insight into what happens to the so-called "lost energy". The fact that the Bernoulli equation can be derived from thermodynamics gives the equation more legitimacy.

In addition to the more formal derivations mentioned above, we will discuss viscosity in some detail. In particular, the variety of units used for viscosity and the resulting potential for error will be pointed out – so error may be avoided.

Laminar flow will be discussed, and the Hagen-Poiseuille[xv] equation will be developed. Laminar flow in circular conduits will be shown to have a parabolic profile. This profile will be used to relate average velocity, U, to point velocity, V.

AI-2: SI AND CUSTOMARY U.S. UNITS

The multiplicity of units used in engineering is a source of much confusion and error. This is especially the case were different units carry the same name, or the same nominal units may have different temperature and pressure bases.

When writing computer programs, it is best to carefully define the units to be used internally to the program and to do all the unit conversions externally to the main program. This practice allows the same program to be used by many users with different requirements. It also lessens the chances of error.

It is fortunate the world is gradually standardizing on the SI system. The U.S. customary system is still very popular, however. Engineers and technicians must be conversant with at least these two systems. In addition to the units of the two main systems, there are a multiplicity of mixed units in use, many of which are falling into disuse. Since these other units usually involve only the use of conversion factors, they will not be discussed. We now will briefly describe some of the engineering units.

Force

Force is that which causes a mass to accelerate. Mass is the substance of a body. Acceleration is the rate at which its velocity changes. Newton's law defines force. We give two forms of the law, one for SI units and one for

Equations of Incompressible Fluid Flow — APPENDIX AI

U.S. customary units:

$$1N = 1kg \cdot 1m \cdot 1s^{-2} \qquad \text{(AI-1)}$$

$$1lb_f = 1lb_m \cdot 1ft \cdot 1s^{-2} \cdot g_c^{-1}$$

The units are newtons, kilograms, meters and seconds in the SI system. The newton is only a name for the group consisting of the three basic units: kg, m and s. In the U.S. customary system, there are four basic units; hence, we need a dimensional constant, g_c, inserted in the denominator on the right or in the numerator on the left. The basic units of the force equation are pound-force, pound-mass, feet and seconds.

The dimensional constant, g_c, used in the customary U.S. system, has a numerical value of 32.17. This value is identical numerically to the acceleration of gravity at sea level at a latitude of 45 degrees. It is frequently called a gravitational constant, but this terminology is incorrect. The units of acceleration are ft/s². Those of the dimensional constant are ft-lb_m/lb_f-s².

The purpose of the dimensional constant is simply to allow one pound force to equal numerically one pound mass when subject to a gravitational field that would give it an acceleration of 32.17 feet per second squared. This acceleration is the normal acceleration of gravity at sea level at a latitude of 45 degrees. It is supposed that since non-technical people could not separate the concepts of weight (force) and mass, then the confusion should be compounded by giving both concepts the same name.

Lest we think users of metric systems are inherently superior, it is good to remember that, until quite recently, the kilogram-force was used with the kilogram-mass for the same purpose. This use also required a dimensional constant. In fact, the principal advantage of a metric system is it contains repeated multiples of ten.

In switching between systems, it is wise always to write the dimensional constant, but to regard it as equal to one and to be dimensionless when using formulae in the SI system. This allows its position to be kept track of when manipulating formulae and when converting back to the U.S. customary system.

APPENDIX AI | Equations of Incompressible Fluid Flow

Example AI-1

An apple, held by its stem, exerts a force of one newton at a place where the acceleration due to gravity is normally 9.81 m-s^{-2} if it were released. Its mass could be computed from

$$1N = x\ kg \cdot 9.81m \cdot 1s^{-2} \qquad \text{(AI-2)}$$

$$x\ kg = \frac{1N}{9.81m \cdot 1s^{-2}} = 0.102 \frac{N \cdot s^2}{m} = 0.102\ kg$$

The apple would have a mass of 102 grams, 0.225 lb$_m$, or 3.60 ounces. We commonly say it "weighs" 102 grams or 3.6 ounces. It actually "weighs" one newton or 0.225 pounds-force. This is the force with which it is attracted toward the center of the earth.

Pressure

Pressure is defined as force per unit area. A more detailed discussion will follow later in this appendix.

$$1Pa = 1N \cdot m^{-2}, \quad 1bar = 10^5\ Pa = 100kPa \qquad \text{(AI-3)}$$

$$1lb_f \cdot ft^{-2}\ (psf) = (1/144)\ lb_f \cdot in^{-2}\ (psi)$$

In the SI system, the units are pascals, newtons and meters. Since the pascal is so small, one hundred thousand pascals are given the name "bar", which is equivalent to one hundred kilopascals. Similarly, psfa units are replaced by psia units with a suitable multiplier of 144 to make the units more maniable.

Psfa and pascals are still the basis of all scientific and engineering computations even though the nominal units are psia and kilopascals.

Example AI-2

At a place where the acceleration of gravity is 9.81 meters per second per second, a mass of 102 grams of water (3.6 ounces) exerts a force of one newton. Poured into a horizontal tray of one meter by one meter, the water exerts a pressure of 1N-m^{-2} or one pascal, on the bottom of the tray.

In customary U.S. units, 102 grams is 0.225 pounds-mass. The equivalent in pounds-force is found from Newton's law.

$$x\ lb_f = \frac{0.225 lb_m \cdot 32.17\ ft \cdot s^{-2}}{32.17 lb_m \cdot ft \cdot lb_f^{-1} \cdot s^{-2}} = 0.225\ lb_f \qquad \text{(AI-4)}$$

Equations of Incompressible Fluid Flow | APPENDIX AI

Since one meter squared is equal to 10.764 feet squared, the pressure exerted would be,

$$P = \frac{0.225\, lb_f}{1m^2 \cdot 10.764\, ft^2/m^2} = 0.0209\, lb_f \cdot ft^{-2} \quad \text{(AI-5)}$$

Dividing this number by 144 gives the pressure in pounds-force per square inch – 0.000145 psi. Multiplying psi by 27.73 gives the pressure in equivalent inches of water column – 0.004 in. WC (but at what temperature?). The water would be spread very thinly – four thousandths of an inch or one tenth of a millimeter. The smallness of the pascal is the reason kilopascals, megapascals and bar (10^5 Pa) are more prevalent than pascals.

Density

Density is mass per unit volume.

$$\rho = kg \cdot m^{-3}$$
$$\rho = lb_m \cdot ft^{-3} \quad \text{(AI-6)}$$

At "normal" temperature and pressure, water has a density of approximately 1,000 kg-m^{-3} or 1.0 g-cm^{-3}. The ASME steam tables[xxxvi] give air-free (more dense) water as having a density of 62.371 lb$_m$-ft^3 at 60°F (15.5 °C). At "normal" temperature and pressure, air has a density of about 1.2 kg-m^{-3}. At 60 F and 14.7 psia, the density of air is 0.0764 lb$_m$-ft^{-3}. A volume of deaerated water contains 816 times the mass of an equal volume of air at the same temperature and at 14.7 psia.

Specific weight

Specific weight is defined as the weight per unit volume or the force exerted by a mass contained in one unit of volume. The dimensions are newtons per meter cubed or pounds-force per foot cubed.

The dimensional constant is used with customary U.S. units.

$$\gamma = \rho g \frac{kg}{m^3} \frac{m}{s^2} = \rho g \frac{kg}{m^2 s^2} = \rho g \frac{Ns^2}{m^3 s^2} = \rho g \frac{N}{m^3} \quad \text{(AI-7)}$$

$$\gamma = \frac{\rho g}{g_c} \frac{\frac{lb_m}{ft^3} \frac{ft}{s^2}}{\frac{lb_m \cdot ft}{lb_f \cdot s^2}} = \frac{\rho g}{g_c} \frac{lb_f}{ft^3}$$

APPENDIX AI Equations of Incompressible Fluid Flow

Hydraulics engineers have a tendency to use the concept of specific weight or weight density. Chemical engineers can get along without it. Mass density is quite adequate for all cases. Conceptually, it is easier to associate a mass with a volume than to associate a force with a volume. Force is always associated with masses.

Specific weight is numerically equal to density in customary U.S. units as long as the gravitational acceleration is close to 32.17 feet per second squared. In SI units, it is not the same. As an example we will take water with a mass density of 1,000 kilograms per meter cubed.

$$\gamma = \rho g = 1000 \frac{kg}{m^3} 9.81 \frac{m}{s^2} = 9810 \frac{kg}{m^2 s^2} \left(\frac{N}{m^3}\right) \quad \text{(AI-8)}$$

For water with a density of 62.371 lb_m-ft^3,

$$\gamma = \frac{\rho g}{g_c} = \frac{62.371 \cdot 32.17}{32.17} \frac{lb_f}{ft^3} = 62.371 \frac{lb_f}{ft^3} \quad \text{(AI-9)}$$

Similarly, for air at normal conditions of temperature and pressure,

$$\gamma = 1.2 \frac{kg}{m^3} 9.81 \frac{m}{s^2} = 11.8 \frac{N}{m^3} \quad \text{(AI-10)}$$

For air at 60°F and 14.7 psia,

$$\gamma = 0.0764 \frac{lb_m}{ft^3} \frac{g}{g_c} = 0.0764 \frac{lb_f}{ft^3} \quad \text{(AI-11)}$$

Again, the weight density is numerically the same as the mass density in customary U.S. units, but not in SI units. The statement that U.S. unit weight density is numerically equal to mass density is only true as long as the acceleration due to gravity is not too different from 32.17 ft/s².

In customary U.S. units, the ratio of acceleration due to gravity and the dimensional constant appears frequently. Since these numbers are equal, they often are cancelled. It is well to remember the ratio has units.

$$\left[\frac{g}{g_c}\right] \equiv \frac{ft \cdot s^{-2}}{lb_m \cdot ft \cdot lb_f^{-1} \cdot s^{-2}} = \frac{lb_f}{lb_m} \quad \text{(AI-12)}$$

All pounds are not the same! Different units cannot be cancelled.

Equations of Incompressible Fluid Flow | APPENDIX AI

In spite of appearances, this section was not meant to confuse the reader. It was meant to make him or her aware units are often used with inadequate definitions. The reader should make it a habit of checking units each time he or she comes across a new formula. In particular, each term of a formula must have the same units as every other term. If it does not, the formula is fundamentally flawed.

It might be thought the "cookbook" approach would resolve the problem of units. However, even this method relies on:
1. the original formula being correct;
2. the formula being applicable to the problem being analyzed; and,
3. the units (including their temperature and pressure bases) being carefully defined.

There is no such thing as a free lunch – even when cookbooks are used.

Energy

Energy is defined fundamentally as the potential to perform work. Work is defined as force moving through a distance, but this is obviously only one aspect of energy. Energy can be stored or changed from one form to another. It cannot be destroyed; it is a conserved quantity. When we talk about lost energy, we mean the energy is no longer available for use as mechanical energy; it has been transformed to internal energy or has passed to the environment as a flow of heat. The term "irreversibility" is a better term than "lost energy" or "friction loss".

If the force is constant over the distance, work energy is simply the product of the force times the distance. If it is variable, then it is integrated over the distance.

$$E = 1J = 1N \cdot 1m \qquad \text{(AI-13)}$$
$$E = 1 ft - lb_f = 1 ft \cdot 1 lb_f$$
$$1J = 1N \cdot m = 1N(0.225)\frac{lb_f}{N} 1m(3.28)\frac{ft}{m} = 0.738 ft - lb_f$$

AI-3: PRESSURE AT A POINT WITHIN A FLUID

It is often stated that pressure at a point within a fluid is equal in all directions. If this be true, how is it that pressure decreases vertically upwards? We will try to clarify this conundrum by a more formal derivation.

APPENDIX AI | Equations of Incompressible Fluid Flow

In order to understand the concept of pressure within a fluid, we can imagine a free body within a stationary fluid. The free body is at equilibrium with the surrounding fluid so all forces acting upon it are balanced. The body does not move.

The body does not have to be different from the fluid. It can be composed of a volume of the fluid. The analysis is simpler for such a volume.

To give as much generality as possible to the analysis, the free body will be chosen as a tetrahedron (following McCabe and Smith[xxxi]). Figure AI-1 shows the body and the forces have components in the vertical direction.

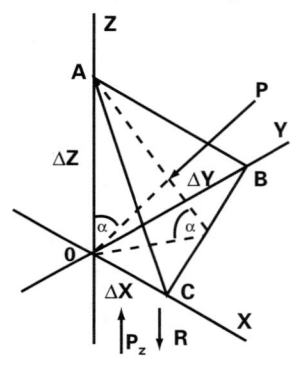

Figure AI-1. Pressure tetrahedron

Forces acting

Fluids (except pseudoplastics) cannot support shear stress without deforming (without moving). Since the fluid is stationary, all forces acting on the body must be perpendicular to the surfaces of the body and must also be at equilibrium, otherwise the body would move. A mathematical statement of this is the algebraic sum of the external forces must equal zero.

In the vertical direction, the forces are:
- the weight of the body acting downwards, $mg/g_c = R$;
- the force due to pressure, P_z, acting upwards on the bottom surface, OBC;
- the vertical (downwards) component of the pressure, P, acting normal to the inclined surface, ABC.

The forces due to the pressures acting on the surfaces, AOC, and, AOB, have no vertical components. They enter only into the analysis of horizontal forces.

From the common statement of Newton's second law, the sum of the external forces equals the mass times the acceleration. The acceleration is zero, so the forces must be balanced. The mathematical statement is,

$$\sum F_Z = \frac{mg}{g_c} - P_Z A_{OBC} + \left(PA_{ABC}\right)_Z = 0 \qquad \textbf{(AI-14)}$$

This is a force balance with the positive direction taken downward. The subscript, Z, on the last term is there to remind us we must find an expression for the vertical component of this force.

Area, A_{OBC}, can be seen from Figure AI-1 to be simply $\Delta x \Delta y/2$. It will be helpful to obtain an expression for area, A_{ABC}, in terms of the same variables, $\Delta x \Delta y$, so they may be cancelled. Figure AI-2 from elementary geometry can be used to develop the relationship. The plane, P´, passes through the base BC of the triangle ABC and the plane, P, is a parallel plane. A perpendicular dropped from the apex, A, through the two planes will cut them at a´ and a. If another perpendicular, a´H, is drawn through a´ and the base, BC, we have three mutually perpendicular lines, Aa´, a´H and BC. A plane through the points A, a´, H is, therefore perpendicular to the base, BC, so the line AH is the height of the triangle BAC. The triangle Ba´C is identical to the triangle bac because it is a parallel projection of triangle bac.

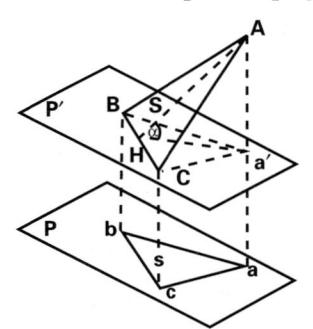

Figure AI-2. Projected areas

From the figure we can see cos α equals length a´H divided by length AH. If S is the area of the larger triangle, ABC, and s is that of the smaller triangle, abc (or a´BC), we have,

APPENDIX AI | Equations of Incompressible Fluid Flow

$$S = \frac{AH \cdot BC}{2} \tag{AI-15}$$

$$s = \frac{d'H \cdot BC}{2}$$

$$d'H = AH \cos \alpha$$

$$s = \frac{AH \cos \alpha \cdot BC}{2}$$

$$s = S \cos \alpha$$

$$S = \frac{s}{\cos \alpha}$$

The two areas are related by the cosine of the included angle. The vertical component of the pressure force, acting on the area BAC, is also related by the cosine of the same angle, as can be seen from Figure AI-1. The force balance of Equation AI-14 can now be written,

$$\sum F_z = \frac{mg}{g_c} - P_z \frac{\Delta x \Delta y}{2} + P \frac{\Delta x \Delta y}{2 \cos \alpha} \cos \alpha = 0 \tag{AI-16}$$

The volume of a pyramid is given by the base area times the height divided by three. The base area is half $\Delta x \Delta y$ so the volume of the pyramid is one sixth $\Delta x \Delta y \Delta z$.

The mass of the free body is its volume times its density, $\rho \Delta x \Delta y \Delta z / 6$. The force balance equation may now be written,

$$\sum F_z = \frac{\rho \Delta x \Delta y \Delta z g}{6 g_c} - P_z \frac{\Delta x \Delta y}{2} + P \frac{\Delta x \Delta y}{2} = 0 \tag{AI-17}$$

$$\frac{\rho \Delta z g}{6 g_c} - \frac{P_z}{2} + \frac{P}{2} = 0$$

If the angle alpha is kept constant while the plane, ABC, (in Figure AI-1), is moved toward the origin, the length Δz becomes dz, a differential quantity. The equation becomes a differential equation,

$$\frac{\rho g}{6 g_c} dz - \frac{P_z}{2} + \frac{P}{2} = 0 \tag{AI-18}$$

When dz becomes negligibly small (vanishes), P_Z equals P. Since the pressure P is normal to an arbitrarily inclined plane, it can be said the point pressures must all be equal to the vertically directed pressure. Furthermore, when analyzing horizontally directed forces, we can simply stand the free body on its side. The first term in the equation no longer applies so all horizontal pressures are equal. The generally accepted statement that pressure at a point is equal in all directions is only true because the point in question is so small as to allow the differential height in Equation AI-18 to be neglected.

AI-4: HYDROSTATIC EQUILIBRIUM

Although pressure at a point is equal in all directions, pressure does change in the vertical direction. This is explained by the following analysis based on Figure AI-3. The figure shows a column of fluid extending in the Z direction. We will take the upward direction as positive. A thin slice of the fluid can be taken as a free body and the development in AI-19 can be made. The sum of the external forces must be zero, otherwise, the body would move.

$$\sum F_Z = PS - (P + dP)S - \frac{g\rho S dz}{g_c} = 0 \qquad \textbf{(AI-19)}$$

$$-SdP - \frac{g\rho S dz}{g_c} = 0$$

$$\frac{dP}{\rho} + \frac{gdz}{g_c} = 0$$

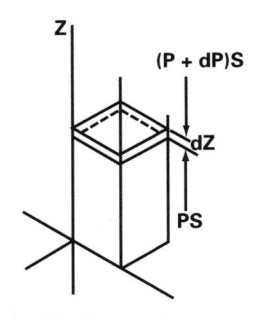

Figure AI-3. Hydrostatic equilibrium

The last equation of the set AI-19 is a differential equation relating change in pressure to change in elevation. It also relates two terms that have units of energy. If the fluid can be taken as incompressible and the gravitational

APPENDIX AI | Equations of Incompressible Fluid Flow

acceleration is constant (if the difference in elevation is not excessive), the equation can be integrated directly to,

$$\frac{1}{\rho}\int_1^2 dP + \frac{g}{g_c}\int_1^2 dz = 0 \qquad \text{(AI-20)}$$

$$\frac{P_2 - P_1}{\rho} + \frac{g}{g_c}(z_2 - z_1) = 0$$

$$\frac{P_2}{\rho} + \frac{g}{g_c}z_2 = \frac{P_1}{\rho} + \frac{g}{g_c}z_1$$

In a constant density static fluid, there is an inverse relationship between pressure and elevation. If one increases, the other decreases.

Equation AI-20 can also be integrated for compressible fluids. However, in this case, the density varies with temperature and pressure and the acceleration of gravity varies with height, so the integration becomes more difficult.

Note we started the analysis by considering forces acting on a free body and ended with an expression in terms of energy per unit mass. The expression could have been derived directly from the Bernoulli[1] equation by realizing that, in static equilibrium, there are no losses ($h_f = 0$), there is no kinetic energy ($U^2/2g_c = 0$) and there is no work transfer (w = 0).

In hydraulics practice, the concept of specific weight (force per unit volume), γ, is used. The above equation is then written, for an incompressible fluid,

$$\frac{P_2}{\rho\frac{g}{g_c}} + z_2 = \frac{P_1}{\rho\frac{g}{g_c}} + z_1 \qquad \text{(AI-21)}$$

$$\left[\rho\frac{g}{g_c}\right] \equiv \gamma$$

$$\frac{P_2}{\gamma} + z_2 = \frac{P_1}{\gamma} + z_1$$

Density has dimensions of pounds-mass per foot cubed or kilograms per meter cubed. Specific weight has units of pounds force per foot cubed or newtons per meter cubed. In the author's opinion, these latter units have resulted in much grief.

AI-5: FRICTION LOSSES EXPLAINED

The term "fluid friction loss" is a logical inconsistency used to describe the conversion of mechanical energy to internal energy and heat flow. Friction is a macroscopic phenomenon occurring when objects such as brake pads are applied to brake drums or disks. When we deal with fluids, especially with gases, physics teaches that the molecules do not touch. The molecules of fluids approach one another until their force fields overlap. When the repulsive force overwhelms the attractive force, the molecules are mutually repelled.

If molecules do not touch, and if total energy is conserved, what is a more logical explanation for the mechanical energy losses involved in fluid flow than that of fluid friction?

A more satisfying explanation of fluid friction losses is found in a knowledge of thermodynamics. Sadi Carnot[ii] showed work (mechanical) energy could be converted completely to thermal energy, but that, even in an ideal cycle, thermal energy could be converted only partially to mechanical energy. In the latter case, even with an ideal heat engine (a concept which Carnot invented), some energy must flow to a cooler sink under the influence of the temperature difference. This valuable concept was the source of many advances in thermodynamics – in particular the idea of entropy as the quantity that governs natural change.

This same concept gives a more satisfactory explanation of fluid mechanical energy (friction) losses than is generally given in hydraulics texts.

Obstructions, compression and turbulence

When portions of a flowing fluid meet an obstruction such as a bend, a valve or even the roughness of a pipe wall, compression takes place. This phenomenon is well described in textbooks under the concept of stagnation – stagnation temperature, stagnation pressure. When compression takes place, the local temperature increases – even with so-called incompressible fluids. The local pressure also increases. Turbulence results as the mass of a portion of the fluid that is at a momentarily higher pressure takes off along the path of least resistance, exchanging static energy for kinetic energy.

In turbulent flow, it is easy to see that as eddies impinge upon irregularities in a pipe wall or even upon other eddies, there is compression and, therefore, a local temperature and pressure increase. The eddies become self-propagating.

APPENDIX AI | Equations of Incompressible Fluid Flow

Lost work really means conversion to internal energy with possible heat energy flow

Carnot clearly identified heat flow due to a temperature difference without conversion of some of the thermal energy to work as a pure loss of mechanically available energy. The energy is still present; it is simply in a form that cannot contribute to bulk movement. Even with an ideal Carnot[ii] engine, there is a flow of unconverted thermal energy to a cooler sink. This energy is what is normally considered a loss. Carnot even showed how to compute these losses. The ideal Carnot engine is simply a mental construct that allows estimation of the maximum amount of mechanical energy that can be obtained under a temperature difference.

Laminar flow and compression

Even in laminar flow, where it is not so obvious compression is occurring, compression does take place. Bulk flow in a conduit is still in the axial direction, but molecular motion is random and mass does move across layers in a direction normal (perpendicular) to the layer. Continuity demands the normal mass flow be equal in each direction. In other words, equal amounts of molecular mass enter as leave a given layer.

The slower molecules always slow the bulk motion of the faster molecules. This phenomenon results in compression and compression results in increased temperature.

Once there are (many) temperature differences in a flowing fluid, there are flows of thermal energy (heat). Some, although very little, of this energy can be recuperated when it causes expansion of a local volume and, therefore, the production of work. However, by the second law of thermodynamics, and as demonstrated by Carnot, not all thermal energy can be converted to work. In the real world, the higher local temperatures associated with turbulence result in heat transmission through the pipe walls to the environment. This is obviously a loss of available mechanical energy – energy that could have been used in moving the fluid down the line.

Another portion of the thermal energy remains in the fluid. The fluid simply becomes warmer, but at a bulk temperature less than the highest local temperature generated by compression.

The many small temperature differences cannot be measured individually, but the overall conversion of mechanical to thermal energy can be measured and correlated as pressure energy that must be supplied externally to

Equations of Incompressible Fluid Flow | APPENDIX AI

maintain a constant flow rate. This correlation is what is done through the Darcy-Weisbach[vii] relationship, the friction factor and the Reynolds[v] number.

The Carnot engine

The reader can get an idea of the loss mechanism by realizing the Carnot[ii] engine was conceived of as one that would give the maximum efficiency of conversion of heat energy flow to work. The efficiency of such an engine is simply given by the temperature difference between the source of heat and the sink divided by the absolute temperature of the source. Note that a real engine does not have to exist. Carnot's purpose was simply to establish the limits of the conversion of heat to work.

We can visualize small volumes of fluid as miniature Carnot engines that convert some heat flow to work and pass on the remainder; the higher temperatures that generate heat energy flow are due to fluid compression as described above. Since we know compression of turbulent fluid only increases its temperature by a few degrees at the most, we can get a feeling for the order of magnitude of the conversion.

Suppose the stagnation phenomenon causes the temperature to increase by five degrees Celsius and the bulk temperature is 298 degrees Kelvin. The efficiency of a Carnot engine operating between these two temperatures would be 100(303-298)/303 or about 1.7%. This represents the most thermal energy that could possibly be converted to work under these conditions. In reality, we do not have ideal Carnot engines conveniently converting heat energy to work energy.

The volumes of fluid do not undergo a uniform cyclical process and much of the heat flow will be directly from the hot source to the cooler sink. So the recovered work will be much less than even 1.7%. In addition, as heat flows, the temperature differences for subsequent cycles will be lower. A one-degree Celsius difference would give only 0.34% efficiency – *even in the ideal case.*

Recapitulation

To recapitulate, a flowing fluid requires outside energy to keep it flowing. Some of this energy is converted to the kinetic energy associated with flow. When some of the fluid meets obstructions, wall imperfections, other eddies or even when there is molecular diffusion in laminar flow, compression takes place. This stagnation results in locally higher pressures and temperatures momentarily. Mechanical energy associated with the moving mass can be converted to thermal energy during this stagnation process. When the volume of fluid moves off into the main stream again, because of a pressure difference, there will be heat transfer to any lower temperature fluid (there will also be heat transfer to a pipe wall or valve or fitting). The volume of fluid may do net work on the surrounding fluid, but, by the second law, the absolute maximum thermal energy that can be recovered as work is less than 1.7% per five degrees temperature difference. It is less than 0.34% for a one-degree difference. Since there is much direct heat transfer, the actual useful mechanical energy recovered from heat energy flow in a flow process is negligible. The conversion of mechanical energy to internal energy and heat energy flow must be made up by the external source keeping the fluid in motion. *The difference between the total mechanical energy at each end of the conduit represents the irreversibilities – the losses.*

AI-6: FORCE-MOMENTUM CONSIDERATIONS FOR VARIABLE MASS SYSTEMS

Newton's second law is usually written for constant mass systems. As such, it is not a sufficiently general equation to serve as a starting point for some analyses. For instance, the analysis of Bernoulli's[i] equation given in Section AI-8 makes use of the momentum equation without showing how it is derived. When computing velocities and thrusts of rockets or forces and flow in pumps and compressors, a different starting point is advisable.

Variable mass systems

Rockets, pumps and compressors may be categorized as variable mass systems. The quantity of mass either changes with time within the system or it is flowing through the system. We will start the analysis with a generalized system then apply it to a few specific ones.

Figure AI-4 depicts a mass, m, moving with a velocity, v overbar, at an initial time, t_0. A short time later, t_0 plus Δt, the mass has ejected a small mass, Δm, at a velocity, u overbar. The velocity of what is left of the main mass has increased by Δv. (This could be a rocket, but it is not specified at this point so we can derive an equation with the greatest generality.) The overbars represent vector quantities for more generality. They have direction as well as magnitude.

We treat the mass as a system and apply to it Newton's law relating an external force to the rate of change of momentum. The variable, P overbar, is the momentum vector. The force vector will have the same direction as the rate of change of momentum. The reader can find a similar treatment in Halliday and Resnick.

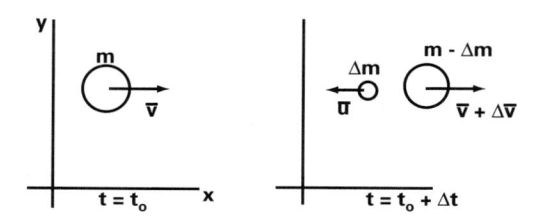

Figure AI-4. Variable mass systems

In what follows, we have adopted the terminology of Halliday and Resnick. The mass is in slugs, i.e. pounds-force divided by the gravitational acceleration, g, or pounds-mass divided by the dimensional constant, g_c. Kilograms can be used directly in the equations.

External force equals the time rate of change of linear momentum

The following development can be made.

$$\bar{F}_{ext} = \frac{d\bar{P}}{dt} \sim \frac{\bar{P}_f - \bar{P}_i}{\Delta t} = \frac{\left[(m - \Delta m)(\bar{v} + \Delta \bar{v}) + \Delta m \bar{u}\right] - m\bar{v}}{\Delta t} \quad \text{(AI-22)}$$

$$\bar{F}_{ext} \sim \frac{m\bar{v} + m\Delta \bar{v} - \Delta m \bar{v} - \Delta m \Delta \bar{v} + \Delta m \bar{u} - m\bar{v}}{\Delta t}$$

$$\bar{F}_{ext} \sim m\frac{\Delta \bar{v}}{\Delta t} - \bar{v}\frac{\Delta m}{\Delta t} - \frac{\Delta m \Delta \bar{v}}{\Delta t} + \bar{u}\frac{\Delta m}{\Delta t}$$

In the limit, as Δt approaches zero, several changes occur:
1. The total mass, m, becomes associated with the instantaneous mass in the body. Effectively, the two states fuse since Δt is so small.

2. The quotient $\Delta m / \Delta t$, a positive mass divided by a positive period, becomes minus dm/dt since the change is associated with loss from a body.

3. The product $\Delta m \Delta v$ becomes the product of two differentially small amounts, dmdv, and may be neglected.

APPENDIX AI | Equations of Incompressible Fluid Flow

Most general equation for variable mass systems

The equation in the limit becomes one of four alternative expressions.

$$\overline{F}_{ext} = m\frac{d\overline{v}}{dt} + \overline{v}\frac{dm}{dt} - \overline{u}\frac{dm}{dt} \quad \text{(AI-23)}$$

$$m\frac{d\overline{v}}{dt} = \overline{F}_{ext} - \overline{v}\frac{dm}{dt} + \overline{u}\frac{dm}{dt}$$

$$m\frac{d\overline{v}}{dt} = \overline{F}_{ext} + (\overline{u} - \overline{v})\frac{dm}{dt}$$

$$m\frac{d\overline{v}}{dt} = \overline{F}_{ext} + \overline{v}_{rel}\frac{dm}{dt}$$

The three terms of the last equation represent the motion of the body. They are all forces. The first term is the mass of the body times its acceleration. This term represents the actual acceleration of the body. If there is no motion, this term is zero. The second term is the sum of all the external forces on the body, and the third is the thrust on the body due to the leaving mass. The thrust is made up of the relative velocity of the mass leaving the body (negative) and the instantaneous rate of change of the mass of the body (also negative).

Equations AI-23 are the most general equations governing variable mass systems obeying the rules of classical mechanics (systems remote from the velocity of light). They cannot be derived from Newton's law in its normal form by treating mass as a variable. If such a derivation is attempted, one term will be missing (the last term of the first equation), because Newton's law in its normal form is a restricted case of a more general law.

Specific example of equations AI-23 – a rocket

Although we are not dealing with rockets in this book, the application of the above equation to a rocket helps fix the terms of the equation clearly. This is the only reason for discussing rockets.

1. The first term of the last equation of the set AI-23 refers to the instantaneous mass of the rocket and to the acceleration of that mass.

2. The second term represents the sum of the external forces acting on the body of the rocket. These forces are variable. They consist mainly of the gravitational attraction acting down and air resistance acting against the motion.

3. The relative velocity of the third term is made up of two components. The first is the velocity of the gases leaving the rocket. The second is the velocity of the rocket.

Equations of Incompressible Fluid Flow | APPENDIX AI

4. The rate of change of mass per unit time is negative for a rocket. When multiplied by the negative relative velocity, the result is a positive thrust in the upward direction. This thrust must overcome the external forces (predominantly gravity) acting on the rocket if the rocket is to move from its launch pad.

Specific example of equations AI-23 – a restrained system

Figure AI-5 represents a restrained system that could be a pipe fitting or a pressure relief valve. The general equation for variable mass systems can be adapted to fittings, pipe bends and relief valves by simply fixing the body within a reference frame. The body consists of an open system through which mass is flowing. Since the body is fixed within the reference frame, its velocity is zero and the first term in the last equation of set AI-23 disappears. We will extend the analysis by considering mass entering as well as mass leaving the system.

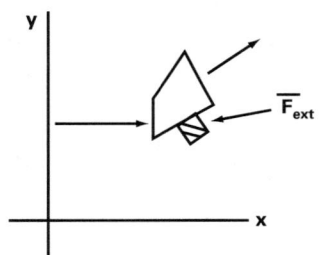

Figure AI-5. Restrained system

For a restrained system, the last term is still the thrust on the system. The mass, m, is the mass of the body and its contents. Since, in Figure AI-5, the velocity vector points in the positive direction and the mass leaving per unit time is negative, the force due to leaving mass points in the opposite direction to flow. The force due to mass entering, however, points in the same direction as the flow.

Steady-state simplification

We can simplify the analysis by considering the steady-state – the most common industrial state for most processes. In the steady-state, the mass flow rate entering equals the negative of the mass flow rate leaving. We can use the signs of the velocities to represent the direction of the resultant force on the system. We have to sum two equations, one for flow out and one for flow in, to get the resultant forces on the system (pipe fitting, valve or other open body).

APPENDIX AI | Equations of Incompressible Fluid Flow

$$0 = \sum \bar{F}_{ext} + \left(\bar{u}\frac{dm}{dt}\right)_{out} + \left(\bar{u}\frac{dm}{dt}\right)_{in} \tag{AI-24}$$

$$\sum \bar{F}_{ext} = -\left[\left(\bar{u}\frac{dm}{dt}\right)_{in} + \left(\bar{u}\frac{dm}{dt}\right)_{out}\right] = -(\bar{u}_{in} - \bar{u}_{out})\dot{m}$$

Note the mass flow rate is always positive. The sign on the leaving velocity has been changed to allow the mass flow rate to be substituted for the rate of change of the leaving mass. If the inlet velocity is greater than the outlet velocity, the reaction will be in the negative direction. The anchor must supply the resistance.

Equation AI-24 applies to a constant mass flow rate. It states the mass flow rate, an inherently positive scalar, can be multiplied by the vector difference between incoming and outgoing velocities to find the sum of the external forces that must be present to keep the system fixed in space. If the incoming velocity is greater than the outgoing one, the force on the system points in the general direction of flow and the system must be restrained by a force (an anchor) in the opposite direction. If the outgoing velocity is greater, the force points against the general direction of flow and the anchor must be on the other side of the fitting. The exact direction of the external force is given by the vector difference of the velocities.

Caveat — Steady-state assumptions

The last equation only applies to steady-state mass flow. This fact is often not clearly stated when the equation is cited. If this fact is not understood, your cookbook will lead you to burn some of your creations. For instance, the initial, instantaneous, impact forces on a pressure relief valve are about twice the forces estimated on the basis of steady-state flow.

The external force is the sum of external forces that balance the momentum forces. This sum includes differential pressure across the body, viscous forces opposite flow, gravitational forces and forces due to internal pressure acting on the walls of the fitting if they are not parallel. We will use this sum when we describe Darcy[vii] and Bernoulli[i] equations.

Because of the form of Equation AI-24, it is often stated that the sum of the external forces on a body equals the differences between time rate of change of momentum being convected in and out of the body. If this statement is felt to be elegant, then it is worth remembering. If not, it is best to simply think of the derivation in terms of relationships between forces, masses and velocities that occur under well-defined circumstances.

AI-7: DERIVATION OF THE DARCY EQUATION

The friction factor and the Darcy equation

The Darcy[vii] or Darcy-Weisbach equation is an empirical equation given about 1850 to express the irreversibilities due to incompressible fluid flowing through a conduit. For a horizontal conduit of round section, these irreversibilities are expressed as,

$$h_f = \frac{\Delta P}{\rho} = f_M \frac{L}{D} \frac{U^2}{2g_c} \qquad \text{(AI-25)}$$

The first two terms in equation AI-25 are derived directly from the Bernoulli[i] equation (I-20) by considering the density to be constant and the pipe to be horizontal and of constant section. This leaves only the mechanical energy loss term and the static energy terms ($\Delta p/\rho$ is the difference between two static energy terms for the incompressible fluid). *Equation AI-25 gives a way of measuring energy losses directly by measuring pressure drop in horizontal, constant diameter pipe.* The pressure drop is the upstream less the downstream pressure, in this case. Once a correlation among the length, diameter, kinetic energy and friction factor is obtained for horizontal pipes, it can be applied to all pipes, vertical, inclined and horizontal, by including all the terms of the Bernoulli equation.

The last term in AI-25 is an empirical correlation of the experimentally measured losses and known quantities. Equation AI-25 is the Darcy equation. The factor of proportionality is called the Moody friction factor. It can be seen to be dimensionless if the units are checked. It is therefore the same number whether SI (g_c equals one, no dimensions) or customary U.S. units are used. The velocity, U, is the average velocity across the conduit section. The dimensional constant, g_c, has units of lb_m-ft/lb_f-s^2.

The relationship is valid completely only when a fully developed flow profile has been established. For turbulent flow, this normally occurs about 30 diameters downstream from a pipe entrance from a vessel. The relationship applies to both laminar and turbulent flows and to smooth or rough pipes.

The friction factor for liquids is found from experiment. The irreversibilities, equivalent to the pressure drop divided by the density in constant diameter pipe, are divided by the terms to the right of the friction factor. Many workers have established correlations among the friction factor, so found, and other parameters.

APPENDIX AI | Equations of Incompressible Fluid Flow

From dimensional analysis and experiment, the parameters that most influence the friction factor, when flow is turbulent, are the Reynolds[v] number and the relative roughness of the pipe. The relative roughness is the absolute roughness of the pipe divided by the diameter. The absolute roughness is the height of the actual asperities on the pipe wall. Fortunately, commercially available pipe is extruded on mandrels that give a fairly reproducible relative roughness.

Laminar flow — the Hagen-Poiseuille equation

G. H. L. Hagen[xv], a German hydraulic engineer, published results of experiments on three brass tubes with diameters of 2.55 mm (0.1 inch), 4.02 mm (0.16 inch) and 5.91 mm (0.23 inch). The lengths were 47.4 cm (18.7 inch), 109 cm (42.9 inch) and 105 cm (41.3 inch).

Louis Marie Poiseuille[xv], a French physician and physicist, shortly afterwards independently presented results on flow in glass capillaries. The flows were clearly in the laminar range. The Reynolds[v] numbers were below 2,000. The equation resulting from the work of these two researchers was,

$$\Delta P = 32 \frac{\mu}{g_c} \frac{UL}{D^2} \qquad \text{(AI-26)}$$

The differential pressure given in Equation AI-26 is that pressure drop associated with losses. It is not recoverable pressure. It might be better to divide both sides of the equation by density and state that the left-hand side represents losses or irreversibilities. The coefficient, 32, is a pure number. It has no dimensions. This can be seen if the appropriate dimensions for all the other terms are substituted into the equation. The viscosity units in AI-26 are $lb_m \cdot ft^{-1} \cdot s^{-1}$.

Equation AI-26 is sometimes transformed from one in velocity to one in volumetric flow.

$$\Delta P = 128 \frac{\mu}{g_c} \frac{QL}{\pi D^4} \qquad \text{(AI-27)}$$

By equating the Darcy-Weisbach[vii] equation to the Hagen-Poiseuille equation over the same permanent pressure drop and solving for the friction factor, the laminar flow friction factor can be obtained as a function of the Reynolds number.

Equations of Incompressible Fluid Flow | APPENDIX AI

$$\Delta P = 32 \frac{\mu}{g_c} \frac{UL}{D^2} = \rho f_M \frac{L}{D} \frac{U^2}{2g_c} \qquad \text{(AI-28)}$$

$$f_M = \frac{64\mu}{DU\rho} = \frac{64}{N_{\text{Re}\,D}}$$

When flow is in the laminar region, it is found that the effect of relative roughness is negligible because of the viscous sublayer of fluid near the wall.

It is to be noted the Darcy-Weisbach[vii] equation maintains the form,

$$h_f = \frac{\Delta P}{\rho} = f_M \frac{L}{D} \frac{U^2}{2g_c} \qquad \text{(AI-29)}$$

for both turbulent and laminar flow. Superficially, this fact seems to contradict previous statements that, in laminar flow, pressure drop is directly proportional to average velocity and not U^2. The apparent inconsistency is resolved when one realizes the laminar flow friction factor (Equation AI-28) contains a velocity term, U, in the denominator. This cancels one of the velocity terms in the numerator. The equation for laminar flow is linear with velocity, in spite of appearances. In industrial situations laminar flow occurs mainly with highly viscous materials – polymers and the like.

The critical zone

The range of flows involving Reynolds[v] numbers between 2,000 and 4,000 is given the name "critical" zone. Again, terms such as critical and absolute are used rather indiscriminately. The friction factors in this range of flows are indeterminate because of the fluctuations that occur between the two principal regimes. There are large uncertainties. The critical zone is the transition zone between two rather well defined zones.

Turbulent flow

In most industrial applications, turbulent flow is the norm.

Smooth pipes

Blasius[xxxvii] (student of Prandtl[xxxviii]), in 1911, was the first to establish the fact that the friction factor followed a functional relationship in Reynolds[v] number below a Reynolds number of 100,000 in hydraulically smooth pipes. This relationship is given as,

$$f_{Blasius} = 0.3164 N_{\text{Re}\,D}^{-1/4} \qquad \text{(AI-30)}$$

APPENDIX AI | Equations of Incompressible Fluid Flow

Blasius[xxxvii] made use of (correlated) available data, including some from Cornell University.

In 1914, it was concluded from limited air and water tests that the relationship between the friction factor and the Reynolds[v] number was independent of the fluid. This permitted exploration of irreversible energy conversions in compressible fluids. The validity of the assumption that friction factors for gases are the same as those for liquids will be explored more fully when the work of D.S. Miller[xvi] is discussed.

Prandtl[xxxviii] developed another equation for the relationship between a friction factor (a wall friction factor, f_s) and the pipe Reynolds number. This equation correlated published experimental data over a larger range of Reynolds numbers.

In 1858, Darcy[vii] tested 22 pipes of cast iron, lead, wrought iron, asphalted iron and glass. Nikuradse[xxxix] correlated Darcy's data. Nikuradse screened sand to produce different grades based on diameter. He carefully glued the sand on the internal walls of pipes. Using this method, he was able to vary the relative roughnesses of pipe walls between 1/500 and 1/15, about 33:1.

Colebrook[xl] empirically arranged the Prandtl equation and one developed by von Kármán to produce a formula that gave remarkably good results in correlating published data.

Louis F. Moody[viii] in 1944, presented the plot that carries his name. His friction factor, the one used in this book, is probably the most common one in engineering circles.

Machine computing became simpler when the Churchill-Usagi[xli] relationship (See Appendix II.) was developed. The Churchill-Usagi equation allows direct computation of the friction factor from basic data. Hard-to-read graphs are no longer required.

Derivation of the Darcy equation

This derivation follows that given in McCabe and Smith, *Unit Operations of Chemical Engineering*, 1967.

Consider steady flow of an incompressible fluid through a horizontal, circular pipe. Consider a disc of fluid as a free body within a section as shown in Figure AI-6.

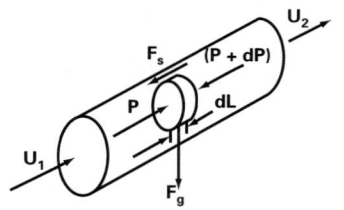

Equations of Incompressible Fluid Flow | APPENDIX AI

Figure AI-6. Derivation of the Darcy equation

The momentum equation gives,

$$\sum Fg_c = \dot{m}(\beta_2 U_2 - \beta_1 U_1)$$
$$\sum F = P_1 A_1 - P_2 A_2 - F_s - F_g$$
(AI-31)

Because the point velocity, v, is not constant across a section, a correction factor, beta, is included to permit the average velocity, U, to be used in place of v. The term, F_s, is the shear force (necessary to prevent acceleration – if it were not present, the fluid would accelerate) due to viscous resistance between the layers of fluid that constitute the outside of the free body and the surrounding fluid. It is equal to the area of the wetted perimeter of the free body times the shear stress, τ. F_g is the gravitational force. This force will be zero in the horizontal direction.

In case the reader has trouble with the origin of the shear force, its presence is arrived at by logical deduction. For a constant density fluid flowing without acceleration in a horizontal pipe of constant section, the velocities at the two sections are equal, as are the correction factors. The momentum term sums to zero. The gravity term in the second equation is zero in the horizontal direction. There is a measurable pressure drop, so we know a force exists in the forward direction. If a shear force, F_s, did not exist to counterbalance the forward force, acceleration would result. The two forces due to static pressure times the areas over which they act have different signs because they act on the body from different directions. The summation gives all of the external forces acting upon the body.

With steady, unaccelerated flow of a liquid in a horizontal segment of constant diameter, the velocities and their correction factors are equal. Since flow is unaccelerated, the external forces cancel. Their sum is zero.

APPENDIX AI | Equations of Incompressible Fluid Flow

$$\sum F = 0 = P_1 A_1 - P_2 A_2 - F_s - F_g \qquad \text{(AI-32)}$$
$$\sum F = 0 = \pi r^2 P - \pi r^2 (P + \Delta P) - 2\pi r \Delta L \tau - 0$$
$$\sum F = 0 = P - P - \Delta P - \frac{2\tau \Delta L}{r}$$
$$0 = \frac{\Delta P}{\Delta L} + \frac{2\tau}{r}$$

At a given cross section, the pressure is constant across the section, otherwise bulk flow would occur in the normal direction. The ratio of the pressure difference, $P_2 - P_1$, between two sections a small distance apart to the incremental length also is constant across the section. Therefore, the first term is independent of r, the radius, which obviously varies across the section. However, the equation must still hold, so the sheer stress, tau, must vary directly with r. Also, because of the flat velocity profile at the center of the pipe, tau is zero when r is zero. In other words, if shear existed at the center line, there would be a noticeable velocity profile at the center.

Using the subscript, w, to indicate a fixed location at the wall, we can write,

$$0 = \frac{\Delta P}{\Delta L} + \frac{2\tau}{r} \qquad \text{(AI-33)}$$
$$0 = \frac{\Delta P}{\Delta L} + \frac{2\tau_w}{r_w}$$
$$\frac{\tau}{r} = \frac{\tau_w}{r_w}$$

The last equation of AI-33 states that the ratio of the shear stress to the radius is constant across the pipe. It says nothing about the linearity of the relationship — the stress is more intense closer to the wall.

From the Bernoulli[1] equation, head loss in steady-state flow of constant density fluid in a constant diameter, horizontal pipe is given by (ΔP is $P_2 - P_1$):

$$-\frac{\Delta P}{\rho} = h_{fs} \qquad \text{(AI-34)}$$

Equations of Incompressible Fluid Flow — APPENDIX AI

The subscript, fs, draws attention to the fact that the losses in mechanical energy are mainly found in the layer of fluid closest to the wall. Note this may be confusing because the term on the left refers to the bulk change between two sections.

Substituting the second equation of the set AI-33, into this equation,

$$0 = \frac{\Delta P}{\Delta L} + \frac{2\tau_w}{r_w} \qquad \text{(AI-35)}$$

$$-\Delta P = \frac{2\tau_w \Delta L}{r_w}$$

$$h_{fs} = -\frac{\Delta P}{\rho} = \frac{2\tau_w \Delta L}{r_w \rho} = \frac{4\tau_w \Delta L}{\rho D}$$

The Fanning friction factor is a parameter used to correlate flow effects. It is defined as,

$$f_F \equiv \frac{\tau_w}{\rho \left(\dfrac{U^2}{2g_c} \right)} \qquad \text{(AI-36)}$$

The Fanning friction factor, by the definition of AI-36, is seen to be the ratio of the shear stress at the wall to the product of the fluid density and kinetic energy. It is the experimentally determined coefficient that relates the shear stress at the wall to the product of density and kinetic energy. The friction factor is dimensionless. When SI units are used, g_c equals one and is dimensionless. The shear stress at the wall can be estimated with the first of the equations in the set AI-35, not from AI-36.

The ratio of shear stress to density that appears in the above definition, AI-36, also appears in the relation between head losses and wall shear, AI-35. So, we can substitute the AI-36 into AI-35 as follows:

$$h_{fs} = \frac{4\tau_w \Delta L}{\rho D} = 4 f_F \frac{\Delta L}{D} \frac{U^2}{2g_c} \qquad \text{(AI-37)}$$

It is more convenient to keep the standard kinetic energy term intact, so the term $4f_F$ becomes f_M, the Moody[viii] friction factor. The equation with this factor becomes,

$$h_{fs} = f_M \frac{\Delta L}{D} \frac{U^2}{2g_c} \qquad \text{(AI-38)}$$

Equation AI-38 is the Darcy[vii] equation. Obviously the Moody[viii] friction factor is four times the Fanning factor. The Moody factor is often called the Darcy or the Blasius[xxxvii] friction factor. The subscript, s, refers to the fact that skin "friction" is conceived of as being the main cause of head loss when no boundary layer separation occurs. *If form friction (irreversibilities) comes in to play, due to boundary layer separation from obstructions or changes in flow directions, h_f becomes greater than h_{fs}. The symbol h_L is often used to designate form friction losses.*

The utility of the Darcy equation is the irreversibility in units of energy per unit mass can be found by experiment for a large variety of situations. The ratio of the irreversibility to the corresponding length to diameter ratio and kinetic energy gives the Moody friction factor. This factor has been correlated with the pipe wall roughness and the fluid densities, viscosities and flow rates. The correlation means the friction factor can be found in the friction factor charts or by the Churchill-Usagi[xli] equations. The friction factor charts can be found in Figures I-4 and II-1. The Churchill-Usagi equation is given in Appendix II.

AI-8: DERIVATION OF THE BERNOULLI EQUATION INCLUDING IRREVERSIBILITIES

This derivation follows that originally given by Weber and Meissner, *Thermodynamics for Chemical Engineers*, 1957[xlii]. The derivation is in customary U.S. units. It was given originally for compressible fluids, but since it is general, it is included here for the incompressible case. Each step of the derivation should be examined to make sure it applies to both incompressible and to compressible fluids.

This equation is for steady-state duct flow (see Figure AI-7). It is derived for an arbitrary expansion whose elevation rises dZ for every dL along the centerline. The cross sectional area increases by dA over the same length.

Equations of Incompressible Fluid Flow | APPENDIX AI

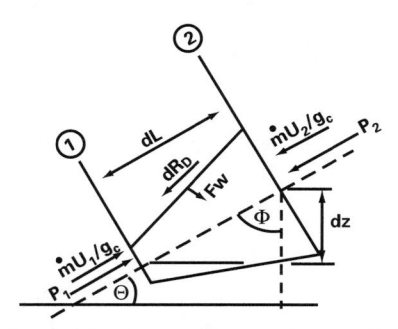

Figure AI-7. Development of the Bernoulli equation

In thermodynamic terms, the analysis is for a control volume over a differential expansion delineated by two normal sections, 1 and 2. Mass flow is into section 1 and out of section 2. This is an open (flow) system.

Force balance

In what follows, letters identify quantities at section 1. The same letters plus the differentials identify quantities at section 2.

A. Forces directed downstream:

1. Pressure times area at section 1, PA

2. Momentum force, because of the steady-state assumption, at section 1, $\dot{m}U/g_c$

3. Axial component of pressure reaction normal to duct wall. This pressure is equal in all directions, so the only force influencing flow is that applied against the difference between the downstream and upstream areas, $(P + dP/2)dA \sim PdA$ (second order differential neglected).

B. Forces directed upstream:

1. Pressure times area at section 2, $(P + dP)(A + dA) \sim PA + PdA + AdP$
2. Momentum force at section 2, $\dot{m}(U + dU)/g_c$
3. Viscous force, R_D. This is the force that is judged to be present because of the lack of acceleration of the fluid. It is responsible for the irreversibilities. It is conceived of as being concentrated at the duct walls due to the steep velocity profile at this location, but this is a simplification.
4. The force of gravity directed downward along dZ has an upstream component along dL.

APPENDIX AI | Equations of Incompressible Fluid Flow

The force due to gravity is $(1/g_c)$(volume/specific volume)(acceleration due to gravity):

$$F_Z = \frac{mg}{g_c} = \frac{g}{g_c} \frac{A_{av}}{v_{av}} dL = \frac{g}{g_c} \left[\frac{A + \dfrac{dA}{2}}{v + \dfrac{dv}{2}} \right] dL \qquad \text{(AI-39)}$$

Since the cosine of the angle, ϕ, that dL makes with dZ is dZ/dL, the upstream force due to gravity along dL is,

$$F_L = F_Z \cos\phi = F_Z \frac{dZ}{dL} = \frac{g}{g_c} \left[\frac{A + \dfrac{dA}{2}}{v + \dfrac{dv}{2}} \right] dZ \qquad \text{(AI-40)}$$

So seven forces have been identified for this steady-state force balance:
- two momentum forces in opposite directions;
- two forces from external pressures in opposite directions;
- one viscous force generated by the presence of the duct walls and fluid motion, directed upstream;
- one wall reaction force directed downstream;
- and, one gravity force directed upstream.

If this were a horizontal parallel duct, the last two forces would have no influence on fluid movement. They would be normal to the flow direction.

At steady-state (constant mass flow rate), the sum of all the forces in the L (centerline) direction is zero – they are all balanced.

$$\sum F_L = 0 = PA + \frac{\dot{m}U}{g_c} - (P + dP)(A + dA) \qquad \text{(AI-41)}$$

$$- \frac{\dot{m}(U + dU)}{g_c} - dR_D - \frac{g}{g_c} \left[\frac{A + \dfrac{dA}{2}}{v + \dfrac{dv}{2}} \right] dZ + \left(P + \frac{dP}{2} \right) dA$$

If all the factors are multiplied and the resulting second order differentials are ignored as being small relative to the first order differentials, the

Equations of Incompressible Fluid Flow | APPENDIX AI

following equation results:

$$-AdP = \frac{\dot{m}\,dU}{g_c} + \frac{g}{g_c}\frac{A}{v+dv/2}dZ + dR_D \qquad \text{(AI-42)}$$

For steady-state mass flow,

$$\dot{m}_1 = \dot{m}_2 = const = \rho_1 A_1 U_1 = \rho_2 A_2 U_2 = \frac{AU}{v} \qquad \text{(AI-43)}$$

The constant term on the right of Equation AI-43 (made up of three variables) may be substituted for the constant mass flow rate in Equation AI-42 and the equation may be manipulated, as follows:

$$0 = A_1 dP + \dot{m}\,dU/g_c + dR_D + \frac{g}{g_c}\frac{A_1}{v_1+dv/2}dZ \qquad \text{(AI-44)}$$

$$0 = A_1 dP + \frac{UA}{v}\frac{dU}{g_c} + dR_D + \frac{g}{g_c}\frac{A_1}{v_1+dv/2}dZ$$

$$0 = v\frac{A_1}{A}dP + \frac{UdU}{g_c} + \frac{v}{A}dR_D + \frac{g}{g_c}\frac{vA_1/A}{v_1+dv/2}dZ$$

As $\quad dP \Rightarrow 0, \quad A \Rightarrow A_1, \quad A_1/A \Rightarrow 1, \quad (v/A) \Rightarrow v_1/A_1, \quad v/(v_1+dv/2) \Rightarrow 1$

$$-vdP = \frac{UdU}{g_c} + \frac{g}{g_c}dZ + (\frac{v}{A})_1 dR_D$$

$$-vdP = \frac{UdU}{g_c} + \frac{g}{g_c}dZ + dh_f$$

The coefficient of the dR_D term in AI-44 is a constant and it can be taken inside the differential. A new differential is thus created which we have called dh_f — the differential losses in mechanical energy due to viscous drag forces or compression. The subscript, 1, shows the values of v and A are those at section 1 — they are constant. All the terms in Equation AI-44 have units of force times length divided by mass. They are energy per unit mass.

Integrating all terms between sections 1 and 2 (definite integral), but leaving the first term under the integral sign because its integration depends on the path (process — adiabatic, isothermal, etc.), the following equation results:

$$\int_1^2 vdP + \frac{\Delta(U^2)}{2g_c} + \frac{g}{g_c}\Delta Z + {}_1h_{f2} = 0 \qquad \text{(AI-45)}$$

APPENDIX AI | Equations of Incompressible Fluid Flow

The units in each equation must be consistent ones. If SI or any other system based on three fundamental units is used, g_c becomes one and is dimensionless. It is advantageous to write the equations with this term so as not to forget the acceleration of gravity term, g, when switching between systems.

Each of the terms of Equation AI-45 except the last represents changes in energy that can be converted to changes in any of the other terms. The last term represents energy that is converted from the other terms. The last conversion is unidirectional. It comes from viscous forces working in the direction opposite the flow and producing local temperature increments due to compression and irreversibilities due to dispersion of heat energy without recovery as useful work.

Viscous forces cause compression of liquids or gases; compression results in a local temperature increase from the work of compression; differential temperature causes heat energy flow; heat energy flow without recuperation of mechanical energy is a loss of mechanical energy (but not of total energy). We have used the subscripts 1 and 2 on the last term just to emphasize the fact the integration was over a length of pipe and the irreversible energy conversion (the irreversibility) belongs to that length. The subscripts will be dropped subsequently.

The first term represents the difference in pressure-volume energy (static) along the conduit. The second one is the difference in kinetic energy. The third is difference in potential energy. The last term represents mechanical energy converted to thermal energy (internal energy or heat flow) due to the irreversibilities. All terms are on a per unit mass basis.

Another way of looking at the above equation is to make the negative of the first term equal to all the other terms. It can then be said that the static energy may be converted to any of the other three forms of energy including mechanical energy losses due to irreversibilities. This is a reasonable explanation because one cannot have movement without a difference in mechanical energy.

Modifying the Bernoulli equation to include a pump or turbine

If a pump or turbine is present, the Bernoulli[i] equation may be modified as,

$$-\int_1^2 vdP + w_n = \frac{\Delta(U^2)}{2g_c} + \frac{g}{g_c}\Delta Z + h_f \quad \textbf{(AI-46)}$$

The second term represents pump or compressor work added to the fluid per unit mass flowing. If a turbine is present the negative sign must be used.

The terms on the right represent the change in energy per unit mass as it flows through the system – kinetic energy, potential energy and irreversible changes to thermal energy.

Restricted forms of the Bernoulli[i] equation
Equation AI-45, the differential form of the Bernoulli equation including irreversibilities, is general. Various restrictions (constraints) can be placed on it to approximate specific processes.

Incompressible fluids
For incompressible fluids, the first term on the left of AI-46 can be integrated directly to $-v(P_2 - P_1)$ or the equivalent $-(P_2 - P_1)/\rho$.

Ideal gases flowing isothermally
For gases that can be represented as ideal (most gases at low pressures), the first term on the left can be integrated directly to $RT\ln(P_1/P_2)$ when there is no temperature change. We will discuss this form more fully in the chapter on compressible flow. This assumption is not applicable to most in-plant situations. It is mostly used for long, uninsulated pipelines.

Comparison of the Bernoulli equation with the first law equation for open systems
We will give two approaches to this comparison. Remember the symbol, h, is used for enthalpy, not for irreversibilities, in the first law equation. The first approach relies on more knowledge of thermodynamics than the reader might possess, but it is given for completeness. The second approach is more readily understandable.

The first thermodynamic approach
The first law statement for open systems has already been given in Chapter I as,

$$\delta q + \delta w_n = dh + \frac{UdU}{g_c} + \frac{gdX}{g_c} \tag{AI-47}$$

$$q + w_n = \Delta h + \frac{\Delta(U^2)}{2g_c} + \frac{g\Delta X}{g_c}$$

The subscript, n, on the work term means "net". This is the work actually transmitted to the fluid. The subscript, f, is a synonym sometimes used for n.

APPENDIX AI | Equations of Incompressible Fluid Flow

The differential form can be rewritten,

$$\delta q - dh + \delta w_n = \frac{UdU}{g_c} + \frac{gdX}{g_c} \qquad \text{(AI-48)}$$

For an ideal system (thermodynamically reversible), dq equals Tds, so Tds can be substituted for dq. The variable, s, is specific entropy. Also, one of the fundamental equations of thermodynamics is,

$$-vdP = Tds - dh \qquad \text{(AI-49)}$$

Substituting into the first law statement, we have,

$$-vdP + \delta w_n = \frac{UdU}{g_c} + \frac{gdX}{g_c} \qquad \text{(AI-50)}$$

The equation, AI-50, is identical to the Bernoulli[i] equation except for the term representing irreversibilities. In other words, it is identical to Bernoulli's original equation. The thermodynamically derived equation assumes ideality (no irreversibilities), as did Bernoulli's original equation.

All the terms are energy per unit mass. When integrated, the two terms on the left can be regarded as driving potentials that produce the two terms on the right. The latter terms are seen to be the kinetic energy and potential energy changes.

For the real (irreversible) case, an arbitrary term can be added which represents the mechanical energy per unit mass that is converted to thermal energy and is no longer available to do work.

$$-vdP + \delta w_n = \frac{UdU}{g_c} + \frac{gdX}{g_c} + dh_f \qquad \text{(AI-51)}$$

The two terms on the left of AI-51 can be regarded as the driving potentials. The three terms on the right are seen as changes resulting from the driving potentials, kinetic energy, potential energy, and lost work (converted to thermal energy) due to irreversibilities.

This is a rather restricted viewpoint. It might be better to regard two of the terms of Equation AI-51 as representing unidirectional changes, dh_f and dw_n. The first of these terms always takes energy from the other terms. The second one either adds energy, if it is associated with a pump or a compressor, or removes energy if it is associated with a turbine (the sign is changed to a negative one).

Second thermodynamic approach

The second approach to the problem is to use the concept of "lost work" or an even worse term, "internal heat". Although the names chosen are dubious, the approach is probably better as it attacks the problem of useful mechanical energy that is degraded to less useful thermal energy more directly. The terms "lost work" and "internal heat" can be replaced by the single term "irreversibility". The valuable piece of insight that follows was obtained from Benedict, R.P., *Fundamentals of Pipe Flow*, Wiley 1980[xliii].

From thermodynamics, the general energy balance, on a unit mass basis, for steady-state flow can be given for an open system as follows:

$$\delta q + \delta w_n = du + Pdv + vdP + \frac{UdU}{g_c} + \frac{g}{g_c}dZ \qquad \text{(AI-52)}$$

Equation AI-52 states that energy, ideal and non-ideal, crossing the closed boundary of an open system, is only recognized as heat energy or work energy. It further establishes an energy balance between changes in energy crossing the boundary and changes inside the system: internal energy, ideal work of the system, flow work, vdP, kinetic and potential energies.

Note equation AI-52 is usually given with the first three terms on the right replaced by dh, the differential enthalpy. The relationship is shown in AI-53.

$$dh = d(u + Pv) = du + d(Pv) = du + Pdv + vdP \qquad \text{(AI-53)}$$

For an ideal closed system consisting of one unit of flowing mass, the work done by the system equals the Pdv term. In other words, if there were no losses, the work done by the closed system on its environment would exactly equal the integrated value of the Pdv term. The variable P is the internal pressure and the differential dv is the change in internal specific volume in the system under study.

$$\delta\left(w_{by}\right)_{closed,rev} = Pdv \qquad \text{(AI-54)}$$

$$0 = Pdv - \delta\left(w_{by}\right)_{closed,rev}$$

The work done by a closed, irreversible system on its environment will always be less than that of a reversible, ideal system. The difference between the ideal Pdv work and the actual work done by the system will always be a positive number, not zero. This is the basis for the concept of entropy (Greek word for "that which gives direction"). This difference is what has been termed lost work, internal heat, or simply "irreversibility".

APPENDIX AI Equations of Incompressible Fluid Flow

It can be written as,

$$\delta F = Pdv - \delta\left(w_{by}\right)_{closed, irrev} \qquad \text{(AI-55)}$$

All terms have units of energy per unit mass.

The first law for a closed system may be written as,

$$\delta q = du + \delta\left(w_{by}\right)_{closed} \qquad \text{(AI-56)}$$

There is no constraint of reversibility or irreversibility on the term for the work performed by the system since the first law is universal (Equation AI-56 holds in either case). The work is done by the system, not on the system.

If we add the two previous equations we can eliminate the work actually performed by the system in favor of the term Pdv. Note the actual work is measured by the change in the system boundary against an external force. This is the same whether the work is reversible or irreversible.

$$\delta q + \delta F = du + Pdv \qquad \text{(AI-57)}$$

This equation brings together most of the important concepts of thermodynamics. It includes first and second law concepts. First law concepts involve energy balances; second law concepts involve the direction of natural processes. The equation states that the sum of the heat crossing a closed system's boundaries and the term that has been variously termed lost work, internal heat generated, or irreversibility is equal to the sum of the changes in internal energy and the product Pdv. The Pdv term would have been equal to the actual work in the ideal case.

Note the equation is written as an equality between functions that depend on the path traversed for their integration and functions that are point functions (reproducible functions of state that only depend on the end points for their integration). The lost work term, being positive always, will always serve to increase the heat energy flow from the system, the internal energy term or the Pdv term. The amount that this term influences the two point terms will depend on the constraints imposed on the system — whether heat is allowed to flow or not.

So now we know what happens to "lost" energy. It is generally transformed to internal energy and Pdv energy or it flows to the environment as heat, or it does both. Heat is defined as being energy flow caused by a temperature difference. If the heat energy remains in the system, it increases the internal energy and the Pdv energy. The fact that the "lost" mechanical energy can be estimated directly from the differences in total mechanical energy between two points and can be correlated to heat flow is further confirmation of Joule's "mechanical equivalent of heat".

When the first law equation with losses is substituted into the general steady-state equation for open systems, AI-52, the following equation results.

(AI-58)

$$\delta q + \delta w_n = (du + Pdv) + vdP + \frac{UdU}{g_c} + \frac{g}{g_c}dZ$$

$$\delta q + \delta w_n = (\delta q + \delta F) + vdP + \frac{UdU}{g_c} + \frac{g}{g_c}dZ$$

$$\delta w_n = \delta F + vdP + \frac{UdU}{g_c} + \frac{g}{g_c}dZ$$

The following observations can be made. The term on the left of the last equation of set AI-58 is the net work (the fluid work) that crosses the control volume and enters the fluid. The first term on the right is the term that has been called lost work, internal heat generated, or irreversibility. It is identical to the term h_f that was originally treated as an arbitrary correction for "losses" of mechanical energy.

The above equation, derived from thermodynamic considerations, is identical to the Bernoulli[i] equation with losses, derived from force-momentum balance considerations. So, the Bernoulli equation may be derived from force balance considerations or from first and second law considerations. This goes a long way towards proving its validity. Practice and experimentation do the rest. Given a well-proven equation, engineers have established correlations that extend its usefulness.

AI-9: LAMINAR FLOW AND THE HAGEN-POISEUILLE EQUATION

Laminar flow velocity profile

In laminar flow the velocity profile across a circular conduit is parabolic. In defining viscosity we stated that velocity varied linearly between the parallel plates. We will now clarify the apparent contradiction between the two statements.

APPENDIX AI | Equations of Incompressible Fluid Flow

Figure AI-8 is a free body diagram representing laminar flow in a circular conduit.

It defines the terms used in the following development. In unaccelerated, steady-state, horizontal flow, a force balance on the free body requires the sum of the external forces be equal to zero.

$$\sum F_x = 0$$
$$\pi r^2 P_1 - \pi r^2 P_2 - R_D = 0$$
(AI-59)

All the terms of equation AI-59, including R_D, have units of force.

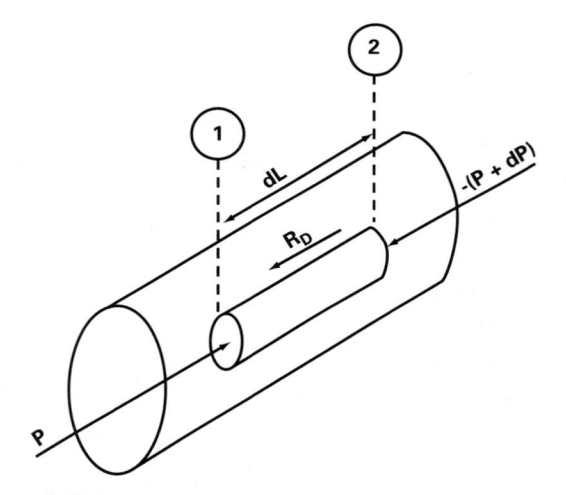

Figure AI-8. Horizontal laminar flow

The last term of AI-59 represents the force acting externally on the periphery of the free body. This force must be present to prevent acceleration. From the definition of viscosity for parallel plates, I-6, we have seen the force is proportional to the viscosity, the area affected by the force and the velocity gradient. The units of viscosity are $lb_m/ft\text{-}s$.

$$R_D = \frac{\mu}{g_c} A \frac{dV}{dZ}$$
(AI-60)

246 Flow of Industrial Fluids—Theory and Equations

Equations of Incompressible Fluid Flow — APPENDIX AI

The area affected is that of the periphery of the cylinder, $2\pi r dL$. We can make these substitutions into the force balance.

$$\sum F_x = 0 \tag{AI-61}$$

$$\pi r^2 P_1 - \pi r^2 P_2 - \frac{\mu}{g_c} 2\pi r dL \frac{-dV}{dZ} = 0$$

$$r(P_1 - P_2) - \frac{\mu}{g_c} 2 dL \frac{-dV}{dZ} = 0$$

$$-r \frac{dP}{dL} - 2 \frac{\mu}{g_c} \frac{-dV}{dZ} = 0$$

$$dZ = dr$$

$$-\frac{dP}{dL} r dr + 2 \frac{\mu}{g_c} dV = 0$$

The negative sign on the point velocity is due to the fact that, unlike the case of the flat plate, the velocity diminishes as the radius increases. The distance between the plates has been replaced by the radius.

Since the pressures are those measured along the cylinder at points separated by a distance, dL, we can call their difference, dP, an infinitesimal pressure difference.

It is to be noted that, at a section, section 1 for instance, the change in pressure with distance is constant. The same change is seen at any point at the same section.

We have a differential equation whose indefinite integral will give us a relation between the radius and the velocity *at a given section*.

$$\frac{dP}{dL} \int r\, dr = 2 \frac{\mu}{g_c} \int dV + C \tag{AI-62}$$

$$\frac{dP}{dL} \frac{r^2}{4} = \frac{\mu}{g_c} V + C$$

To find the constant, C, we note V is equal to zero at the wall, so,

$$C = \frac{dP}{dL} \frac{r_w^2}{4} \tag{AI-63}$$

Flow of Industrial Fluids — Theory and Equations

APPENDIX AI | Equations of Incompressible Fluid Flow

We can substitute the value of this constant into equation AI-62.

$$\frac{dP}{dL}\frac{r^2}{4} = \frac{\mu}{g_c}V + \frac{dP}{dL}\frac{r_w^2}{4}$$ (AI-64)

$$-\frac{1}{4}\frac{dP}{dL}\left(r_w^2 - r^2\right) = \frac{\mu}{g_c}V$$

To simplify the relationship, we note that at the center of the pipe the velocity is a maximum and the radius is zero. These facts give,

$$-\frac{1}{4}\frac{dP}{dL}\left(r_w^2\right) = \frac{\mu}{g_c}V_{max}$$ (AI-65)

We can now divide the last equation into the previous one to find that,

$$\frac{r_w^2 - r^2}{r_w^2} = \frac{V}{V_{max}}$$ (AI-66)

In mathematical terms, for laminar flow, the velocity profile across a circular pipe is parabolic. The point velocity, V, is a function only of the maximum velocity, the location within the pipe and the pipe diameter.

Relation between pressure drop and average velocity in laminar flow

Poiseuille's law

To find the relationship between point velocity and the average velocity in laminar flow we will start with Figure AI-9. This figure shows a hollow cylinder whose wall thickness may be imagined as small as we wish. It is differentially thick.

The circumference of the hollow cylinder is $2\pi r$ feet. The cross sectional area is $2\pi r dr$ feet squared. The volumetric flow through the hollow cylinder is $Q = V(2\pi r dr)$ feet cubed per second. The mass flow is $m = rV(2\pi r dr)$ pounds-mass per second.

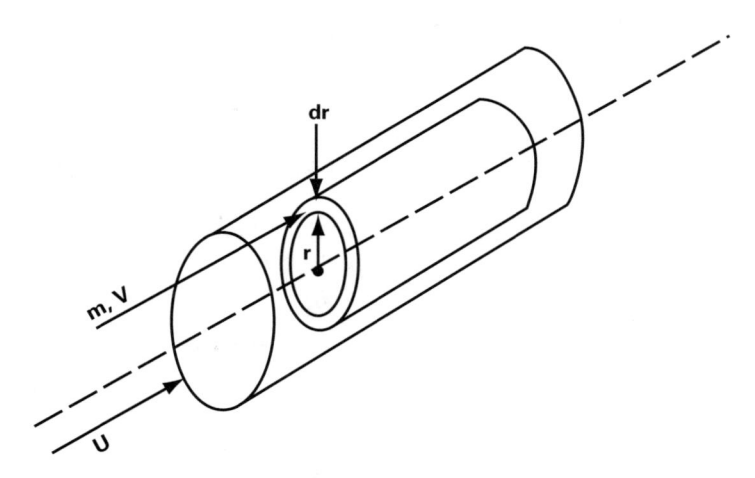

Equations of Incompressible Fluid Flow | APPENDIX AI

Figure AI-9. Hollow cylinder model for Poiseuille's law

To relate the average velocity, U, to the point velocity, V, we find two different expressions for the same quantity. That quantity is the mass flowing through a pipe constituted of an infinite series of concentric hollow cylinders. It is the steady-state mass flow. The relationship developed in Equation AI-65 is substituted for the point velocity, V.

$$\rho U \pi r_w^2 = 2\pi \rho \int_0^{r_w} V r dr \tag{AI-67}$$

$$U = \frac{2}{r_w^2} \int_0^{r_w} V r dr$$

$$U = \frac{2}{r_w^2} \int_0^{r_w} \frac{-dP}{dL} \frac{g_c}{4\mu} \left(r_w^2 - r^2 \right) r dr$$

$$U = \frac{2}{r_w^2} \frac{g_c}{4\mu} \frac{-dP}{dL} \int_0^{r_w} \left(r_w^2 - r^2 \right) r dr$$

$$U = \frac{2}{r_w^2} \frac{g_c}{4\mu} \frac{-dP}{dL} \left[\int_0^{r_w} r_w^2 r dr - \int_0^{r_w} r^3 dr \right]$$

$$U = \frac{2}{r_w^2} \frac{g_c}{4\mu} \frac{-dP}{dL} \left[r_w^2 \frac{r_w^2}{2} - \frac{r_w^4}{4} \right]$$

$$U = \frac{2}{r_w^2} \frac{g_c}{4\mu} \frac{-dP}{dL} \left[\frac{r_w^4}{4} \right]$$

$$U = \frac{1}{8} \frac{g_c}{\mu} \frac{-dP}{dL} r_w^2$$

APPENDIX AI | Equations of Incompressible Fluid Flow

The negative sign on the pressure derivative refers to the fact that the pressure decreases in the direction of L. The pressure derivative with length is constant across a section.

The last equation can be transformed as follows:

$$U = \frac{1}{8}\frac{g_c}{\mu}\frac{-dP}{dL}\frac{D^2}{4} \qquad \text{(AI-68)}$$

$$\frac{-dP}{dL} = 32\frac{\mu}{g_c}\frac{U}{D^2}$$

If all the terms on the right are constant over a finite length of pipe, as they would be in steady-state, laminar flow of an incompressible fluid, in a horizontal, constant diameter pipe, we can integrate the last equation directly.

$$-\int_1^2 dP = 32\frac{\mu}{g_c}\frac{U}{D^2}\int_1^2 dL \qquad \text{(AI-69)}$$

$$P_1 - P_2 = 32\frac{\mu}{g_c}\frac{U}{D^2}L$$

The last equation states that the pressure drop associated with a fluid undergoing laminar flow in a circular pipe is directly proportional to the viscosity, the average velocity and the length of the pipe. It is inversely proportional to the diameter squared (also to the area).

This equation is sometimes written in terms of the volumetric flow rate, Q, as follows,

$$Q = UA = U\frac{\pi D^2}{4} \qquad \text{(AI-70)}$$

$$U = \frac{4Q}{\pi D^2}$$

$$P_1 - P_2 = 32\frac{\mu}{g_c}\frac{U}{D^2}L = 128\frac{\mu}{g_c \pi}\frac{Q}{D^4}L$$

The constants 32 and 128 are non-dimensional in the above equations. If the term, g_c, is treated as also being non-dimensional and as being equal to one, the equations may be used with SI units.

250   Flow of Industrial Fluids—Theory and Equations

Equations AI-70 and AI-71 give a means of experimentally determining viscosity by forcing liquids through capillary tubes of fixed lengths and diameters.

Under the same assumptions of horizontal flow, incompressible fluid, straight pipe, the Bernoulli[i] equation reduces to,

$$-\int_1^2 vdP = \int_1^2 dh_f \qquad \text{(AI-71)}$$
$$-v(P_2 - P_1) = h_f$$
$$h_f = v(P_1 - P_2) = \frac{32\mu L U v}{g_c D^2}$$

This is Poiseuille's[xv] equation for the energy per unit mass transformed from mechanical energy to thermal energy due to fluid irreversibilities in laminar flow.

Poiseuille's equation for mechanical energy losses in laminar flow should be compared with Darcy's[vii] equation for the mechanical energy losses due to turbulent flow

$$h_f = f_M \frac{L}{D} \frac{U^2}{2g_c} \qquad \text{(AI-72)}$$

Clearly in laminar flow, the losses (and therefore the pressure drop in straight pipe) are linear with the average velocity across a section. In turbulent flow, they are proportional to the average velocity squared.

Relationship of a common form of equation for N_{ReD} to AI-71

Frequently, the following equation is found for the pipe Reynolds[v] number:

$$N_{ReD} = 6.31 \frac{W}{d\mu_{cP}} \qquad \text{(AI-73)}$$

The units used in AI-73 are lb_m/h for W, inches for d and centipoise for μ. This really is an example of mixed units, not customary U.S. units, although the formula is very commonly used in North America.

APPENDIX AI | Equations of Incompressible Fluid Flow

The equation can be derived as follows:

$$N_{Re\,D} = \frac{\rho U D}{\mu} = \frac{\rho U D \pi (D/4)}{\mu \pi (D/4)} = \frac{\rho U A}{\mu \pi (D/4)} = \frac{w}{\mu \pi (D/4)} = \frac{4(12)W}{3600\mu\pi d} \quad (AI\text{-}74)$$

$$[\mu_{cP}] \equiv \frac{1\,g}{100\,cm \cdot s}$$

$$\mu = \frac{g}{cm \cdot s} \frac{1}{453.59}\frac{lb}{g}\frac{30.48\,cm}{1\,ft} = P \cdot \frac{30.48}{453.59}\frac{lb_m}{ft \cdot s} = \frac{cP}{100} \cdot \frac{30.48}{453.59}\frac{lb_m}{ft \cdot s}$$

$$N_{Re\,D} = \frac{4(12)\dfrac{in}{ft}W\dfrac{lb_m}{h}}{3600\dfrac{s}{h}\pi d(in) \cdot \dfrac{cP}{100} \cdot \dfrac{30.48}{453.59}\dfrac{lb_m}{ft \cdot s}} = \frac{6.316}{d\mu_{cP}}\frac{h}{lb_m}W\frac{lb_m}{h} = 6.316\frac{W}{d\mu_{cP}}$$

AI-10: SUMMARY OF APPENDIX AI

Appendix I has given some detailed definitions of the concepts and some derivations of the formulae presented in Chapter I. In particular:
- force, pressure, density and specific weight were defined;
- point pressures were analysed mathematically, as were changes in pressure with elevation;
- Newton's law in terms of force versus rate of change of momentum (AI-23) was discussed as being more fundamental than the more familiar force versus mass and acceleration. It serves as the basis for many of the later derivations in the book;
- detailed derivations of the Darcy[vii] equation and the Bernoulli[i] equation were given;
- restricted forms of the Bernoulli equation were discussed;
- a comparison of the Bernoulli equation derived from momentum-force balance considerations with one derived from thermodynamic considerations was made to help reinforce the equation's legitimacy;
- laminar flow was described, mathematically;
- a very common equation for the Reynolds[v] number involving mixed units was derived.

Most of the tools needed to analyse fluid flow problems have now been presented. Their further application requires nothing more than practice and extension.

APPENDIX **AII**

Losses In Incompressible Fluid Flow

AII-1: PURPOSE — PROVIDING CHAPTER II DETAILS

This appendix covers in much greater detail some of the information given in Chapter II regarding loss computations. The reader is reminded that, in the context of this book, the terms "mechanical energy losses", "losses" and "irreversibilities" are synonymous. Permanent losses and pressure drop are not synonymous. Permanent losses are the irreversible conversions of mechanical energy to thermal energy.

Relationships are described in detail for the following:
- the relationship of the valve coefficient, C_v, to the loss coefficient, K;
- the relationship between head units and energy/mass units;
- the Churchill-Usagi[xli] equations for friction factor computations;
- the difference between irreversibilities and pressure drop;
- the different types of K factors and their usefulness;
- generalized loss coefficients and negative loss coefficients.

Fairly extensive tabulations and formulae will be given for the loss coefficients, the K factors. The peculiarity of negative loss coefficients will be explained.

The valve coefficient, C_v, is readily available for all control valves from vendors' data or it can be obtained from Driskell's[xvii] table of values for typical valve types. The valve coefficient can be used to obtain an equivalent K factor. The K factor can then be used to estimate irreversibilities. The relationship between the valve coefficient and the loss coefficient will be derived.

APPENDIX AII | Losses In Incompressible Fluid Flow

The relationship between head units and energy per unit mass units will be clearly established. The Bernoulli[i] equation will be shown in two forms; one form will use head units, the other energy per unit mass units.

The great utility of the Churchill-Usagi[xli] equations is in their use during automated computations – simulations for instance. However, they are reasonably simple and they can be programmed into a hand-held calculator. It is much easier to read a digital scale than to interpolate a graph.

AII-2: RELATION OF VALVE COEFFICIENT, C_v, TO LOSS COEFFICIENT, K

The valve coefficient is useful for making estimations of flow rates through valves. The loss coefficient is useful in computations involving flow through piping systems. It is convenient to have a relationship between the two coefficients so published data on specific valves may be used in estimations of losses for the total system.

Definition of C_v

By definition, the valve coefficient, C_v, is the proportionality factor between the flow rate of water at 60°F and the square root of the differential pressure across the valve. The valve coefficient was an American invention prior to the days of SI units, so the units of the coefficient are gpm per psi$^{1/2}$.

$$C_v = q / \sqrt{\frac{\Delta P}{G}} = q / \sqrt{\Delta P} \qquad \text{(AII-1)}$$

The specific gravity of water at 60°F is taken as one in the above definition.

It can be seen that the valve coefficient can also be defined as the amount of 60°F water, in U.S. gpm, flowing through a valve with a one-psi pressure drop across it.

Since a gallon is a unit of volume, we can convert it to cubic feet and estimate an average velocity for a one-psi drop across a valve.

$$\frac{0.13368 \; ft^3 / gal}{60 \; s/min} q \; gpm = \frac{0.13368}{60} C_v \sqrt{\Delta P = 1} = UA \; ft^3/s \qquad \text{(AII-2)}$$

$$UA = U \frac{\pi d^2}{4(144)} = \frac{0.13368}{60} C_v \sqrt{\Delta P = 1}$$

$$U = 0.4085 \frac{C_v}{d^2}$$

Losses In Incompressible Fluid Flow — APPENDIX AII

We now have an expression for the average velocity at the inlet to a valve in terms of its coefficient and the *diameter of the inlet* under a pressure drop of one psi. The velocity is in ft/s. The diameter is in inches. The coefficient has units of gpm divided by the square root of pressure in psi. This is a dimensional equation; it uses mixed units.

K as a function of C_v

The generalized loss equation in terms of the original resistance coefficient, K, is often used to establish the relationship between the loss coefficient, K, and the valve flow coefficient, C_v. See Equation AII-3. If not defined otherwise, variables are referred to the valve inlet.

The velocity at the valve inlet, U, with a pressure drop of one psi is obtained from the previously developed equation, AII-2. Also, the pressure drop across a valve is defined as that due to the permanent losses in mechanical energy, the irreversibilities — recovery is not considered to be a factor. The irreversibilities are given by the pressure drop divided by the density if the inlet and outlet diameters are equal. The average velocity in Equation AII-3 is that of the inlet (or outlet if the diameters are equal).

The permanent loss relationship and the K factor equation can be equated as follows:

(AII-3)
$$h_f = \frac{\Delta P}{\rho} = K \frac{U^2}{2g_c}$$

$$h_f = \frac{144(1.0)}{62.365} = 2.309 \; ft - lb_f / lb_m$$

$$2.309 = K \frac{\left[0.4085 \frac{C_v}{d^2}\right]^2}{2g_c}$$

$$890.3 = K \left(\frac{C_v}{d^2}\right)^2$$

Given a valve flow coefficient established by the manufacturer, the resistance coefficient can be computed from last of equations AII-3. The permanent losses across the valve at its wide open position (its nominal C_v) may be found by computing K from the nominal C_v by using the last of equations AII-3 and by making use of the first of equations AII-3 with the actual inlet velocity. Irreversibilities at fractional openings can be found by

APPENDIX AII | Losses In Incompressible Fluid Flow

using the inlet velocity at the associated C_v. The C_v at a given opening may be found from manufacturers' data. The fractional C_v depends on the valve characteristic. In the set of AII-3, h_f and ΔP should be computed as projected values, to the valve faces.

AII-3: RELATIONSHIP BETWEEN ENERGY PER UNIT, MASS UNITS, HEAD UNITS AND PRESSURE UNITS

Hydraulics engineers often use "weight density" in place of mass density and feet or meters of fluid in place of ft-lb$_f$/lb$_m$ or N-m/kg (J/kg). This practice can be explained formally by the following transposition of the Bernoulli[i] equation with irreversibilities:

$$\frac{P_1}{\rho_1} + \frac{U_1^2}{2g_c} + \frac{gZ_1}{g_c} + w_n = \frac{P_2}{\rho_2} + \frac{U_2^2}{2g_c} + \frac{gZ_2}{g_c} + h_L \qquad \text{(AII-4)}$$

$$\frac{P_1}{\rho_1(g/g_c)} + \frac{U_1^2}{2g} + Z_1 + \frac{w_n}{(g/g_c)} = \frac{P_2}{\rho_2(g/g_c)} + \frac{U_2^2}{2g} + Z_2 + \frac{h_L}{(g/g_c)}$$

The units of each individual term in the last equation are now feet or meters of fluid. The units of the group, $\rho_1(g/g_c)$, are force/length^3 (lb$_f$/ft^3 or N/m^3). This is the "weight density", γ, lb$_m$/ft^3.

In the customary U.S. system of units, the acceleration due to gravity and the dimensional constant are numerically equal at sea level at a latitude of 45 degrees. The mass density is, therefore, equal numerically to the weight density in U.S. units at this latitude. In SI units, the dimensional constant is replaced by one and the mass density does not equal the weight density.

If the denominator of the static energy terms is replaced by γ, the last form of the Bernoulli equation becomes,

$$\frac{P_1}{\gamma_1} + \frac{U_1^2}{2g} + Z_1 + \frac{w_n}{(g/g_c)} = \frac{P_2}{\gamma_2} + \frac{U_2^2}{2g} + Z_2 + \frac{h_f}{(g/g_c)} \qquad \text{(AII-5)}$$

All terms have units of length (height, hence head). In hydraulics practice, Z and P/γ are often combined and are called the hydraulic (piezometric) head. They are then written as H_1 and H_2. The above equation becomes,

$$H_1 + \frac{U_1^2}{2g} + \frac{w_n}{(g/g_c)} = H_2 + \frac{U_2^2}{2g} + \frac{h_f}{(g/g_c)} \qquad \text{(AII-6)}$$

This latter form of the Bernoulli equation is often used to develop the equation for flow through an ideal, horizontal restriction. The elevations are equal. There is no energy added or subtracted. The last term is zero (ideal). Section 1 is upstream; section 2 is at the vena contracta.

$$H_1 + \frac{U_1^2}{2g} = H_2 + \frac{U_2^2}{2g}$$

$$U_2^2 - U_1^2 = 2g(H_1 - H_2)$$

(AII-7)

Going one step further, pressure units can be developed by multiplying the Bernoulli[i] equation with irreversibilities by (the constant) density.

$$\Delta P_{lost} = \rho h_f = (P_1 - P_2) + \left(\frac{\rho U_1^2}{2g_c} - \frac{\rho U_2^2}{2g_c}\right) + \left(\frac{\rho g Z_1}{g_c} - \frac{\rho g Z_2}{g_c}\right)$$

(AII-8)

In hydraulics practice (constant weight density),

$$\Delta P_{lost} = \gamma h_f = (P_1 - P_2) + \left(\frac{\gamma U_1^2}{2g} - \frac{\gamma U_2^2}{2g}\right) + (\gamma Z_1 - \gamma Z_2)$$

(AII-9)

It can be seen that both of the above equations represent lost pressure as being the sum of the differences between each of the three terms. The first difference is the actual measured pressure drop, the second a velocity pressure drop and the third and pressure change due to a change in elevation.

These equations are particularly prone to errors in interpretation since it is sometimes difficult to keep (mentally) a non-measurable "pressure drop" separate from a measurable one.

Caveat — Hydraulic engineers' simplification

The hydraulic engineers' approach generates simple equations for restricted uses. The similarity of the equations to the more formal equations leads to confusion among terms such as ρ and γ, Z and H, and g and g_c. For generalists it is best to use the formal form of the Bernoulli equation with mass density.

AII-4: CHURCHILL-USAGI FRICTION FACTOR EQUATIONS

For computer computations, equations are much more convenient than charts and tables for correlations. Stuart W. Churchill[xli] of the University of Pennsylvania presented a convenient method of estimating the friction factor from the Reynolds[v] number and the relative surface roughness

APPENDIX AII | Losses In Incompressible Fluid Flow

(*Chemical Engineering*, Nov. 7, 1977). This method has been successfully used in simulation programs.

The method is claimed to be good over all regimes: laminar, transitional and turbulent. The author has used it and checked it over the laminar and turbulent regimes. It gives excellent results.

It must be remembered the transitional regime is unstable and should be avoided. Numerical stability does not equate to physical stability. So, even if the Churchill-Usagi[xli] equations give specific numerical results between a Reynolds[v] number of 2,000 and 4,000, the results do not represent reality.

The friction factor computed by the Churchill-Usagi method is one half the Fanning[xiv] friction factor and one-eighth the Moody[viii] friction factor. The final result must be multiplied by two or eight to obtain the corresponding Fanning or Moody factors. We will do this at the end of this section.

The equations can not only be computerized; they can be programmed into a hand-held calculator. They reproduce all of the friction factor plots in the literature, including the ones given in this book (from Crane[xiii]).

The equations are,

(AII-10)

$$A = \left[2.457 \ln \frac{1}{\left(\frac{7}{N_{Re}}\right)^{0.9} + \frac{0.27\varepsilon}{D}} \right]^{16}$$

$$B = \left(\frac{37530}{N_{Re}}\right)^{16}$$

$$f = \left[\left(\frac{8}{N_{Re}}\right)^{12} + \frac{1}{(A+B)^{3/2}}\right]^{1/12}$$

$$f_M = 8f$$

It can be seen that the only input terms are the Reynolds number and the relative roughness (absolute roughness divided by the pipe internal diameter in the same dimensions). The absolute roughness of various materials has been given in Table II-1. Additional values can be found in many publications such as Perry[x].

The dimensions of e and of D in equation AII-10 must be consistent. Both must be in feet, meters, millimeters or inches.

AII-5: PRESSURE DROP VERSUS 'FRICTION LOSSES'

There is room for confusion between the concepts of pressure drop and irreversibilities. The literature is full of examples where the two concepts seem to be equated. To avoid inadvertent error, it is good to have a clear concept of the difference.

Pressure drop in a circular pipe flowing full is due to a conversion of static energy into other forms: potential, kinetic and thermal (internal energy) when there is no producer or converter of energy in the line. The producer would be a pump or compressor. The converter would be a turbine. From a mechanical point of view, the irreversibilities associated with flow constitute a loss of useful mechanical energy. In the ideal case, irreversibilities do not exist and kinetic, potential or static energy can be interconverted within physical constraints. Irreversibilities constitute a conversion of mechanical energy that results first in a temperature increase and then in heat flow to the environment.

The Bernoulli[1] equation with no pump or turbine in the section can be written,

$$-\int_1^2 vdP = h_f + \frac{\Delta(U^2)}{2g_c} + \frac{g\Delta Z}{g_c} \quad \text{(AII-11)}$$

The first term is left under the integral sign when compressible fluids are involved. When incompressible fluids are involved, it can be integrated directly to,

$$\frac{P_1 - P_2}{\rho} = \frac{g}{g_c}\Delta Z + \frac{1}{2g_c}\Delta(U^2) + h_f \quad \text{(AII-12)}$$

Written in the above fashion, the Bernoulli equation states that the available static energy difference between two points can be converted to a potential energy difference, a kinetic energy difference or it can be used to overcome viscous resistance. It can increase the elevation of a fluid mass, increase its velocity or it can become "lost". The last term is considered a loss of mechanical energy (conversion to internal energy) because it cannot be converted back to one of the other three forms. This observation resulted in the concept of entropy – that which gives direction.

In a straight, horizontal, equal diameter pipe *with an incompressible fluid*, there is no potential energy change and no kinetic energy change. The static energy change equals the loss term, in this case. By multiplying through by the density, the irreversibilities can be expressed as a pressure drop.

$$P_1 - P_2 = \rho h_f \quad \text{(AII-13)}$$

If there exists a change in elevation, density, or velocity, the simple relationship of Equation AII-13 is no longer true and *the irreversibilities times mass density do not equal the pressure drop.*

AII-6: 'K' FACTORS – LOSS COEFFICIENTS

The K factors are a convenient means of computing the losses in mechanical energy across a commercially available fitting, orifice, valve or other device. A K factor is the proportionality factor between the irreversibilities and the kinetic energy of the fluid at some known section. For incompressible fluids, the K factor can be looked upon as the fraction of available kinetic energy that is converted to mechanically unusable internal energy by some impediment to flow. The subscript, i, is meant to remind the reader to use the average velocity at the section corresponding to the experimental K factor.

$$h_L = K_i \frac{U_i^2}{2g_c} \quad \text{(AII-14)}$$

These factors have been correlated with the average velocity squared and the measured losses for many devices by many workers. They have been tabulated, put in graphical form and put into the form of equations. It is extremely difficult to give credit to the original worker so we will simply give the source from which we obtained the information. Crane[xiii] is an excellent general source. Others will be given as each factor is discussed.

Equation AII-14 is so simple it warrants a few caveats:

Caveats — Velocity used with K factor, Suffixes, K factors, C_v

When using K factors, be sure to use the computed average velocity specified as being associated with the value of K. Do not use a point velocity. The average velocity is usually that in the smaller diameter, but not always. For instance, in the square-edged orifice equation, the velocity is usually that computed for the pipe, not that for the orifice.

Losses In Incompressible Fluid Flow — APPENDIX AII

Be aware of how suffixes are used in equations. Some authors use 1 for upstream and 2 for downstream. Crane[xiii] uses 1 for the smaller diameter and 2 for the larger diameter.

Some authors list K factors for various types of fittings as being constant for all sizes. When geometric similarity does not hold, and it does not in most cases, this cannot be true. Use correlations from reputable sources, such as the Hydraulics Institute[xliv] and Crane.

When a C_v is available for a specific valve, use it to compute a K factor, do not use a general correlation. The manufacturer has tested his valve, so the results will be more accurate.

Although we have adopted the usual convention of using the symbols h_L or h_f for the losses (conversion) of mechanical energy associated with a K factor, remember the true losses can only be established over a length of piping that allows normal velocity profiles at each point of measurement. This is the reason h_L was replaced by $\Delta(\Delta H)$ in Chapter II – as a reminder.

Geometric similarity and dissimilarity

Some K factors remain reasonably constant for all sizes of fitting and some are a function of geometry. If complete geometric similarity holds, as in the case of pipe entrances and exits, the K factor is a constant independent of average velocity. If geometric dissimilarity holds, the K factor is given as a function of the beta ratio.

Table of K factors

Note that, in what follows, all K factors are given for the smaller diameter except for the square edged orifice. The velocity in the smaller pipe must be used in the head loss equation, except for the orifice. If the velocity in the larger pipe is known or is more easily computed, dividing the K factor computed for the smaller pipe by β^4 will give a new, larger, K factor that can be used with the lower velocity in the larger pipe (this does not apply to cases 1, 2, 10 and 11, which are geometrically similar).

APPENDIX AII | Losses In Incompressible Fluid Flow

Table AII-1. K factors

Cause of Loss	K Factor				Case	Source
Pipe entrance, square edged	0.5				1	1
Pipe exit, projecting, square or rounded	1.0				2	1
Sudden enlargement	$K = (1-\beta^2)^2$				3	1
Sudden contraction	$K = 0.5(1-\beta^2)$				4	1
Square edged orifice	$K = 2.8(1-\beta^2)\left[\left(\dfrac{1}{\beta}\right)^4 - 1\right]$				5	2
Gradual enlargement, $\theta \leq 45°$ θ = included angle	$K = 2.6\sin\dfrac{\theta}{2}(1-\beta^2)^2$				6	1
Gradual enlargement, $\theta > 45°$ (same as sudden enlargement)	$K = (1-\beta^2)^2$				7	1
Gradual contraction, $\theta \leq 45°$	$K = 0.8\sin\dfrac{\theta}{2}(1-\beta^2)$				8	1
Gradual contraction, $\theta > 45°$	$K = 0.5\left[\sin\dfrac{\theta}{2}\right]^{1/2}(1-\beta^2)$				9	1
Inward projecting entrance	0.78				10	1
Rounded entrance	r/d 0.02 0.04 0.06	K 0.28 0.24 0.15	r/d 0.100 >0.15	K 0.09 0.04	11	1
Pipe bends and flanged or butt-welded elbows, 90°	r/d 1.0 1.5 2.0 3.0 4.0 6.0	K $20f_T$ $14f_T$ $12f_T$ $12f_T$ $14f_T$ $17f_T$	r/d 8.0 10.0 12.0 14.0 16.0 20.0	K $24f_T$ $30f_T$ $34f_T$ $38f_T$ $42f_T$ $50f_T$	12	1

Source: 1)©Crane Co. All rights reserved. 2) Simpson

Generalized loss coefficients and negative loss coefficients

This section will give a little more theory on loss coefficients. It will also explain the peculiarity of negative loss coefficients.

The use of equivalent pipe lengths as a measure of the irreversibilities caused by the presence of a component in a piping system is not satisfactory for other than approximations. A component can be a bend, miter, tee, orifice plate, valve or any other obstruction to smooth flow. The use of K factors that are established from careful experiments is the ideal approach.

Losses In Incompressible Fluid Flow

A component in a piping system can be preceded and followed by long straight lengths of pipe. This type of experimental set-up allows the flow profiles to be well established upstream and downstream of the disturbance. It is at these sections that measurements are taken.

The pressure measurements are the total pressure measurements. By "total" is meant the sum of the static and velocity pressures. It is not the pressure measured by a pressure gauge.

When the component is removed and the pipes are rejoined linearly, the same flow rate with the same fluid conditions is re-established and measurements are taken at the same sections (see Figure II-3 and the surrounding discussion). The difference between the two total pressure differences, $\Delta(\Delta P_T)$, is a measure of the loss caused by the component. This method of measurement allows inclusion of the influence of irreversibilities that occur upstream and downstream due to the presence of the component — an influence that would be missed if pressure drop were measured directly across the component.

Total pressure loss

For horizontal pipe plus a component, the total pressure loss for incompressible flow can be obtained from the Bernoulli[i] equation.

$$\frac{P_1}{\rho} + \frac{U_1^2}{2g_c} = \frac{P_2}{\rho} + \frac{U_2^2}{2g_c} + h_f \qquad \text{(AII-15)}$$

$$\Delta P_{total} = \rho h_f = \left(P_1 + \frac{\rho U_1^2}{2g_c} \right) - \left(P_2 + \frac{\rho U_2^2}{2g_c} \right)$$

The above equation holds for customary U.S. and for SI units, if the dimensional constant is treated appropriately.

Note carefully the total pressure loss is not the measured pressure drop, $P_1 - P_2$. It is the theoretical pressure loss based on density times the estimated mechanical energy loss.

The difference between total pressure differences due to the component's presence and absence is $\Delta(\Delta P_T)$, as mentioned above. Note this notation is not usual, it is more often written as ΔP. However, we are talking about the difference between two differences. To be more exact we will use the double delta and the subscript, T. If there were no extra losses due to the presence of the component, the difference, $\Delta(\Delta P_T)$, would be zero, whereas ΔP would still reflect the influence of the irreversibilities due to the piping.

The above methodology allows the use of a technique whereby irreversibilities are estimated for the straight pipe and additional irreversibilities are estimated for the presence of the component in that pipe. These irreversibilities can simply be added as part of the system irreversibilities.

K factors

To further increase the utility of the method, the irreversibilities are made non-dimensional by dividing the total pressure drop of an incompressible fluid by $\rho U^2/2g_c$. The velocity, U, is the average velocity at the section to which K is referred.

$$K = \frac{\Delta(\Delta P_{total})}{\rho U^2 / 2g_c} \quad \text{(AII-16)}$$

This equation again applies to both customary U.S. and to SI units when one is substituted for the dimensional constant. Hydraulics engineers use a similar equation where the units are feet or meters and the acceleration of gravity is used in feet per second squared or meters per second squared.

$$K = \frac{\Delta(\Delta H_{total})}{U^2 / 2g} \quad \text{(AII-17)}$$

The advantage of the K factors is they can be tabulated for a given type of component at a fixed Reynolds[v] number. They can then be adjusted for an actual Reynolds number and multiplied by the average kinetic energy ($U^2/2g_c$) at the location for which the average velocity was established to give the actual irreversibilities due to the presence of the component.

About 50 pipe diameters are required to establish reliable K factors. If only 30 pipe diameters are available, the value of the K factor usually deviates by plus or minus 2.0% from that of the longer runs.

The loss coefficients (K factors) appear to become independent of Reynolds number once the Reynolds number is greater than about 10^6. It is unlikely that they become completely independent of Reynolds number.

Negative K factors

The differential pressure across a fitting is taken as a positive number in establishing the K factor. It is the upstream pressure less the downstream pressure. Under certain conditions (combining flows), it is possible to have negative K factors. This peculiarity can be understood by considering a tee whose branch flow is much smaller than the flow through the run. There are two K factors to be established: one for the branch and one for the run.

By examining the Bernoulli[i] equation from the branch to the combined junction, it can be seen that when the flow through the run is much greater than that through the branch, energy is added from the main stream to the branch stream. The downstream specific energy of the combined flow will be greater than that of the branch inlet flow. The pressure will be greater downstream than upstream and a negative K factor results.

This phenomenon has been observed with surprise by many and has even been denied by some. When both streams are considered, there is a net loss (conversion to internal energy) as would be expected.

Classification of loss coefficients by source

It is quite difficult to obtain reliable loss coefficient data for some components. D.S. Miller[xvi] established the following categories in order to judge the reliability of the data:

I. The data are well-established experimental data from two or more sources. The data are cross checked.

II-1. The data are experimentally derived, but have not been cross checked.

II-2. The data are estimated from two or more research programs.

II-3. The data are based on Class I sources, but are outside the experimental range. They are assumed reasonable.

III-1. The data are experimental but the source is thought less reliable.

III-2. The coefficients are based on extrapolations outside the range of data from Class I and Class II sources. There is no information that allows prediction of the effects of being outside the range. In other words, the data are not trustworthy, but there is nothing available that is more reliable.

D.S. Miller also established a classification of components that is useful conceptually. It is given in Table AII-2.

APPENDIX AII | Losses In Incompressible Fluid Flow

TABLE AII-2. Classification of Components

Low Loss Components	Moderate Losses	Large Losses
CONDUIT		
Smooth	Transitional	Rough
TURNING CONDUITS		
Bends and elbows	Miters	90's
DIFFUSING COMPONENTS		
Diffusers	Swages - expanders	Sudden expansions
ACCELERATING JOINTS		
Smooth contraction	Swages - contractions	Sudden contraction
COMBINING FITTINGS		
Branch with rounded approach	Angled branch	90 branch
DIVIDING FITTINGS		
Branch with rounded path	Angled branch	90 branch
OBSTRUCTIONS		
Multiple, parallel	Rotors (turbines)	Convoluted valves

Concept of adverse gradients

When studying flow around bends and through fittings, the concept of an adverse gradient is useful. Potential energy (gZ/g_c) and static energy (P/ρ) can be converted into kinetic energy ($U^2/2g_c$) without excessive loss when the flowing stream is restricted smoothly. The opposite process of converting kinetic energy to one of the other two forms usually involves significant irreversibilities. The reason, according to D.S. Miller[xvi], is to be found in the non-uniformity of the kinetic energy across the section at a bend or fitting versus the uniformity of pressure energy (static) across the same section.

The kinetic energy near the wall is usually very low due to the viscous drag force at the wall. When the fluid near the wall enters a region of rising static pressure (in a diffuser, for example), the fluid must be supplied with energy from the higher energy fluid away from the wall, or it comes to rest. If the static pressure energy downstream becomes sufficiently great, reverse flow can occur locally. This phenomenon causes separation of the stream from the wall and contraction of the area through which the main body of fluid flows. After the contraction, which has a variation in energy distribution across the section, large scale mixing occurs. The mixing, which is thermodynamically irreversible and involves conversion of mechanical energy to thermal energy, has the effect of equalizing the energy distribution. Reattachment of the stream to the conduit wall follows.

If the flow is turned, static pressure gradients exist due to centrifugal forces. Low energy fluid can escape (can be pushed) from a high-pressure boundary towards a region of lower static pressure. Secondary flows can be imposed on the main flow.

Example of combining flows with a negative loss coefficient

To reinforce the statement that it is possible to have a negative loss coefficient (an energy gain per unit mass of one stream) let us analyze the next example. Take a pipe tee used to combine flows to the exit from the run. If the branch flow into the run is less than about one third of the total flow, energy is transferred from the main run flow to the flow from the branch. This results in a higher *specific* energy in the combined flow than would be present if there were no flow through the run.

The loss coefficient for incompressible fluids is defined in terms of subscripts. Subscript 1 represents the fluid entering the main run, subscript 2 is the fluid entering the branch, and subscript 3 is the combined flow leaving the fitting.

$$K_{13} = \frac{\Delta(\Delta P_T)_{13}}{\rho U_3^2 / 2g_c} = \frac{\Delta(\Delta H)_{13}}{U_3^2 / 2g} \qquad \textbf{(AII-18)}$$

If we perform computations on the tee in the case cited, K_{23} will be negative. If we sum the irreversibilities, $\Sigma(\Delta(\Delta P_T))/\rho$, we will find the combined loss in mechanical energy is a positive number, in spite of one member being negative.

Example AII-1

Water is flowing through a tee with a 6-inch Sch. 40 branch and a 12-inch Sch. 40 run. The total flow is 2,000 gpm. The branch flow is 200 gpm. The flow is combining. The pressure downstream is 100 psig and the temperature is 60°F. The orientation can be taken as horizontal to simplify the computation, but this is not necessary if the potential energy terms are added in the Bernoulli[i] equation.

The problem is to compute the irreversibilities from the inlet to the branch to the downstream section and from the inlet to the run to the same downstream section. Remember these irreversibilities are those due to the presence of the fitting including irreversibilities slightly upstream and many pipe diameters downstream caused by this presence (mainly due to the destruction of eddies).

APPENDIX AII | **Losses In Incompressible Fluid Flow**

Caveat — Conversion of irreversibilities to pressure drop

When the irreversibilities are converted to pressure drops, the computed pressures might not be identical to those read on pressure gauges immediately upstream and downstream of the fitting. They will correspond, through the Bernoulli[i] equation, when measured at points sufficiently upstream and downstream and projected to the faces of the component.

D.S. Miller[xvi] gives loss coefficients for various types of fittings in the form of graphs. He gives the loss coefficients as a function of the ratio of flow through the branch to the main flow and of the area of the branch to the main area. In our case,

(AII-19)
$$\frac{A_2}{A_3} = \left(\frac{D_2}{D_3}\right)^2 = \left(\frac{d_2}{d_3}\right)^2 = \left(\frac{6.065}{11.938}\right)^2 = 0.258$$

$$\frac{Q_2}{Q_3} = 0.1$$

$$K_{23} = -0.43$$

$$h_{f23} = K_{23}\frac{U_3^2}{2g_c} = -0.43\frac{(5.73)^2}{2g_c} = -0.219\frac{ft-lb_f}{lb_m}$$

The velocity was obtained by looking up in Crane[xiii] the velocity associated with 2,000 gpm in a 12 inch schedule 40 pipe. The K factors were found from Miller. Since the irreversibilities from the branch to the run are negative, it can be assumed the fluid flowing through the branch has gained more energy due to momentum transfer from the fluid flowing through the run than it has lost due to the irreversibilities.

For the run, using the same approach, K_{13} equals approximately 1.6, so,

(AII-20)
$$h_{f13} = 1.6\frac{(5.73)^2}{2g_c} = 0.816\frac{ft-lb_f}{lb_m}$$

For the branch and the run, these sum to positive irreversibilities, as expected.

Caveat — Authors' definitions of loss coefficients

Be careful to note how the loss coefficients are quoted. Some authors are not too careful in clearly stating which velocity is to be used. If the coefficient was established on the basis of combined flow in the run, the velocity is that of the combined flow. If it was established on the basis of flow in the branch, the velocity is that of the branch flow only.

The reference position in example AII-1 is position 3, downstream. The datum can be taken as the center line of the pipes. The total mechanical energy at position 3 is given by,

$$E_3 = \frac{P_3}{\rho} + \frac{U_3^2}{2g_c} + \frac{gZ_3}{g_c}$$ (AII-21)

$$E_3 = \frac{144(100+14.7)}{62.371} + \frac{(5.73)^2}{2g_c} + 0 = 265.31 \frac{ft - lb_f}{lb_m}$$

The pressure at the position with subscript 2, the branch inlet, is obtained from manipulating the Bernoulli[i] equation with the potential energy term held at zero because of the horizontal arrangement of the fitting.

$$P_2 = \rho \left[\frac{P_3}{\rho} + \frac{U_3^2}{2g_c} + h_f - \frac{U_2^2}{2g_c} \right]$$ (AII-22)

$$P_2 = 62.371 \left[264.8 + 0.51 - 0.219 - \frac{(2.22)^2}{2g_c} \right]$$

$$P_2 = 16529 \frac{lb_f}{ft^2} = 114.79 \frac{lb_f}{in^2} = 100.1 \, psig$$

The pressure entering the branch is only a tenth of a psi higher than the downstream pressure, but it is higher in this case.

We can perform a similar exercise for the run.

$$P_1 = \rho \left[\frac{P_3}{\rho} + \frac{U_3^2}{2g_c} + h_f - \frac{U_1^2}{2g_c} \right]$$ (AII-23)

$$P_1 = 62.371 \left[264.8 + 0.51 + 0.816 - \frac{(5.16)^2}{2g_c} \right]$$

$$P_1 = 16573 \frac{lb_f}{ft^2} = 115.09 \frac{lb_f}{in^2} = 100.4 \, psig$$

The pressure entering the run is four tenths of a psi higher than the downstream pressure.

APPENDIX AII | Losses In Incompressible Fluid Flow

It is instructive to look at the energy changes around the tee and to compute the total mechanical energy at each section and the irreversibilities (mechanical energy converted to internal energy).

The flow rate of total mechanical energy at each position is given by the mass flow rate at that position multiplied by the specific mechanical energy at that point. Position 1 is the run inlet, position 2 is the branch inlet, and position 3 is the combined outlet. Note that even though the symbol, h, is used for specific mechanical energy (head), this is not enthalpy.

$$\dot{m}_1 h_1 = 4.010 \frac{ft^3}{s} 62.371 \frac{lb_m}{ft^3} (265.31 + 0.816) \frac{ft-lb_f}{lb_m} = 66560.2 \frac{ft-lb_f}{s}$$ (AII-24)

$$\dot{m}_2 h_2 = 0.4456 \frac{ft^3}{s} 62.371 \frac{lb_m}{ft^3} (265.31 - 0.219) \frac{ft-lb_f}{lb_m} = 7367.5 \frac{ft-lb_f}{s}$$

$$\dot{m}_3 h_3 = 4.456 \frac{ft^3}{s} 62.371 \frac{lb_m}{ft^3} (265.31) \frac{ft-lb_f}{lb_m} = 73736 \frac{ft-lb_f}{s}$$

The rate at which total mechanical energy is being converted to thermal energy is given by the sum of the rates of entering mechanical energy less the leaving mechanical energy.

$$Losses = 66560.2 + 7367.5 - 73736.0 = 191.7 \frac{ft-lb_f}{s} = 0.246 \frac{Btu}{s} = 260 \, watts$$ (AII-25)

Even though one of the loss coefficients was negative, which indicates an energy gain for that stream, the overall effect was a loss. If these irreversibilities could have been recuperated, the energy could have been used to power a low wattage light bulb.

AII-7: SUMMARY OF APPENDIX AII

Appendix AII has expanded upon the information given in Chapter II. It has concentrated only on passive components. Pumps and turbines will be considered in Chapter III and Appendix III. The essential points covered were:

- the relationship of the valve flow coefficient, C_v, to the loss coefficient, K, was developed and discussed;
- the relationship of head units and energy per unit mass units was developed and the limitations of the hydraulic engineering approach were pointed out;

Losses In Incompressible Fluid Flow — APPENDIX AII

- the very useful Churchill-Usagi[xli] equation that is easily implemented in computer programs was given with a warning about its use in the transitional zone;

- the easily confused relationship of measured pressure drop, which includes recoverable and non-recoverable pressure difference, and loss when expressed in pressure units was, hopefully, clarified;

- much of Appendix AII focused on the use of K factors, on the different types of K factors, on the relationship to a specific velocity at a given location, and on the fact that there exist negative K factors;

- it was shown there is a constant loss of power across a typical fitting in spite of an apparent gain in energy across one of legs.

APPENDIX AIII

Computations Involving Pumps For Liquids

AIII-1: PURPOSE — PROVIDING CHAPTER III DETAILS

Appendix AIII discusses in much greater detail, and more theoretically, some of the information given in Chapter III regarding pumps. The intent is to supply information and knowledge about pumps, including jet pumps, in sufficient detail so those interested in the flow of fluids may integrate both static and dynamic data into their everyday activities.

We will begin with the general theory of centrifugal pumps and then discuss the categories of pumps defined in Chapter III. Only pumps that handle liquids are considered in this appendix.

AIII-2: THEORY OF CENTRIFUGAL PUMPS

Figure AIII-1 is a cross section of a typical single-stage centrifugal pump showing the essentials. The casing, the inlet flange and volute are shown. The impeller turns with the shaft and the shaft is sealed against leakage by packing in a stuffing box. Mechanical seals (not shown) have largely replaced packing. Many pumps are designed so they may use either packing or mechanical seals in a stuffing box according to the end user's wishes.

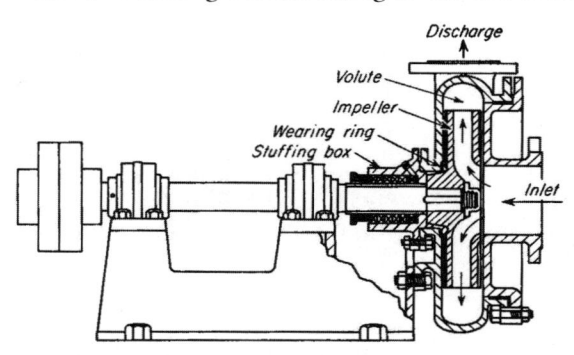

Figure AIII-1. Typical single-stage centrifugal pump

Source: L. McCabe, Unit Operations in Chemical Engineering, 1967, reprinted by permission McGraw-Hill Companies.

APPENDIX AIII | Computations Involving Pumps for Liquids

The arrows show the direction of flow. Sufficient pressure has to be available at the pump suction to force the liquid into the pump impeller. The liquid is then subject to the centrifugal force of the rotating impeller. It is projected radially to the volute where it is collected and directed to the pump discharge.

Total mechanical energy of the liquid is increased by the centrifugal action

In essence, the total mechanical energy of the liquid is increased by the centrifugal action. Energy is transferred to the shaft by the driver and to the fluid by the impeller. The pump is simply the means of energy transfer. The total mechanical energy at the inlet to the pump is the sum of the static (pressure) energy, the kinetic energy and the potential energy. A similar summation gives the total mechanical energy at the discharge. The difference between the two summations gives the mechanical energy transferred to the fluid. Some of the mechanical energy transferred to the shaft is converted to internal thermal (associated with a temperature increase) energy by shock and turbulence. This internal energy is not useful mechanically. The ratio of the mechanical energy transferred to the fluid to that transferred to the shaft gives the pump efficiency.

The rate at which energy is transferred to the shaft is the shaft power, frequently called brake horsepower. The rate at which it is transferred to the fluid is the fluid power or fluid horsepower. The ratio of the two is also equal to the pump efficiency (excluding that of the driver). The difference between the two is the amount of power converted to internal energy (commonly, and incorrectly, called heat) by various causes, all of which will be discussed shortly.

The centrifugal pump is probably the most common pump. The simple centrifugal pump as shown in Figure AIII-1 is most frequently directly connected to an electric motor operating at constant speed. In those areas where 60-Hertz power is available, that speed is often 3600 rpm or 1800 rpm less the slip (resulting in 3,550 rpm or 1,750 rpm). We will start our analysis with this pump in mind.

Pump engineering is a very mature technology

Pump engineering is a very mature technology. Pump manufacturers, therefore, can guarantee the performance of their pumps with a great degree of accuracy – much more than can manufacturers of other types of process equipment.

Computations Involving Pumps for Liquids APPENDIX AIII

When the author was a very young engineer, he heard stories from older engineers in which they would complain about horsepower ratings of pumps. It seemed that, in the early days, since quality control was not too tight, when an engineer bought a five horsepower pump, a 7 ½ h.p. pump was actually delivered due to the fact the manufacturer had to build in a tolerance on the plus side. When the technology matured, the manufacturers were able to tighten their specifications and they started to deliver five horsepower pumps. Older engineers, who had gotten used to taking advantage of the situation, did not appreciate the change.

Analysis of an ideal pump to which corrections will be applied

Figure AIII-2 shows the essentials of a volute pump. The flow through the pump may be analyzed as follows:

1. Liquid enters axially at point (a). It must be under sufficient pressure to prevent vapor cavities from forming within the inlet as the pressure falls due to acceleration.

2. Liquid turns from an axial direction to a radial one from point (a) to point (1) where it enters the channel formed by the impeller and the casing.

3. Liquid flows through the impeller channels from point (1) to point (2). It is accelerated by centrifugal force. This acceleration requires work to be supplied continuously to the impeller shaft. The work is transformed to kinetic energy of the fluid by the impeller.

4. Liquid leaves the impeller at point (2) and moves around the volute to the discharge flange at point (b). Its direction has been changed again, from a radial to a tangential one.

Pump data is usually referenced to the inlet and outlet flange centerlines. Analysis requires that the pump be divided into three volumes: inlet, channels and outlet. The heart of the pump is the impeller, containing the channels, so we will begin the analysis with it.

Impeller

Figure AIII-3 depicts an impeller, its direction of rotation, and the velocity vectors involved. Only one blade is shown.

At steady state, fluid enters the channel at an angle governed by the blade angle, β_1. This is the angle between the (usually) backward leaning blade and the tangent to the circle described by its leading edge. The liquid has an instantaneous velocity, relative to the blade, of v_1. At the same time the

APPENDIX AIII | Computations Involving Pumps for Liquids

inner tip of the blade is moving with a tangential velocity of u_1. A fluid particle entering a channel has a resultant velocity at this point of V_1. The resultant velocity is given the unfortunate name "absolute" velocity. The angle between the absolute velocity vector, v_1, and the tangent to the circle at the inner blade tip is α_1.

The absolute velocity is simply the velocity that a particle has relative to the frozen reference frame of the observer. In other words, if we could take high speed photos of the fluid particles separated by milliseconds, a particle would appear to be moving in the direction of the absolute velocity vector and the instantaneous absolute velocity could be computed.

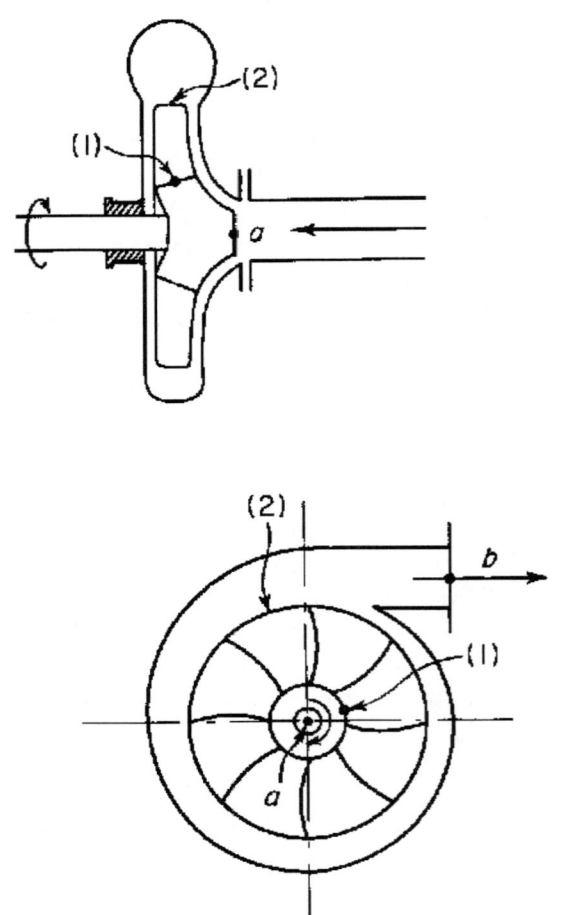

Figure AIII-2. Centrifugal pump showing Bernoulli[i] stations

Source: L. McCabe, Unit Operations in Chemical Engineering, 1967, reprinted by permission McGraw-Hill Companies.

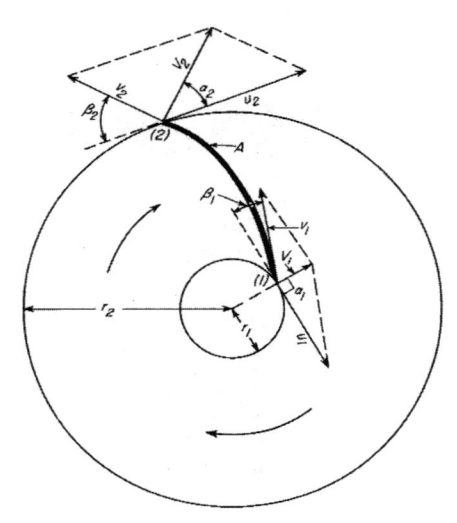

Computations Involving Pumps for Liquids | APPENDIX AIII

Figure AIII-3. Velocity vectors at entrance and discharge of centrifugal pump vane.

Leaving the channel, the fluid particles have been accelerated; they have gained kinetic energy from the work of the impeller. Since they are flowing along the curved blade, the direction of the relative velocity, v_2, is controlled by the angle at the tip of the blade. The angle between the tangent to the circle at the tip and the relative velocity is β_2. The tangential velocity of the tip is u_2. Since the particle is moving relative to the blade and the tip is moving tangentially, the resultant "absolute" velocity can be obtained by the parallelogram law. The vector, V_2, is the result.

Vector V_2 is the instantaneous velocity that a particle has as photographed by a high-speed camera taking pictures in a parallel plane to that of the plane of the impeller. The angle between the absolute velocity and the tangent to the circle at the tip is α_2.

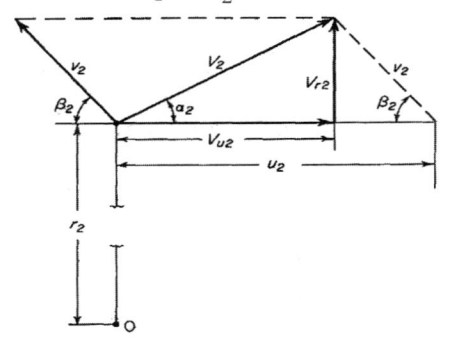

Figure AIII-4. Vector diagram at tip of centrifugal pump vane

Flow of Industrial Fluids—Theory and Equations

APPENDIX AIII | Computations Involving Pumps for Liquids

Simplifying assumptions

For the sake of the analysis, simplifying assumptions (to be corrected later) are made. These are:

1. The point velocity within all channels is identical across a cross section of the channel. This assumption is equivalent to assuming no slip between layers of fluid and no recirculation within the channel.

2. The angle β is the actual vane angle. This assumption is equivalent to assuming what is called perfect guidance. It would be true only if there were an infinite number of curved vanes. In truth, the fluid leaves a channel at varying angles because of varying velocities in the same channel.

For fluid leaving the periphery, the velocity vectors and the tip radius can be depicted as in Figure AIII-4. The absolute velocity vector, V_2, has been resolved into a radial vector, V_{r2}, and a tangential one, V_{u2}. We are now ready to proceed with the analysis.

Analytical principle

In physics and engineering, much economy of effort can be made by finding a universal principle that is always true and then by applying that principle directly without concern for its development.

Newton's law

In linear motion, such a universally true principle is Newton's law, which relates the acceleration of a body (a mass) to the sum of the external forces applied to it. The simplified version of this law is stated as force equals mass times its acceleration, Equation AIII-1. The more fundamental version, which also applies to systems of changing mass, states that the sum of the external forces is equal to the time rate of change of linear momentum, Equation AIII-2. The dimensional constant may be ignored if SI units are used.

$$\sum F_{ext} = \frac{1}{g_c} ma \qquad \text{(AIII-1)}$$

$$\sum F_{ext} = \frac{1}{g_c} \frac{d}{dt}(mv) \qquad \text{(AIII-2)}$$

It should be noted force is intimately associated with mass through these equations; we recognize the presence of forces by their influence on masses.

Rotary analog of Newton's law

In rotary motion, the analog of Newton's law relates torque to the time rate of change of angular momentum, Equation AIII-3.

$$\sum \tau_{ext} = \frac{1}{g_c} \frac{d}{dt}(mrV_{\tan}) \quad \textbf{(AIII-3)}$$

The units of torque are foot-pounds-force or newton-meters. The angular momentum is the term in parentheses. The mass of a particle is m. The radius, r, is that of the instantaneous circle of revolution of the particle of mass. The tangential velocity is $V_{\tan}$ (which is called V_{u2} in Figure AIII-4), the tangential component of the absolute velocity. The torque required to effect a time rate of change of the terms in parenthesis is often given without the summation sign; summation of all external forces being understood.

Note torque is not energy in spite of having the same units. Torque represents force applied at a distance, not force travelling through a distance.

If we draw a control boundary so as to include the volume within a channel of Figure AIII-2, mass is seen to cross the boundary at two sections, (1) and (2). No mass crosses the boundaries other than at the inlet and outlet. The rotary speed of the channel and the mass flow rate will be assumed constant.

The external torque associated with the passage of a particle at each of the sections is given by Equations AIII-4. These equations are the rotary analogs of equations AI-24.

$$\sum \tau_{ext} = \frac{1}{g_c}\frac{d}{dt}(mr_2 V_{2\tan}) - \frac{1}{g_c}\frac{d}{dt}(mr_1 V_{1\tan}) \quad \textbf{(AIII-4)}$$

$$\sum \tau_{ext} = \frac{1}{g_c}\dot{m}(r_2 V_{2\tan} - r_1 V_{1\tan})$$

The tangential velocities and the radii are constant. The quantity of mass changes as it enters (disappears into) and leaves (emerges from) the control volume. This change is equal to the mass flow rate and is the same for both stations under the steady state assumption. If the rate of change of angular momentum is greater at the exit than at the inlet, the external torque is positive. Torque has to be applied to cause the change. If the torque were negative, we would be dealing with a turbine, and energy would be extracted.

APPENDIX AIII | Computations Involving Pumps for Liquids

On the assumption of perfect guidance, all particles have the same mass flow rates, so all of the mass flowing in a channel, and indeed all mass flowing in the impeller, can be lumped into the mass flow rate term. The sum of the external torque applied to each particle is the total torque applied to all the mass flowing through the pump.

Simplification

By design of the blade at the inlet, *for a given flow rate*, the absolute velocity vector at the inlet can be made to point radially, at ninety degrees to the tangential vector. This means its tangential component is zero and Equation AIII-4 reduces to Equation AIII-5.

$$\tau_{ext} = \frac{1}{g_c} \dot{m} r_2 V_{2\tan} \qquad \text{(AIII-5)}$$

The power (energy per unit time) required to maintain a body in circular motion against a resisting torque is given by equation AIII-6. We have dropped the subscript, ext, for convenience.

$$P = \tau \omega \qquad \text{(AIII-6)}$$

The units of torque are foot-pounds-force or newton-meters. The angular velocity, ω, has units of inverse time. So, the units of power are foot-pounds-force per second or newton-meters per second or joules per second, depending on the quantities of measure chosen. Equation AIII-7 is used with customary U.S. units to express power as horsepower.

$$P = \frac{\tau \omega}{550} \qquad \text{(AIII-7)}$$

There are 550 ft-lb$_f$/s in one horsepower (the average rate at which a theoretical horse can do work).

The work done per unit mass is obtained by dividing power by the mass flow rate (energy per unit time divided by mass per unit time equals energy per unit mass).

$$w = \frac{P}{\dot{m}} = \frac{\tau \omega}{\dot{m}} = \frac{r_2 V_{2\tan} \omega}{g_c} \qquad \text{(AIII-8)}$$

The units are foot-pounds-force per pound-mass or, if SI units are used and g_c equals one and is dimensionless, they are newton-meters per kilogram or joules per kilogram. Note the work and power are the work and power imparted to the fluid. Keep in mind the same names are used to express quantities that may differ by efficiency factors. The brake horsepower or the shaft horsepower will be greater than the fluid horsepower. Equation AIII-8 applies to the case when the inlet tangential velocity of a particle is zero.

Relating channel hydraulics to inlet and outlet hydraulics

The above equations refer to flow through the channels from points (1) to (2) of Figure AIII-2. We have to relate the inlet (a) and outlet (b) flows to these equations. The Bernoulli[1] equation can be written between points (a) and (1) and between points (2) and (b). Elevation (potential energy) changes can be neglected and there is no pump work in these sections.

$$\frac{P_a}{\rho} + \frac{\alpha_a U_a^2}{2g_c} = \frac{P_1}{\rho} + \frac{U_1^2}{2g_c}$$
$$\frac{P_2}{\rho} + \frac{U_2^2}{2g_c} = \frac{P_b}{\rho} + \frac{\alpha_b U_b^2}{2g_c}$$
(AIII-9)

The alpha correction factors are applied to the flows in the conduits because of the turbulent profile within these conduits and the use of average velocity, U. They are not applied to the channels because of the assumption of perfect guidance (parallel flow). The velocities represented by U_1 and U_2 are those perpendicular to the entrance to and exit from the ideal channels.

Within the channel, when the inlet tangential velocity is zero, a Bernoulli equation can be written linking theoretical work per unit mass flowing to the fluid energy terms, temporarily neglecting elevation changes and losses but including the work per unit mass done on the fluid.

$$w = \frac{r_2 V_{2\tan} \omega}{g_c} = \left(\frac{P_2}{\rho} + \frac{U_2^2}{2g_c}\right) - \left(\frac{P_1}{\rho} + \frac{U_1^2}{2g_c}\right)$$
(AIII-10)

We can now substitute equations AIII-9 into equation AIII-10 to give,

$$w = \frac{r_2 V_{2\tan} \omega}{g_c} = \left(\frac{P_b}{\rho} + \frac{\alpha_b U_b^2}{2g_c}\right) - \left(\frac{P_a}{\rho} + \frac{\alpha_a U_a^2}{2g_c}\right)$$
(AIII-11)

So, the external work per unit mass of fluid passing through the pump can be measured by the radius of the pump impeller at the tip, the tangential velocity component of the flowing fluid at the impeller tip and the rotational speed in radians per second. It can also be measured by the differ-

APPENDIX AIII | Computations Involving Pumps for Liquids

ences of the sums of the static (pressure) energies and the kinetic energies measured at the outlet and inlet flanges. From a practical point of view, the latter form is more useful. The kinetic energies have a correction factors, alpha, because the average velocities, U, were used for the pipe flow. The inlet tangential velocity is zero.

The first sum in parenthesis on the right of AIII-11 is called the "discharge head" and the second one is the "suction head". The difference is called the total dynamic head or developed head. The individual terms are often called pressure heads and velocity heads. It is important to keep the distinctions between pressure and energy clear and not to fall into the trap of using the word "head" indiscriminately and to assume it simply means discharge pressure. In our terms, head is energy per unit mass. It is the author's contention that it is easier to make the distinctions among the various head terms when energy per unit mass is the conceptual tool rather than feet or meters. This is especially true when the fluid is compressible.

Expressed in terms of head in foot-pounds-force per pound-mass or in newton-meters per kilogram, the total dynamic head is given as,

$$\Delta H_{ideal} = h_b - h_a = \frac{\omega r_2 V_{2\tan}}{g_c} = w_{ideal} \qquad \text{(AIII-12)}$$

The subscript, (ideal), is to remind us we have not yet applied correction factors for the assumptions that were made. The suction and discharge heads can be computed from measurable pressure, flow rate and pump inlet and outlet internal diameters. They give the ideal total dynamic head directly.

Relation of total dynamic head to volumetric flow in an ideal pump

The volumetric flow through a pump is given by the following equation.

$$q_{ideal} = V_{2rad} A_p \qquad \text{(AIII-13)}$$

The units are either cubic meters per second or cubic feet per second. The velocity is the velocity perpendicular to the circumference of the impeller. The area is the sum of the areas of the channel cross sections at the periphery.

From Figure AIII-4, the tangential component of the absolute velocity is given in Equation AIII-14.

$$V_{2tan} = u_2 - \frac{V_{2rad}}{\tan \beta_2} \quad \text{(AIII-14)}$$

From Equation AIII-12 combined with AIII-13 and AIII-14 and the assumption of zero inlet tangential velocity, we can derive the following relationship.

$$\Delta H_{ideal} = \frac{\omega r_2 V_{2tan}}{g_c} = \frac{\omega r_2}{g_c}\left(u_2 - \frac{V_{2rad}}{\tan \beta_2}\right) \quad \text{(AIII-15)}$$

$$= \frac{u_2}{g_c}\left(u_2 - \frac{V_{2rad}}{\tan \beta_2}\right) = \frac{u_2}{g_c}\left(u_2 - \frac{q_{ideal}}{A_p \tan \beta_2}\right)$$

It can be inferred from Equation AIII-15 that the ideal total dynamic head, TDH, is proportional to the tangential kinetic energy. The equation also shows the ideal pump, at a constant rotational speed, has a linear relationship between the total dynamic head (energy transferred to fluid by the pump per unit of flowing mass) and the volumetric flow rate. More importantly, it shows this relationship is independent of the fluid. Since we have not made any assumptions regarding the nature of the fluid, the above equation is also valid for real fluids (in an ideal pump).

The relationship between head and volumetric flow rate given above is the main reason pump manufacturers use head and not pressure on their curves. An economy of description is gained. One curve fits all fluids pumped through the same pump. If manufacturers used pressure on their curves, they would have to specify the fluid density.

Caveat — Conversion of 'head' to pressure

Another word of warning is warranted here. To obtain pressure units, it is simply necessary to multiply head as defined above by density in compatible units. However, total dynamic head includes both the energy per unit mass associated with pressure and that associated with velocity. So multiplying by density does not give the differential pressure across the pump. It gives a number that differs from the differential pressure by the difference in velocity pressures (the pressure generated in reducing the velocity to zero). Since this number is not greatly different from the differential pressure, it is often assumed the two numbers are one and the same, but this is not so.

APPENDIX AIII | Computations Involving Pumps for Liquids

The blades on pumps are usually curved backwards to the direction of rotation so the blade angle, β, is less than 90 degrees and the tangent of the angle is positive. With some fans and blowers, the blade angle may be curved in the forward direction. This is not often done with pumps because it produces problems of flow instability.

Equation AIII-15 shows that if the tangent of the angle is positive, the (ideal) head curve will slope downwards with increasing flow. If it is infinite, the slope will be horizontal for all flows. If the tangent is negative, the slope of the TDH curve versus q will be positive. These characteristics correspond to blades whose leaving angles are less than 90 degrees, equal to 90 degrees and greater than 90 degrees when measured in the clockwise direction from the tangent line.

AIII-3: PERFORMANCE OF REAL CENTRIFUGAL PUMPS

The developed head (the total dynamic head) of a real centrifugal pump is much less than that of its ideal model. The rate at which mechanical energy is transmitted to the fluid is less than the rate at which it is transmitted to the shaft. The fluid horsepower is less than the brake horsepower. The efficiency of the pump excluding the driver, η, is less than one. Efficiency here is being defined as mechanical energy per unit mass flowing transmitted to the fluid divided by energy per unit mass flowing transmitted to the shaft.

Head losses versus power losses

"Losses" are frequently classified as head losses or power losses. The distinction comes about because some deviations from the ideal pump model cause a lower total dynamic head without causing a decrease in the energy per unit mass pumped. This fact is of interest theoretically and for didactic reasons.

Head losses: Correcting the previous assumptions – imperfect guidance

In the real pump, guidance is not perfect. Flow is not parallel through the channels. There may even be reverse flow in portions of the channel causing a circulating pattern to be imposed on the overall flow.

Two vector diagrams are given in Figure AIII-5. They may help explain the real situation. These diagrams represent a fixed volumetric flow rate and a fixed pump speed. The right hand diagram (dashed lines) represents the case of perfect guidance; the left-hand diagram (solid lines) represents a real case.

The bases of the diagrams are identical in length because the peripheral velocity, u_2, is identical in both cases. Similarly, the altitudes of the triangles represent the ideal radial velocity on the right and the actual radial velocity on the left. Both of these velocities are related to the volumetric flow rate and are equal.

The so-called absolute velocity of the fluid decreases in the real case due to the lack of perfect guidance. The stream leaves the impeller at a more backward angle so that the angle, β, is decreased. The tangential component of the absolute velocity, V_{2tan}, also decreases ($V_{u2} \rightarrow V'_{u2}$). Since by Equation AIII-15, the total dynamic head is proportional to the tangential velocity, the total dynamic head decreases in the real case.

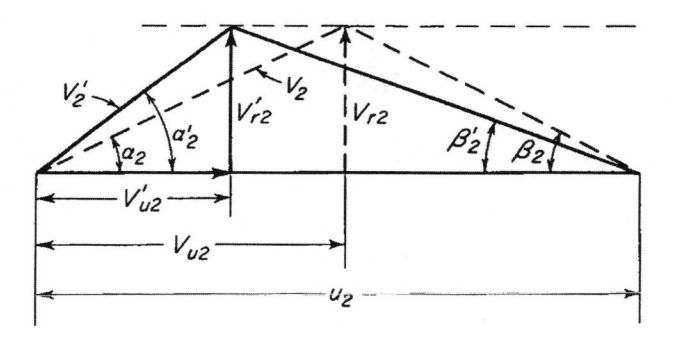

Figure AIII-5. Vector diagrams describing perfect and imperfect guidance

Source: L. McCabe, Unit Operations in Chemical Engineering, 1967, reprinted by permission McGraw-Hill Companies.

The energy requirement per unit mass is proportional to the tangential velocity by Equation AIII-8. Therefore, power decreases with the decrease in tangential velocity and the efficiency is not affected by this isolated cause (of imperfect guidance).

The reduction in total dynamic head due to less-than-perfect guidance is quite severe. The reduction is greatest at low flow rates and least at the design point when the flow streams become more nearly parallel. If we plot ideal head versus volumetric flow as in Figure AIII-6, we can depict a new theoretical head discounted for guidance losses by a line below the ideal head curve. This line will have a slightly less negative slope. The area between the two lines represents the losses.

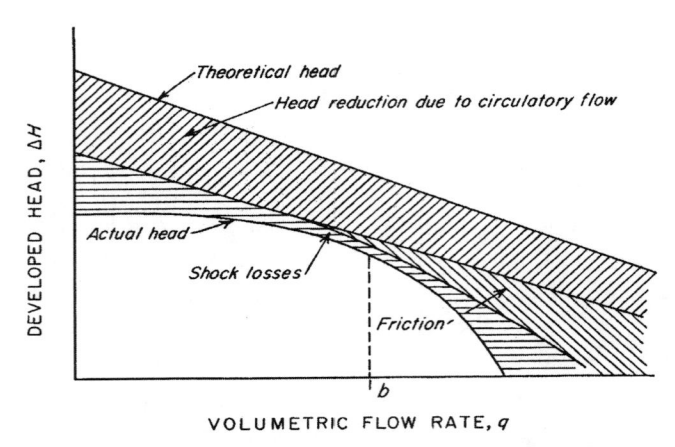

APPENDIX AIII | Computations Involving Pumps for Liquids

Figure AIII-6. Reduction in theoretical 'head' due to various causes

Source: L. McCabe, Unit Operations in Chemical Engineering, 1967, reprinted by permission McGraw-Hill Companies.

Head losses: Correcting the previous assumptions – irreversibilities

According to the complete Bernoulli[i] equation, fluid irreversibilities decrease the developed head (TDH). Useful mechanical energy is converted into unrecoverable internal energy, which manifests itself by a temperature rise. According to the Darcy[vii] equation, these losses increase with the square of the flow rate. Figure AIII-6 shows the line representing head discounted for so-called friction losses. It has the same origin as the second line above, but it is concave downward. It diverges increasingly with higher flow rates from the previous curve.

Head losses: Correcting the previous assumptions – shock losses

The father of thermodynamics, Sadi Carnot[ii], had a father, General Hypolite Carnot. The grandfather of thermodynamics not only was an army general under Napoleon, he was an important politician (member of the Directorate) during the French revolution and he was a hydraulician who wrote extensively on such things as shock losses in turbomachinery.

Hypolite Carnot was well aware that if a fluid did not enter a channel at the same angle as the blade or if it left at an angle different from that of the surrounding fluid, a loss in available mechanical power occurred due to "shock". The maximum mechanical power could not be generated unless shock was minimized. Some have suggested the concern of Hypolite Carnot for efficiency and its relationship to shock losses was the stimulus that lead Sadi Carnot to develop his theories on ideal thermodynamic engines and on the second law of thermodynamics.

286   Flow of Industrial Fluids—Theory and Equations

Computations Involving Pumps for Liquids — APPENDIX AIII

For our purposes, it should be realized a sudden change in flow direction causes turbulence, loss of head and conversion of mechanical energy into internal thermal energy. This change in direction can be due to a mismatch between the incoming fluid velocity and that of the leading edge of the impeller blades. It can also be due to the sudden introduction of a leaving stream at an angle to a circulating stream. It results in a lower mechanical efficiency, η.

Corrective measures – diffusers

Large volute pumps have huge losses due to shock. Frequently, stationary diffusers are added to guide the fluid more smoothly to the pump exit as shown in Figure AIII-7. Diffusers, as well as guiding the fluid smoothly, reduce its velocity and increase its pressure by providing a smoothly expanding path.

Diffusers perform two functions. They reduce shock losses and they convert kinetic energy to static (pressure) energy.

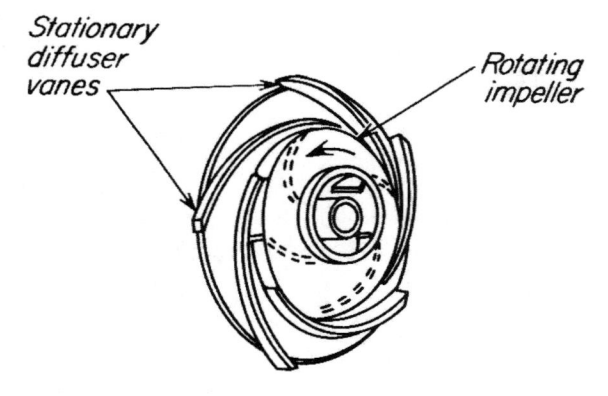

Figure AIII-7. Impeller and stationary vanes in diffuser pump

Source: L. McCabe, Unit Operations in Chemical Engineering, 1967, reprinted by permission McGraw-Hill Companies.

Best efficiency point

A fixed speed centrifugal pump is designed to be most efficient at a fixed flow rate. Moving away from this flow rate in either direction causes the efficiency to decrease. More energy is consumed per unit mass being pumped. The point at which the efficiency is greatest, the best efficiency point, corresponds to the design flow of the pump and the design head (TDH).

The shock losses increase on either side of the best efficiency point. In Figure AIII-6, the shock losses are shown to discount the head versus flow curve. The combination of mechanical energy losses produces the curve that is typical for a particular pump design.

Other power losses in a pump

Internal circulation
The pump impeller has clearance between it and the casing. The high pressure at the discharge and low pressure at the suction causes liquid to circulate internally between the impeller and the casing. This results in "disk friction" (again, mechanical energy converted to internal energy). The internal circulation is maintained at the expense of energy that would otherwise appear as energy within the fluid.

Bearing losses
Mechanical "friction" is present in pump shaft bearings. "Friction" is a force against the direction of motion that must be overcome by a force supplied by the driver. Additional energy per unit mass must be supplied to the shaft to overcome all of the above mechanical energy losses. The rate at which energy is supplied to the shaft constitutes power. Theoretically, the following equation is true.

$$P = \frac{1}{g_c} \dot{m} \omega r_2 V_{2\tan} \qquad \textbf{(AIII-16)}$$

So, the theoretical power is close to a straight-line function of the mass flow rate or, since liquids can be considered constant density fluids, of the volumetric flow rate. The rate at which energy is transmitted to the fluid is the hydraulic power. The rate it is transmitted to the shaft of the pump is the shaft or brake power. The ratio of the fluid power to the shaft power is the pump efficiency exclusive of any driver's efficiency, η.

Power versus volumetric flow – a schematic description
Figure AIII-8 is a plot of power versus volumetric flow in a pump.

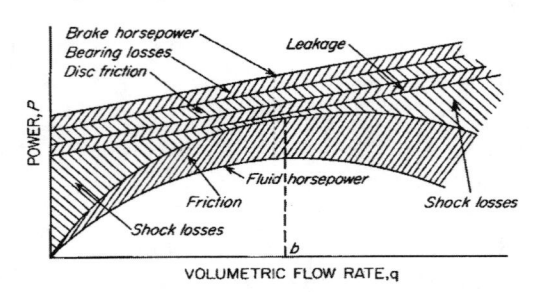

Figure AIII-8. Power versus volumetric flow in a centrifugal pump

Source: L. McCabe, Unit Operations in Chemical Engineering, 1967, reprinted by permission McGraw-Hill Companies.

Computations Involving Pumps for Liquids | APPENDIX AIII

The uppermost line represents theoretical power supplied to the shaft — frequently termed brake horsepower. The line starts at some positive value corresponding to all of the losses occurring when the pump is dead-headed (operating against a closed valve). As flow increases, the power to the shaft increases more or less linearly.

Several lines are drawn parallel to and below the brake horsepower line. The vertical distance between each line represents the power lost for various reasons: bearing friction, disc friction, leakage and internal circulation.

Below the lowest straight line is drawn a curved line. The intervening space represents irreversibilities due to shock. The line is convex upward. This line starts at zero flow where all of the remaining irreversibilities are those due to shock. It comes close to the lowest straight line at the design operating point where the shock losses are minimized. It diverges from the lower straight line from there onwards as the shock losses increase again.

Below the curved line representing power discounted for shock, leakage, disc friction and bearing losses, there is another curved line that represents power actually transmitted to the fluid as mechanical power. The space between this curve and the previous one represents losses due to flow through the pump. The two lower curves diverge at a rate roughly proportional to the flow rate squared. This causes the fluid power curve to have a definite maximum at its design operating point.

Head, efficiency and power versus volumetric flow

In customary U.S. units, the head in feet (which is really foot-pounds-force per pound-mass), the efficiency and the brake horsepower are frequently plotted versus volumetric flow. Each plot will be read against a different ordinate scale. The horsepower scale usually starts above zero.

The brake horsepower curve usually slopes upward from its minimum value in a fairly linear fashion. The efficiency curve curves upward from zero and peaks at the design operating point. On most process pumps, the head curve slopes downwards following a curve that is concave downward. The curve can drop off quite radically past the design operating point giving rise to the expression, "the pump ran out on its curve".

Pumps with high heads and high speeds often have a peaked head curve. There is usually a definite hump close to shutoff. This type of pump has limited rangeability. If the pump is throttled too far back on its curve it

APPENDIX AIII | Computations Involving Pumps for Liquids

can become unstable because there are two flow rates that can be in equilibrium with the same total dynamic head. The pump will hunt between the two equilibrium states and the surging flow can cause damage to the driver and the pump.

Maximum head per stage and multistage pumps

The tip speed of a centrifugal process pump limits the maximum practical total dynamic head to between 70 and 100 foot-pounds-force per pound-mass or in terms of feet of fluid 70 to 100 feet. This is about 20 to 30 meters or 210 to 230 newton-meters per kilogram (or J/kg).

If we take 85 foot-pounds-force per pound-mass or 85 "feet" as the average, such a pump working on 60°F water will produce 85 ft-lb$_f$/lb$_m$ times 62.4 divided by 144 or 36.8 psi of differential pressure. Increasing the pressure substantially requires multistaging, or a different design, or both.

In multistaging, the developed heads per stage add to give the overall head. The overall efficiency is found by multiplying the stage efficiencies. If they are all equal, then the overall fractional efficiency is equal to the single stage fractional efficiency raised to a power equal to the number of stages. A seven-stage pump with an efficiency per stage of 90% would have an overall efficiency of 47.8%. Better than half the energy input to the shaft is wasted in the conversion to internal energy whose first manifestation is an increase in the temperature of the fluid.

Use of the Bernoulli[i] equation with pumps

We will repeat the most general form of the Bernoulli equation and then apply it to a system containing a pump. Several examples of computations will be given. A simple system consists of an open tank, a pump and associated piping. The discharge piping from the pump feeds an open receiver. The surface of the liquid of the first tank is designated as point 1. The open discharge pipe above the receiver is point 2. The following can be written:

$$\frac{P_1}{\rho} + \frac{g}{g_c} Z_1 + \frac{\alpha_1 U_1^2}{2 g_c} + w_{12} = \frac{P_2}{\rho} + \frac{g}{g_c} Z_2 + \frac{\alpha_2 U_2^2}{2 g_c} + h_{12} \qquad \text{(AIII-17)}$$

The term, w_{12}, is the energy the pump transmits to the fluid per unit mass of flowing fluid. The term, h_{12}, is the energy per unit mass of flowing fluid converted to internal energy. The suffix, 12, on each term is meant to show these terms refer to mechanical energy per unit mass that is added or converted to thermal energy within the system somewhere between sections 1 and 2. The subscripts, 1 and 2, indicate the energy per unit mass is associated with the specific sections.

Computations Involving Pumps for Liquids | APPENDIX AIII

The term, h_{12}, is usually called a loss term. These irreversibilities occur in the fluid in the piping, not in fluid in the pump. The term, w_{12}, refers only to energy transmitted to the fluid. The pump also has losses not included in the piping loss term. These losses occur due to friction in stuffing boxes and bearings and in fluid irreversibilities within the pump. The mechanical energy transmitted to the fluid is equal to that transmitted to the shaft less the sum of the losses within the pump.

$$w_{12} = w_s - h_p \equiv \eta w_s \tag{AIII-18}$$

$$\eta = \frac{w_s - h_p}{w_s} = 1 - \frac{h_p}{w_s}$$

$$\eta < 1.0$$

The fluid energy per unit mass transmitted by the pump is given in terms of the shaft energy per unit mass multiplied by an efficiency factor that is always less than one.

The Bernoulli[i] equation with pump losses and pump shaft work taken into consideration takes on the following form.

$$\frac{P_1}{\rho} + \frac{g}{g_c} Z_1 + \frac{\alpha_1 U_1^2}{2g_c} + \eta w_s = \frac{P_2}{\rho} + \frac{g}{g_c} Z_2 + \frac{\alpha_2 U_2^2}{2g_c} + h_{12} \tag{AIII-19}$$

This equation is applicable to customary U.S. units and to SI units if the dimensional constant is taken as one with no units. The subscript, s, means the work is that transmitted to the shaft.

Note equation AIII-19 describes energy changes between sections 1 and 2 external to the pump. The shaft work is transformed to hydraulic work by the efficiency factor which takes into consideration pump irreversibilities. The mechanical energy losses given by h_{12} are those external to the pump.

Total heads

The concept of "total" head refers to a summation of individual heads at the pump inlet flange and at the discharge flange. These locations are designated in the following equation by points 1 and 2. These locations are not the same as those in equation AIII-19. The difference in head is also given the designation "total" (total dynamic head), so the adjective "total" becomes meaningless through overuse. The total dynamic head, TDH, is the mechanical energy imparted to the fluid by the pump impeller per unit mass of flowing fluid.

APPENDIX AIII | Computations Involving Pumps for Liquids

When analyzing a pump, we draw an imaginary boundary around it cutting through the inlet and outlet flanges. The rest of the physical system is considered only through its influence at the boundaries. Only the external influences at the flanges (pressure, velocity and elevations) are considered. Since we have reduced the system to what exists between these two points, the piping loss term, h_{12}, drops out of the Bernoulli[i] equation. The efficiency factor remains with the shaft work term.

$$\eta w_s = \left(\frac{P_2}{\rho} + \frac{g}{g_c}Z_2 + \frac{\alpha_2 U_2^2}{2g_c}\right) - \left(\frac{P_1}{\rho} + \frac{g}{g_c}Z_1 + \frac{\alpha_1 U_1^2}{2g_c}\right)$$

$$\eta w_s = h_2 - h_1 = \Delta H = (TDH)$$

$$w_s = \frac{\Delta H}{\eta}$$

(AIII-20)

The first line of Equation AIII-20 is a specific energy balance. It states that the specific (per unit mass) energy transmitted to the fluid is added to the total incoming specific energy, subscript 1. The sum equals the outgoing specific energy, subscript 2. The arrangement of the equation shown allows computation of the energy requirement at the shaft of a pump in order to effect the change shown on the right.

Analysis of terms

It is worth analyzing the above set of equations. The first term on the left in the first and second equations is the energy per unit mass flowing delivered to the fluid by the pump. In the last equation, it is the energy per unit mass of flowing fluid delivered to the shaft. The units are foot-pounds-force per pound-mass or, if g_c is treated as one, no units, and SI units are used elsewhere, they are joules per kilogram or newton-meters per kilogram.

The first term on the right in parenthesis of AIII-20 is the total discharge head, which can be replaced by the sum, h_2. The units of both terms in parenthesis are the same as those of the input term. The second term parenthesis on the right is the total suction head, which can be replaced by the sum, h_1.

The only arbitrary terms in the equations are the elevations, Z_1 and Z_2. They are arbitrary because they depend on a datum selected for convenience. Since the results of the Bernoulli equation depend upon differences, the datum selected will not change the results. The values do depend upon the system being analyzed. The subscripts, 1 and 2, must be used consistently in a given problem.

Computations Involving Pumps for Liquids | APPENDIX AIII

The difference between the total discharge and the total suction heads, ΔH, is called the total dynamic head. It is the actual energy per unit flowing mass transmitted to the fluid at a particular flow rate. It will change over the operating range of the pump.

The three components of each of the total heads are also called "heads". The first component is the static or pressure head, pressure divided by mass density in the above equation – pressure divided by weight density in the hydraulic equations. The second component is elevation head. The third component is the velocity head. The first component is numerically the largest contributor to total head. The other two are relatively small when their differences are considered.

The use of the word "head" to describe at least six different components, the differences between them and the differences of their "total" sums obviously can lead to confusion and error. The author has made such errors himself and has observed that, even in very respected publications, such conceptual errors have been made. It is very easy to use the word "head" in a general sense and then to fall into the conceptual trap of mis-applying it in specific computations. If you use the word, you are advised to qualify it always with words such as "total discharge", "total suction", "total dynamic", "velocity", "static suction", and "static discharge". The point has been belabored sufficiently.

Power to the shaft

The power to the shaft in terms of horsepower is called the "brake" horsepower. Presumably, the term grew out of widespread use of the Prony brake to measure power. The horsepower unit was a natural development at the time when horses were a common means of motive power. It has little meaning today. It, nevertheless, retains its popularity in many sectors, but is a confusing term when conversions are made. For instance one horsepower is not exactly equal to one cheval vapeur.

Power to the shaft is the rate at which energy is transmitted to the shaft. Fluid power is the rate at which it is transmitted to the fluid. The ratio is the efficiency – power in divided by power out or energy in divided by energy out. Power to the shaft is distinct from power to the driver. The input power to a shaft is the output power from a driver. The input power to the driver will be somewhat greater. The driver has its own efficiency.

The set of equations in AIII-21 describes shaft (subscript s) and fluid (subscript f) power and horsepower. The latter is in U.S. units only.

APPENDIX AIII | Computations Involving Pumps for Liquids

$$P_f = \dot{m}\Delta H = \dot{m}w_f = \dot{m}\eta w_s \qquad \text{(AIII-21)}$$

$$P_s = \dot{m}w_s = \dot{m}\frac{\Delta H}{\eta}$$

$$\eta = \frac{P_f}{P_s}$$

$$P_s = \frac{\dot{m}w_s}{550} = \frac{\dot{m}\Delta H}{550\eta}\,hp$$

The first three equations in the set AIII-21 may use SI or customary U.S. units. In both cases the units are energy transmitted per second — foot-pounds-force per second or joules per second. Eta is the fractional efficiency. The last equation uses customary U.S. units only. The coefficient, 1/550, is the conversion between foot-pounds per second and horsepower.

AIII-4: REAL CENTRIFUGAL PUMPS – SUCTION LIFT, CAVITATION AND NPSH

Equation AIII-19 gives the general relationship describing systems containing pumps. The pump work term and the piping loss term are not point functions. The loss term is distributed across the system external to the pump and the pump term does not take into account its specific location. Within limits, the pump can be anywhere within the piping system. In this section we will establish the limits.

The suction pressure can be below atmospheric pressure, but the total mechanical energy at the suction cannot be below the energy associated with the vapor pressure of the liquid. If it is, flashing may result and perhaps cavitation. Flashing will not occur if the sum of the velocity head and the pressure head, in other words if the sum of the kinetic energy and the static energy, is sufficiently greater than the energy associated with the vapor pressure of the liquid at the flowing temperature.

The criterion to avoid flashing is stated more specifically in equation AIII-22.

$$\frac{\alpha_f U_f^2}{2g_c} + \frac{P_f}{\rho_f} > \frac{P_{fv}}{\rho_f} \qquad \text{(AIII-22)}$$

In this equation the subscript, f, is used simply to indicate the location is a specific one anywhere in the fluid stream. The subscript, fv, indicates vapor pressure at the temperature of the fluid at the location. The velocity profile coefficient, α_f, can be ignored in most cases (can be equated to one). The equation applies to the SI system if the dimensional constant is made equal to one and is dimensionless. It is to be noted the pressure at the location can be below the vapor pressure as long as the kinetic energy term is sufficiently great.

An explanation of equation AIII-22 is vapor pressure is an inherent characteristic of matter and is a function of temperature. The tendency to vaporize that a liquid possesses is inhibited by externally imposed forces. The term on the right of the inequality represents the static energy the fluid has inherently, because of its temperature. This energy tends to separate the molecules.

If there is no motion, there is no bulk kinetic energy and only the external pressure energy opposes the internal inherent energy. When the sum on the left dips below the term on the right, equilibrium is disturbed and vaporization occurs. The temperature drops to a new equilibrium temperature. A portion of the liquid is vaporized.

NPSH

The excess of the sum of the kinetic energy and the static energy over the inherent energy represents a measure of safety in the avoidance of flashing and potential cavitation. It is given the name "net positive suction head". This head is computed from conditions measured at the inlet flange of a pump. Equation set AIII-23 describes the concept.

$$NPSHA \equiv H_{sv} = \frac{\alpha_s U_s^2}{2g_c} + \frac{P_s}{\rho_s} - \frac{P_{sv}}{\rho_s} = \frac{\alpha_s U_s^2}{2g_c} + \frac{P_s - P_{sv}}{\rho_s} \quad \textbf{(AIII-23)}$$

The subscripts reflect the fact we are referring to conditions at the suction (inlet) flange of the pump. The letter A added to NPSH means available.

Suction lift

The concept of net positive suction head applies to pumps, even when they are operated in a suction lift mode. Pumps operating in this mode must be primed before they are started (see Section III-5, Priming Centrifugal Pumps Operating in a Suction Mode).

APPENDIX AIII | Computations Involving Pumps for Liquids

The Bernoulli[1] equation can be written for the suction piping of a system operating under suction lift and the NPSHA equation, AIII-23, can be combined as follows:

$$\frac{P_1}{\rho} + \frac{g}{g_c}(0) + \frac{(0)_1^2}{2g_c} + (0) = \frac{P_s}{\rho} + \frac{g}{g_c}Z_s + \frac{\alpha_s U_s^2}{2g_c} + h_{1s} \quad \text{(AIII-24)}$$

$$\frac{P_1}{\rho} = \frac{P_s}{\rho} + \frac{g}{g_c}Z_s + \frac{\alpha_s U_s^2}{2g_c} + h_{1s}$$

$$\frac{P_1}{\rho} - \frac{g}{g_c}Z_s - h_{1s} = \frac{\alpha_s U_s^2}{2g_c} + \frac{P_s}{\rho}$$

$$H_{sv} = \frac{\alpha_s U_s^2}{2g_c} + \frac{P_s}{\rho_s} - \frac{P_{sv}}{\rho_s} = \frac{P_1}{\rho} - \frac{g}{g_c}Z_s - h_{1s} - \frac{P_{sv}}{\rho_s}$$

$$H_{sv} = \frac{P_1 - P_{sv}}{\rho} - h_{1s} - \frac{g}{g_c}Z_s$$

Assumptions

The assumptions in the above development are: the datum elevation is the liquid level in the feed tank or sump, the velocity at that point is essentially zero, there was no additional pump in the pipe being considered and the pump was already primed and ready to go when started. The terms involving elevation and kinetic energy at point 1, and pump energy between point 1 and the pump suction are all zero. The density is assumed constant between point 1 and the pump suction.

Analysis

The development started with the general Bernoulli equation, it eliminated the terms that were made equal to zero by the stated assumptions and then rearranged the equation so it could be substituted into the NPSHA equation, AIII-23. This allowed the NPSHA equation to be written in terms of variables that are known to be true for the stated problem.

The last equation in the set AIII-24 states that the NPSHA, or the theoretical safety factor, in energy units per unit mass, is equal to the algebraic sum of three terms.

The first term is difference between the pressure on the liquid surface in the feed tank and the vapor pressure of the liquid at the suction flange divided by the liquid density. This is the most important term. Vapor pres-

Computations Involving Pumps for Liquids | APPENDIX AIII

sure and the pressure on the surface are expressed in absolute units. The vapor pressure and the density are measured at the flowing temperature at the suction flange.

The second term represents the irreversibilities in the inlet section. It must be in the same units of energy per unit mass and it must include all losses – pipe entrance, any foot valves, other valves, pipe bends or strainers. It is subtracted from the first term because it represents a loss of safety factor.

The last term represents the loss of safety factor due to the elevation of the pump suction being higher than the surface of the liquid. It is also subtracted from the first term. If the pump is below the surface, this factor will become positive. The sign will be changed automatically if distance below the datum is considered to be negative.

The last equation of the set AIII-24, as written, has units of foot-pounds-force per pound-mass or joules per kilogram depending on the units chosen and the use of the dimensional constant. If the same equation is multiplied through by the density, the units of each term become those of pressure. If the original equation is divided through by the ratio g/g_c, the units of each term become those of manometric head, feet or meters.

Utility of NPSHA and NPSHR

As long as a pump is operating within its design constraints, little attention need be paid to the net positive suction head available or required. In addition, operations and maintenance personnel have little say in the initial choice of a system, so they tend to assume a lack of interest is justified.

It is only when a pump has been misapplied, or the fluid temperature has changed radically and troubleshooting is necessary, do operating or control personnel become involved with NPSH problems. Problems involving NPSH occur when the NPSH required by the pump is greater than that available. When this happens, it is usually accompanied by symptoms such as unusual noise at the pump or loss of pumping ability. If it is allowed to continue, cavitation damage can occur in the impeller channels and in the casing.

Troubleshooting involves making sure there is no unnecessary blockage in the pump suction lines caused by, for example, plugged strainers or partially closed valves. The next step is usually to check the pump data versus the operating conditions. The pump data should give the design condi-

APPENDIX AIII | Computations Involving Pumps for Liquids

tions and an NPSHR curve. The actual operating condition can be used to compute the NPSHA, at least at one point. It is important the data be exact, not old nor incorrect due to measurement errors. Temperature is especially important, or more exactly the vapor pressure is important.

We will give some examples of computed NPSHA following the summary of equations to follow.

Summary of equations using different units for NPSHA

The equation set AIII-25 summarizes transformations between units used to compute NPSHA.

(AIII-25)

$$H_{sv} = \frac{P_1 - P_{sv}}{\rho} - h_{1s} - \frac{g}{g_c} Z_s \; ft-lb_f/lb_m$$

$$H_{sv} = \frac{P_1 - P_{sv}}{\rho} - h_{1s} - \frac{g}{1} Z_s \; J/kg$$

$$\rho H_{sv} = P_1 - P_{sv} - \rho h_{1s} - \rho \frac{g}{g_c} Z_s \; psf$$

$$\rho H_{sv} = P_1 - P_{sv} - \rho h_{1s} - \rho \frac{g}{1} Z_s \; Pa$$

$$H_{sv} \frac{g_c}{g} = \frac{P_1 - P_{sv}}{\rho \frac{g}{g_c}} - h_{1s} \frac{g_c}{g} - Z_s \; ft$$

$$H_{sv} \frac{1}{g} = \frac{P_1 - P_{sv}}{\rho \frac{g}{1}} - h_{1s} \frac{1}{g} - Z_s \; m$$

In hydraulic practice, the denominators of the pressure terms in the last two equations are called specific weight, γ. The units are force per unit volume.

AIII-5: POSITIVE DISPLACEMENT PUMPS

We have already discussed the principal difference between a positive displacement pump and a centrifugal pump. This difference is due to the incompressible nature of the liquids and the volumetric displacement of positive displacement pumps. It is not possible to throttle these pumps as a means of controlling flow. Some path for fluid flow must always be open. The size of the section of this path depends on the backpressure that can be tolerated by the pump and piping system.

Suction lift and cavitation

Positive displacement pumps, if their sealing mechanism and check valves are adequate, can elevate liquid to close to the theoretical limit of static lift without the presence of a foot valve, once primed. It has already been pointed out in Chapter III that a reciprocating pump requires an additional acceleration head in order to make sure the liquid is always in contact with the receding piston.

Positive displacement pumps are not immune to cavitation. The amount of damage done will depend on their construction and speed.

Discharge characteristics

The discharge characteristics of positive displacement pumps vary from very smooth with screw pumps to pulsating with reciprocating pumps. The initial concern for pulsating pumps is purely mechanical. Pulsations are damped in order to reduce the consequences to the piping and connected equipment. Spring hangers and adequate supports are used on piping. Flexible couplings and shock absorbing supports are used on equipment.

Positive displacement pumps are not normally damped with the measurement or control of the flowing fluids in mind. The primary concern of the pump engineer or equipment engineer is the longevity of the equipment. Damping for the purpose of measurement and control usually is left to control systems personnel.

AIII-6: THEORY AND ANALYSIS OF JET PUMPS

Jet pumps appear simple, but are not. They consist of what looks like a tee whose branch is the suction and whose run consists of two connections, the driving fluid connection and the discharge connection. The names given to the different types of jet pumps in an attempt to describe the various possible combinations of fluids and the various operating configurations, in particular, lead to confusion. In this appendix we are only concerned with jet pumps that have liquids at each connection.

Principle of operation

The way a jet pump functions is as follows:
1. The motive fluid is accelerated to a high velocity in the nozzle. Some of the static (pressure) energy is converted to kinetic energy. The pressure at the vena contracta drops radically, most often below atmospheric pressure.

APPENDIX AIII | **Computations Involving Pumps for Liquids**

2. The driven fluid is forced into the entrance to the mixing section by the difference in pressure between the source and the low pressure at the nozzle discharge. This point is common to both fluid streams.

3. The two fluids mix in a highly irreversible manner and momentum is transferred between them. The mixed fluid leaves the mixing section at an intermediate pressure, higher than the suction inlet pressure but lower than the motive pressure.

4. A diffusing section decelerates the mixed fluids and converts some of their kinetic energy back to static (pressure) energy.

If we compare a jet pump to a centrifugal pump, we see one essential difference is the motive fluid is added to the driven fluid. More fluid reaches the downstream sink than left the suction source. This extra fluid is not present in the case of a centrifugal pump. The centrifugal pump simply adds energy, not mass, to the suction fluid. Therefore, in comparing efficiencies, only the mass flow to the suction of the jet pump should be considered. The motive liquid could have been let down directly to the sink without passing through the pump – it had sufficient pressure. When comparing efficiencies, we should only consider the energy necessary to move suction fluid from the source to the sink.

Theory of the mixing section – force and rate of change of momentum balance

The following analysis is based on that originally given by Weber and Meissner[xlii]. Figure AIII-9 is a simplified longitudinal section of a jet pump. We can create a control volume for the purpose of the analysis by taking a section across the entrance and one at the exit of the mixing chamber. For simplicity, consider the mixing chamber to have parallel sides.

The external forces acting upon the control volume are due either to static pressure acting on the cross sectional area, to the forces generated by the rate of change of momentum associated with the flowing streams or to viscous drag.

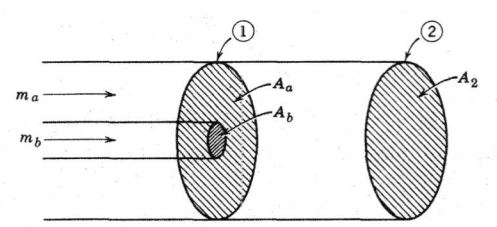

Figure AIII-9. Ejector analysis

Source: Weber & Meissner

Computations Involving Pumps for Liquids APPENDIX AIII

The pressure, P_1, at the entrance to the control volume can be taken as constant across the entire section. It is common to both the fluid discharging from the nozzle and the fluid entering from the suction inlet. The force is the product of this pressure and the total area at section 1. It acts to the right. The pressure, P_2, at the discharge from the control volume also acts across the entire area. The associated force is the product of this pressure and the cross sectional area at section 2. It acts to the left. The viscous drag force, R_D, resists fluid motion and acts to the left.

The sum of the inlet areas can be taken to be equal to the outlet area if we ignore the nozzle wall thickness. The sum of the inlet mass flow rates is exactly equal to the discharge mass flow rate.

A steady state force balance can be set up as follows:

$$P_1 A_a + P_1 A_b + \frac{\dot{m}_a U_a}{g_c} + \frac{\dot{m}_b U_b}{g_c} = P_2 A_c + \frac{(\dot{m}_a + \dot{m}_b) U_c}{g_c} + R_D \quad \text{(AIII-26)}$$

$$P_1 A_c + \frac{\dot{m}_a U_a}{g_c} + \frac{\dot{m}_b U_b}{g_c} = P_2 A_c + \frac{(\dot{m}_a + \dot{m}_b) U_c}{g_c} + R_D$$

Using customary U.S. units, we will assume both streams contain water at 60°F, 62.5 lb_m/ft^3. There is no viscous drag (R_D is to be neglected) and the process is adiabatic (no heat transfer to or from the boundary).

Data
$A_a = 0.25$ ft²
$A_b = 0.10$ ft²
$A_c = 0.35$ ft²
$U_a = 10.0$ ft/s
$U_b = 50.0$ ft/s
$P_1 = 40.0$ psia

The velocity at the discharge from the control volume can be computed from the continuity equation.

$$(UA\rho)_a + (UA\rho)_b = (UA\rho)_c \quad \text{(AIII-27)}$$

$$U_c = \frac{(UA)_a + (UA)_b}{A_c}$$

$$U_c = \frac{10.0(0.25) + 50.0(0.1)}{0.35} = 21.4 \, ft/s$$

APPENDIX AIII | **Computations Involving Pumps for Liquids**

The mass flow rates can be obtained from the velocities, densities and cross sectional areas.

$$\dot{m}_a = (UA\rho)_a = 10.0(0.25)62.5 = 156.25 \ lb_m/s$$

$$\dot{m}_b = (UA\rho)_b = 50.0(0.1)62.5 = 312.50 \ lb_m/s$$

$$\dot{m}_c = \dot{m}_a + \dot{m}_b = 156.25 + 312.5 = 468.75 \ lb_m/s$$

(AIII-28)

If we assume the viscous drag force to be negligible in the force balance equation, AIII-26, the only unknown is the discharge pressure, P_2.

$$40.0(144)0.35 + \frac{312.50(50.0)}{32.17} + \frac{156.25(10.0)}{32.17}$$

$$= 0.35(144)P_2 + \frac{468.75(21.4)}{32.17}$$

$$P_2 = 44.41 \ psia$$

(AIII-29)

Note the increment in pressure over the mixing section is 4.41 psi. This is another example of pressure recovery. This increment is before the diffuser where the pressure will be increased even more. A real jet pump will be analyzed shortly in order to compare jet pump efficiency with that of a centrifugal pump.

Effect of temperature on density

In the above example, an assumption of constant density was made. It is worthwhile checking to see if the assumption was warranted. In order to do this, we will make use of the first law of thermodynamics for open (flowing) systems. The expressions, dh, h, and Δh refer to specific enthalpy, not irreversibility; H is total enthalpy and J is the conversion factor to mechanical energy units from thermal units (778.16 foot-pounds-force per Btu – one in SI units).

$$\delta q + \delta w_s = dh + \frac{UdU}{g_c} + \frac{g}{g_c}dZ \qquad \text{(AIII-30)}$$

Assume $dZ = \delta q = \delta w_s = 0$

$$0 = \int_1^2 dh + \frac{1}{g_c}\int_1^2 UdU = \Delta h_{12} + \frac{1}{g_c}\frac{\Delta U_{12}^2}{2}$$

$$\Delta h_{12} = -\frac{\Delta U_{12}^2}{2g_c}, \quad \Delta \dot{H}_{12} = -\dot{m}\frac{\Delta U_{12}^2}{2g_c}$$

$$\dot{m}_{a+b}\left(h_2 - h_{a+b}\right) = \frac{\dot{m}_a U_a^2 + \dot{m}_b U_b^2 - \dot{m}_c U_c^2}{2g_c J}$$

$$= \frac{156.25(10.0^2) + 312.5(50.0^2) - 468.75(21.4^2)}{2(32.17)778} = 11.63\,Btu/s$$

$$\left(h_2 - h_{a+b}\right) = 11.63/468.75 = 0.02\,Btu/lb_m$$

$$\Delta h = C_p dT \sim (1)\Delta T$$

$$\Delta T \sim 0.02°F$$

This very small change in temperature along the mixing section was due to the change in enthalpy that, in turn was due to the change in velocities. The actual change in a real case would be slightly greater because irreversibilities were ignored in the above analysis. However, one can see the order of magnitude is quite small. The assumption of constant density, therefore, is not bad. Note that, if the temperatures of the suction and driving streams are radically different, the density will change along the mixing section.

APPENDIX AIII | **Computations Involving Pumps for Liquids**

Analysis of jet pump efficiency

We will establish the equations necessary to compute the efficiency of a jet pump on the same basis as that of a centrifugal pump so we may compare the relative efficiencies. The affects of fluid irreversibilities, suction and discharge elevations and motive pressures will be investigated.

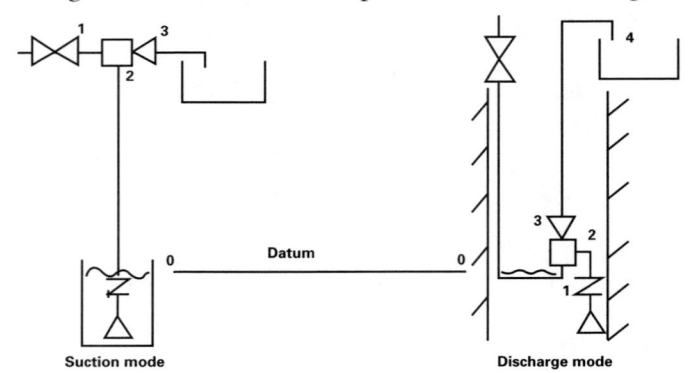

Figure AIII-10. Jet pump configurations

Figure AIII-10 gives examples of the two common modes of jet pump operations.

The following assumptions will be made: 1) A strainer and poppet foot valve are used, adding to the losses in the suction piping. This is done to prevent back flow of water to the suction tank and to limit the size of solids so as not to plug the jet pump. 2) The suction line must be at least 0.5 m longer than the suction lift. 3) The suction tank and the discharge line are at atmospheric pressure. This simplifies the analysis and it is quite a common situation. (When jet pumps are used as injectors, however, they do operate with different suction and discharge pressures.)

Definition of efficiency

In the case of a centrifugal pump, efficiency is the rate at which mechanical energy is transferred to the fluid divided by the rate at which it is transferred to the shaft. An efficiency of one would mean one hundred percent of the mechanical power to the shaft was transferred to the fluid. Efficiency is sometimes stated equivalently as the mechanical energy per unit mass flowing that is transmitted to the fluid divided by the mechanical energy per unit mass flowing that is transmitted to the shaft.

Note a distinction has been made between the two forms of energy transmitted. One form is mechanical energy and the other is thermal energy. Only the former is considered useful from the point of view of producing work. Note Carnot[ii] efficiency is defined as the ratio of the mechanical energy obtained from an engine to the total thermal energy delivered to it. In contrast, pump efficiency is defined in terms only of mechanical energy, fluid energy divided shaft energy.

A jet pump has additional mechanical energy in the discharge stream because of the presence of liquid from the motive stream. This additional energy should not be counted in the efficiency computation. The suction stream is the stream we wish to elevate to a higher energy level. Therefore, in order to compare a jet pump to a centrifugal pump, we should ratio the mechanical energy difference projected to the discharge and the suction flanges of a mass flow equivalent to the suction stream to the mechanical energy present at the motive fluid inlet.

The rate of transfer of energy from the driving stream is the energy per unit mass at the motive inlet times the mass flow rate of the motive stream. In what follows, the subscripts are 1) motive, 2) suction and 3) discharge. Mass flows 2) and 3) are equal if we discount the presence of the motive stream in the discharge.

$$\dot{E}_3 = \dot{m}_3 \left[\frac{P_3}{\rho} + \frac{U_3^2}{2g_c} + \frac{g}{g_c} Z_3 \right]$$

$$\dot{E}_2 = \dot{m}_2 \left[\frac{P_2}{\rho} + \frac{U_2^2}{2g_c} + \frac{g}{g_c} Z_2 \right]$$

$$\dot{E}_1 = \dot{m}_1 \left[\frac{P_1}{\rho} + \frac{U_1^2}{2g_c} + \frac{g}{g_c} Z_1 \right]$$

(AIII-31)

Note the definition of efficiency given here does not correspond to any of the definitions previously given for centrifugal pumps. Our definition is simply a way of measuring the ratio of the useful energy output to energy input for purposes of comparison with the efficiencies of centrifugal pumps. We have assumed the alphas equal to one.

The purpose of establishing the comparison is to try to put an objective number on efficiency from the point of view of energy used in accomplishing a specific task. We have already pointed out the choice of a jet pump is made usually on grounds other than efficiency.

APPENDIX AIII | Computations Involving Pumps for Liquids

The ratio of the increment in suction stream mechanical energy to the motive stream mechanical energy gives the relative efficiency as follows:

(AIII-32)

$$\eta = \frac{\dot{E}_3 - \dot{E}_2}{\dot{E}_1}$$

$$\eta = \frac{\dot{m}_2 \left[\dfrac{P_3}{\rho} + \dfrac{U_3^2}{2g_c} + \dfrac{g}{g_c} Z_3 - \dfrac{P_2}{\rho} - \dfrac{U_2^2}{2g_c} - \dfrac{g}{g_c} Z_2 \right]}{\dot{m}_1 \left[\dfrac{P_1}{\rho} + \dfrac{U_1^2}{2g_c} + \dfrac{g}{g_c} Z_1 \right]}$$

$$Z_2 \sim Z_3$$

$$\eta = \frac{\dot{m}_2}{\dot{m}_1} \frac{\left[\dfrac{P_3}{\rho} + \dfrac{U_3^2}{2g_c} - \dfrac{P_2}{\rho} - \dfrac{U_2^2}{2g_c} \right]}{\left[\dfrac{P_1}{\rho} + \dfrac{U_1^2}{2g_c} + \dfrac{g}{g_c} Z_1 \right]}$$

The development in equation set AIII-32 shows the relative efficiency of a jet pump for comparison with a centrifugal pump is equal to the ratio of the mass flow rates of the suction fluid to motive fluid times the ratio of the difference in specific energy across the pump to the specific energy in the motive stream. The elevation, Z_1, in the case of a jet pump operating in the discharge mode, is practically zero. The elevation enters into the efficiency computation only when the suction mode is used.

Computing the terms of the efficiency equation

Since the discharge pressure and the inlet pressure at the jet pump are rarely given in manufacturers' data sheets, they must be computed from the flow rates that are usually given versus motive pressures. The pressure at the discharge of the jet pump, P_3, must be back computed from known conditions at the end of the line. The pressure at the suction inlet to the jet pump, P_2, must be computed from the known conditions at the surface of the liquid in the suction tank.

The average sectional velocities can be computed from the given flow rates and internal dimensions of the connected piping. We have to make certain assumptions from manufacturers' recommendations. These recommendations are to minimize the number of elbows, especially in the discharge piping, and to include a strainer and check valve in the suction piping.

Computations Involving Pumps for Liquids | APPENDIX AIII

We will write the Bernoulli[i] equation with irreversibilities using subscripts (u) for upstream and (d) for downstream so as to fix our thinking. The loss coefficients will be identified by the summation sign. The terms that do not apply to the specific case will then be eliminated and the equation will be applied to the suction and discharge lines of the jet pump.

The result is

$$_u W_d = \frac{P_d}{\rho} - \frac{P_u}{\rho} + \frac{U_d^2}{2g_c} - \frac{U_u^2}{2g_c} + \frac{g}{g_c} Z_d - \frac{g}{g_c} Z_u + {}_d(h_L)_u \qquad \text{(AIII-33)}$$

$$_u W_d = \frac{P_d}{\rho} - \frac{P_u}{\rho} + \frac{U_d^2}{2g_c} - \frac{U_u^2}{2g_c} + \frac{g}{g_c} Z_d - \frac{g}{g_c} Z_u + \sum_i K_i \frac{U_i^2}{2g_c}$$

The last term in each equation is the energy per unit mass transformed to internal energy. These terms are identities. When they are equated, the Darcy[vii] equation is recovered.

The subscripts on the loss coefficients, K_i, refer to the diameter in which the velocity is measured. The velocity in the kinetic energy term must be associated with the same diameter. Equation AIII-33 is a general equation that would allow computations for work done by a jet pump to be equated to differences in static energy, kinetic energy, potential energy and to conversion of energy caused by piping components.

We will use the equation only for the suction or the discharge piping, so the work terms drop out of the equations.

For the two piping configurations shown in Figure AIII-10, the differences are one additional elbow in each of the suction and the discharge lines when in the discharge mode configuration plus the different lengths. The configurations differ by the numerical values of the various terms and by the fact the suction lift becomes discharge lift.

We will perform computations on a hypothetical case based on the second sketch of Figure AIII-10, the discharge mode, in order to compare approximate efficiencies of centrifugal pumps and jet pumps.

Suction piping in the discharge mode

For the suction piping in the discharge mode, the losses can be identified with the suction strainer and poppet foot valve, one standard elbow and the length of suction pipe. The loss coefficient for a typical poppet foot

APPENDIX AIII | Computations Involving Pumps for Liquids

valve can be obtained from Crane[xiii] as $420f_T$, where f_T is the standardized friction factor for nominal piping. For 25 mm (1 inch) nominal piping, the standardized friction factor is 0.023, so the K factor for the check valve-strainer combination is 9.66. That for the standard elbow is 0.023(30) or 0.69. The K factor for the piping is given by f_M (L/D). The Moody[viii] friction factor must be obtained from the fluid velocity, viscosity, pipe diameter and roughness. It can be looked up in charts or obtained from the Churchill-Usagi[xli] relationship. For the purpose of this exercise, we will use the standardized friction factor assuming fully turbulent flow.

For the suction piping between points 0 and 2, there is no pump. The liquid simply is forced by pressure differential from approximately zero velocity at the liquid surface to the velocity at the inlet to the jet pump. When the zero level is taken as the liquid level in the sump, the Bernoulli[i] equation reduces to

$$0 = \frac{P_2}{\rho} - \frac{P_0}{\rho} + \frac{U_2^2}{2g_c} + \frac{g}{g_c}Z_2 + \sum_i K_i \frac{U_i^2}{2g_c}$$

$$\frac{P_2}{\rho} = \frac{P_0}{\rho} - \frac{U_2^2}{2g_c} - \frac{g}{g_c}Z_2 - \sum_i K_i \frac{U_i^2}{2g_c}$$

(AIII-34)

A habit of many is to use kilopascals for pressure units and to divide all the other terms by 1,000 in order to obtain consistency. This leads to the possibility of forgetting the conversion unit.

Units in AIII-34 are pressure in pascals or psf, density in kilograms per meter cubed or lb_m/ft^3, velocity in meters per second or ft/s, acceleration, g, in meters per second per second or ft/s^2, elevation in meters or feet, pipe linear length in meters or feet and pipe diameter in meters or feet. The pipe length is not identical to the elevation change. In this case it is one half a meter longer. With SI units, the dimensional constant, g_c, is one and dimensionless. The K factor for the check valve is dimensionless.

Discharge piping in the discharge mode

The presence of an additional elbow in the discharge mode configuration should be noted. The upstream section is point 3 and the downstream section is taken through the stream of water as it emerges from the piping, section 4.

The loss coefficients are those associated with flow through the length of pipe, f_M (L/D), with the two elbows, $2(30)f_T$, (once more from Crane[xiii]) and the coefficient of the exit loss from the pipe, $K_{EXIT} = 1$.

Computations Involving Pumps for Liquids | APPENDIX AIII

The velocities at points 3 and 4 are identical, so the two kinetic energy terms drop out of the equation. The specific Bernoulli[i] equation with $Z_4 \neq Z_3$ becomes

$$0 = \frac{P_4}{\rho} - \frac{P_3}{\rho} + \frac{g}{g_c}Z_4 - \frac{g}{g_c}Z_3 + [f_M(L/D) + 2(30)f_T + 1]\frac{U_4^2}{2g_c} \quad \text{(AIII-35)}$$

$$\frac{P_3}{\rho} = \frac{P_4}{\rho} + \frac{g}{g_c}(Z_4 - Z_3) + [f_M(L/D) + 2(30)f_T + 1]\frac{U_4^2}{2g_c}$$

Equation AIII-35 can be made more general by replacing the bracketed term by the sum of the K_i for the discharge piping. The Moody[viii] friction factor will be replaced by f_T for convenience.

General expression for comparative jet pump efficiency

Starting with AIII-32, we can substitute the last of equations AIII-34 and AIII-35 into the efficiency relationship to eliminate the unknown pressures, P_2 and P_3.

$$\eta = \frac{\dot{m}_2}{\dot{m}_1} \cdot \frac{\frac{P_3}{\rho} + \frac{U_3^2}{2g_c} - \frac{P_2}{\rho} - \frac{U_2^2}{2g_c}}{\left[\frac{P_1}{\rho} + \frac{U_1^2}{2g_c} + \frac{g}{g_c}Z_1\right]} \quad \text{(AIII-32)} \quad \text{(AIII-36)}$$

$$\frac{P_3}{\rho} = \frac{P_4}{\rho} + \frac{g}{g_c}[Z_4 - Z_3] + [f_M\frac{L}{D} + 2(30)f_T + 1]\frac{U_4^2}{2g_c} \quad \text{(AIII-35)}$$

$$\frac{P_2}{\rho} = \frac{P_0}{\rho} - \frac{U_2^2}{2g_c} - \frac{g}{g_c}Z_2 - \sum_2 K_2 \frac{U_2^2}{2g_c} \quad \text{(AIII-34)}$$

$$\eta = \frac{\dot{m}_2}{\dot{m}_1} \cdot \frac{[\frac{P_4}{\rho} + \frac{g}{g_c}[Z_4 - Z_3] + [f_M\frac{L}{D} + 60f_T + 1]\frac{U_4^2}{2g_c} + \frac{U_3^2}{2g_c} - \frac{P_0}{\rho} + \frac{g}{g_c}Z_2 + \sum_2 K_2 \frac{U_2^2}{2g_c}]}{\left[\frac{P_1}{\rho} + \frac{U_1^2}{2g_c} + \frac{g}{g_c}Z_1\right]}$$

incompressible fluids, discharge mode $P_4 = P_0, U_4 = U_3, Z_3 \approx Z_2 \approx 0$

$$\eta = \frac{\dot{m}_2}{\dot{m}_1} \cdot \frac{\frac{g}{g_c}Z_4 + [f_M\frac{L}{D} + 60f_T + 2]\frac{U_3^2}{2g_c} + \sum_2 K_2 \frac{U_2^2}{2g_c}}{\left[\frac{P_1}{\rho} + \frac{U_1^2}{2g_c} + \frac{g}{g_c}Z_1\right]}$$

APPENDIX AIII | Computations Involving Pumps for Liquids

It is worthwhile to analyze the terms of the last equation of the set AIII-36. The first term on the right, the ratio of the mass flow rates, suction flow to driving flow, reflects the fact we are only interested in the fraction of mass equivalent to the suction flow. The second term, $g/g_c(Z_4)$, is the energy per unit mass corresponding to the height from zero to Z_4 through which a unit mass will be raised in the discharge pipe; the third term is the mechanical energy necessary to make up for discharge line irreversibilities; the fourth is the mechanical energy converted to thermal energy in the suction line.

In case there should be confusion regarding the multiplication of the suction mass flow rate by the discharge irreversibilities, the discharge losses *per unit mass* are a function of the actual velocity in the pipe. The friction factor, diameter and Reynolds[v] number must be based upon the actual velocities in the discharge and suction piping. This is in accord with our intention to compare useful output with input in both jet pumps and centrifugal pumps. We are trying to compare apples and apples.

Equation AIII-36 is a model for a working equation for computing the efficiency of a jet pump in the configurations given in Figure III-10. It can be modified for other situations.

Example of a real case based on a jet pump operating in the discharge mode

This is an example of a deep well ejector. We will use a manufacturer's published data for a 25-mm jet pump in our example. The size in millimeters is the nominal orifice size corresponding to a nominal one inch in diameter. For the piping we will use U.S. nominal Sch. 40, 1 inch pipe for the suction and discharge and U.S. nominal Sch. 40, ¾ inch pipe for the driving fluid. We will use nominal factors, f_T, from Crane[xiii] for both the Moody[viii] factor and the factor at complete turbulence to simplify the computations. The data is summarized in the table that follows.

Pipe, sch. 40 nominal diameters	Driving	Inlet	Discharge
	18.75 mm, 3/4 inch	25 mm, 1 inch	25 mm, 1 inch
Pipe length		1 m, 3.28 feet	6 m, 19.68 feet
Fittings		1 el, 1 strainer, 1 check valve	2 els, 1 exit
ΣK		10.52	7.56
Flow rate	2.2 m³/h, 9.69 gpm	2.1 m³/h, 9.23 gpm	4.3 m³/h, 18.92 gpm
Flow velocity	5.83 fps	3.43 fps	7.03 fps
Pressure	3.5 bar(g), 65.4 psia		

Computations Involving Pumps for Liquids — APPENDIX AIII

When a jet pump is operating in the discharge mode shown by the right hand sketch of AIII-10, the discharge piping and suction piping coefficients become

$$\sum_i K_i = \left(f_M \frac{L}{D}\right)_{DISCH} + 2(30)f_T + 1$$

$$\sum_j K_j = \left(f_M \frac{L}{D}\right)_{SUCT} + 9.66$$

The subscript, i, refers to discharge and the subscript, j, refers to suction. For clean commercial, one inch, schedule 40 steel pipe, f_T is 0.023. The suction length is one meter (3.28 ft) and the discharge pipe length is 6 meters (19.68 ft), so the total lift is 19.68 + 3.28/2 = 21.32 feet (one half the suction piping is submerged).

The nominal K factors for the suction and discharge piping are respectively (0.023)3.28/(1.049/12) + 9.66 =10.52 and (0.023)19.68/(1.049/12) + 2(30)0.023 +1 = 7.56. The number 9.66 is the K factor for the suction strainer-foot valve combination in the suction pipe. It may come as a surprise to find that the suction piping (called inlet piping) has a greater mechanical energy loss than the discharge piping when the discharge piping is longer. This is because the form factor influence of the strainer, elbow and foot valve is greater than that of the discharge fittings.

With a driving gauge pressure of 3.5 bar (65.4 psia), the flows at the suction (1 inch Sch. 40), at the driving fluid source (3/4 inch Sch. 40) and at the combined discharge (1 inch Sch. 40), from vendor's data, are 2.10 m³/h (9.23 gpm, 3.43 fps), 2.20 m³/h (9.69 gpm, 5.83 fps) and 4.30 m³/h (18.92 gpm, 7.03 fps).

$$\eta = \frac{2.1}{2.2} \frac{\left[\frac{g}{g_c}(21.32) + \frac{7.03^2}{2g_c} + 7.56\frac{7.03^2}{2g_c} + 10.52\frac{3.43^2}{2g_c}\right]}{\left[\frac{65.4(144)}{62.365} + \frac{5.83^2}{2g_c} + \frac{g}{g_c}1.64\right]}$$

$$\eta = 0.955 \frac{21.320 + 0.768 + 5.807 + 1.924}{151.123 + 0.528 + 1.64}$$

$$\eta = 0.955(29.219/153.291) = 0.182 = 18.2\%$$

The mass flow rate ratio has been replaced by the volumetric flow rate ratio because the densities are the same in this case.

APPENDIX AIII | Computations Involving Pumps for Liquids

It can be seen that the efficiency of a centrifugal pump which is normally around 60% compares with 18.2% in this case. This confirms the statement that the decision to use a jet pump is always made on the basis of practicality, not efficiency.

AIII-7: WORKED PROBLEMS

Examples of computed NPSHA in different units

Figure AIII-11 shows a centrifugal pump operating in a suction lift mode. The elevation change from the water surface to a pump center line is 20 feet. There is a foot check valve at the suction pipe inlet that is 3 feet below the water surface. The suction fittings include two screwed elbows and a screwed tee. Plug valves are used to valve in one sump or another. We will assume the worst case (greatest height, most resistance). The suction pipe size is 4 inch Sch. 40. We will assume clean commercial steel pipe. NPSHA is compared to NPSHR to see if the chosen pump will function.

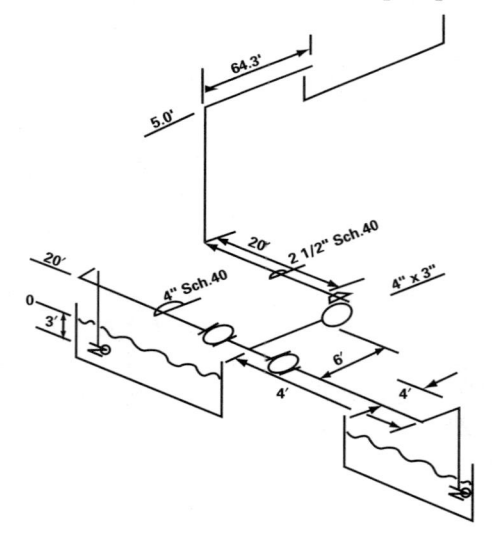

Figure AIII-11. Suction lift

The problem of priming is taken care of by filling the suction piping and pump casing with liquid, before starting the pump. We will consider this to have been done. We will do the first computation twice, at two different temperatures, to show the importance of the temperature variable. The first computation will then be repeated using SI units.

Computations Involving Pumps for Liquids — APPENDIX AIII

DATA, Case 1

Fluid = water
q = 100 U.S. gpm
T = 60°F
Vapor pressure = 0.257 psia
μ = 1.1 cP
ρ = 62.37 lb$_m$/ft^3
N_{Re} = 50.6qρ/dμ = 50.6(100)62.37/4.026(1.1) = 71,262 (Common mixed unit formula from Crane[xiii])

From the friction factor charts or, more accurately from the Churchill-Usagi[xli] relationships, the friction factor for the above Reynolds[v] number in 4-inch schedule 40 pipe, is obtained.

f = 0.0242   (From Churchill-Usagi, relative roughness = 0.00015)

We will use Crane's fully turbulent coefficients for the fittings and valves, since we are dealing with a hypothetical case. The first thing to do is to list all the components and to compute the resistance coefficients and to sum them.

Component	Computations		K
Pipe	0.0242(37/4.026)12	=	2.669
Els	2 x 30(0.017)	=	1.020
Tee (branch)	60(0.017)	=	1.020
Foot valve	420(0.017)	=	7.140
Plug valve	18(0.017)	=	0.306
	Sum K	=	12.155

Losses h_L = (Sum K)U^2/(2g$_c$) = 12.155(2.52^2)/(2 x 32.17)
                                          = 1.200 ft-lb$_f$/lb$_m$.

Using the last equation in the set AIII-24, the NSPHA can be computed as,

H_{sv} = 144(14.7 - 0.257)/62.37 - 1.200 - (g/g$_c$)20 = 33.346 - 1.2 - 20 = 12.146 ft-lb$_f$/lb$_m$.

The numbers show that, for cold water, the first term dominates and the second term is very small. The NPSHA depends on the piping configuration. If the NPSHR of the pump chosen is greater than the NPSHA, the pump is unsuitable for that elevation.

APPENDIX AIII | Computations Involving Pumps for Liquids

Comparison of the first and fifth equations of AIII-25 shows them to give numerically identical results. Only the units change. Hydraulicians would say we have computed net positive suction head available of 12.146 feet. This is the number we must compare with the stated NPSHR of the pump.

DATA, Case 2

In Case 2, we will change the temperature. This changes the viscosity, the density, the Reynolds^v number and the vapor pressure.
Fluid = water
q = 100 U.S. gpm
T = 120°F
Vapor pressure = 1.695 psia
μ = 0.53 cP
ρ = 61.713 lb_m/ft^3
N_{Re} = 50.6qρ/dμ = 50.6(100)61.713/4.026(0.53) = 145,345
f = 0.0228
K_{pipe} = fL/D = 0.0228(37/4.026)12 = 2.514
Sum K = 11.944
h_L = 12.0(2.52^2)/(2x32.17) = 1.184 ft-lb_f/lb_m.
H_{sv} = 144(14.7 - 1.695)/61.713 - 1.184 - (g/g_c)20 = 30.346 - 1.184 - 20 = 9.162 ft-lb_f/lb_m.

Note for a temperature change from 60°F to 120°F we have lost about 3 ft-lb_f/lb_m (or 3 feet) of available NPSH. If the computation is repeated for 180°F water, a not uncommon water temperature in pharmaceutical and biotechnology work, it will be found the computed NPSHA becomes negative. No pump would function at that elevation and temperature. The piping layout would have to be changed.

DATA, Case 3

In Case 3, we will repeat Case 1 with SI units.
Fluid = water
q = 100 U.S. gpm = 6.309 x 10^{-3} m^3/s
T = 60°F = 288.7°K
Vapor pressure = 0.257 psia = 1.772 kPa
μ = 1.1 cP = 1.1x10^{-3} Pa-s
ρ = 62.37 lb_m/ft^3 = 999.04 kg/m^3
D = 4.026/12 ft = 0.3355 ft = 0.10226 m
U = 2.52 fps = 0.768 m/s
N_{Re} = DUρ/μ = 0.10226(0.768)999.04/1.1 x 10^{-3} = 71 328
(The 0.1% discrepancy is due to rounding errors)

Computations Involving Pumps for Liquids — APPENDIX AIII

f = 0.0242   (Same number from Churchill-Usagi[xli]. Absolute roughness and Reynolds[v] numbers are pure numbers)

Computations	Components		K
Pipe	0.0242(11.2776/0.10226)	=	2.669
Els	2 x 30(0.017)	=	1.020
Tee (branch)	60(0.017)	=	1.020
Foot valve	420(0.017)	=	7.140
Plug valve	18(0.017)	=	0.306
		Sum K =	12.155

(pure numbers are the same as Case 1)

Losses $h_L = (\text{Sum K})U^2/2$ = 12.155(0.768²)/2
            = 3.586 J/kg

H_{sv} = (101 350 − 1 772.0)/999.04 − 3.586 − 9.805(6.096) = 99.674 − 3.586 − 59.771 = 36.317 J/kg

To obtain meters of manometric head from energy per unit mass units, divide by g = 9.805 m/s. This gives an equivalent metric manometric head of 3.704 meters or 12.152 feet. The minor discrepancy with Case 1 is due to rounding errors.

A few typical problems

We will use the data associated with Figure AIII-11 and the data of Case 1 to solve a few typical problems.

Problem statement 1

Given the piping configuration of Figure AIII-11, compute the shaft work to the pump, the fluid work, the pressure drop across the pump and the suction and discharge pressures. We take the general Bernoulli[i] equation of AIII-19 and simplify it based on known facts. The first problem will consider only fluid work. We make the following assumptions:

Assumptions

1. Point 2 is the cross section of the stream as it emerges from the pipe.
2. The mechanical energy "losses", h_{12}, are the combined suction and discharge piping irreversibilities, 1.2 + 8.8 = 10.0 ft-lb$_f$/lb$_m$.
3. The pressure at point 1 is the atmospheric pressure as is that at point 2.
4. The fluid densities can be taken as the same at both points (the thermal energy generated by the pump as conversion of mechanical energy will change the density a little, but engineering judgement will be used).

APPENDIX AIII | Computations Involving Pumps for Liquids

5. The velocity profile correction, alpha, will be approximate one because the flow is fully turbulent and we are not looking for ultimate accuracy.
6. The fluid velocity at point 1 is essentially zero (the tank is sufficiently large that the velocity at this point can be taken as zero).
7. The elevation at point 1 will be taken as zero (the elevation is not important per se, the difference in elevation is important; therefore, the elevation can be chosen to satisfy the requirements of the problem).
8. The centrifugal pump has an efficiency of 60%. This is a common guess when the real number is not known.

The Bernoulli[1] equation can be simplified, under the above assumptions to,

$$\eta w_s = \frac{g}{g_c} Z_2 + \frac{U_2^2}{2g_c} + h_{12}$$

$$\eta w_s = \frac{g}{g_c} 50 + \frac{6.7^2}{2g_c} + 10_{12} = 60.698 \; ft - lb_f / lb_m$$

$$w_s = \frac{60.698}{\eta} = \frac{60.698}{0.6} = 101.162 \; ft - lb_f / lb_m$$

This is the work transmitted to the shaft of the pump by the driver per unit of flowing mass. The driver will require more energy than this due to its own efficiency rating.

Problem statement 2

Find the differential pressure developed by the pump at the operating point given by the data of Figure AIII-11 and Case 1.

Assumptions

1. The discharge is on the same center line as the inlet.
2. Since only the pump is being considered we can set points 1 and 2 as being the inlet and discharge flanges of the pump. The losses are those of the pump itself, not of the piping.
3. The alpha correction factors for kinetic energy are again equal to one.

Computations Involving Pumps for Liquids | APPENDIX AIII

The Bernoulli[i] equation can be simplified for this case to,

$$\eta w_s = \frac{P_2}{\rho} - \frac{P_1}{\rho} + \frac{U_2^2}{2g_c} - \frac{U_1^2}{2g_c}$$

$$P_2 - P_1 = \rho\left[\eta w_s - \frac{U_2^2 - U_1^2}{2g_c}\right] = 62.37\left[60.698 - \frac{6.7^2 - 2.52^2}{2(32.17)}\right]$$

$$P_2 - P_1 = 3748.4 \, psf = 26.0 \, psi$$

Note the difference between kinetic energy terms in the above equation has a value of only 0.539. This fact leads many to ignore the terms altogether and to state that the shaft work is equal to the difference in pressure divided by the fluid density. This habit can lead to serious error when compressible fluid problems are considered.

Problem statement 3

Compute the power supplied to the pump (not the motor) in foot-pounds-force per second and in horsepower using the above-developed data.

There are no assumptions. The power supplied to the pump shaft is equal to the shaft work per unit mass of flowing fluid times the mass flow rate.

$$P_s = w_s \dot{m} = w_s A_1 U_1 \rho$$

$$P_s = 101.162(0.0884)2.52(62.37) = 1405.7 \, ft-lb_f/s$$

$$P_s = \frac{1405.7}{550} = 2.556 \, hp$$

Pump efficiency is not considered. This is not the power supplied to the driver or to the fluid. The power to the driver is this number divided by the driver efficiency. The power to the fluid is this number times the pump efficiency.

APPENDIX AIII | Computations Involving Pumps for Liquids

AIII-8: SUMMARY OF APPENDIX AIII

Appendix III has concentrated on giving the basic theory and equations of pumps in order to provide an understanding of their interactions with fluid systems. This knowledge will be expanded in later chapters and appendices.

- The centrifugal pump was treated in greater detail than any other pump, and the turbine was not treated. The turbine has a similar theory to the pump and is less prone to such problems as cavitation.

- It was pointed out the essence of any pump is the transfer of energy per unit mass to a fluid. It was also recommended to think in terms of these units instead of feet and meters of head units. Not only are energy units more logically satisfying, their use can help avoid pitfalls for the unwary.

- Enough theory was developed so the fairly simple governing equations could be accepted with some degree of comfort. This statement applies in particular to centrifugal pumps. Positive displacement pumps are best understood by the limitations imposed upon their application by the incompressible nature of liquids.

- Once the governing equations were developed, the differences between some of the different types of centrifugal pumps were more easily discussed. The importance of efficiency in determining the characteristics of some of the pumps was pointed out. Efficiency is not normally a concern to operations or control personnel, but it is worth considering because it does affect dramatically pump performance in more ways than just an economic one.

- The commonly used measures of performance were given with a number of examples. This was done in an attempt to clear away some of the misconceptions due to the use (and misuse) of units.

- The all important Bernoulli[i] equation was again stressed as being one of the most important tools for the solution of fluid flow problems – including those associated with pumps and turbines.

- The net positive suction head problem was dealt with in a fair amount of detail. Samples were given in different units. This was done for two reasons. The first reason is operations or control personnel do not normally spend too much time thinking about NPSH, and yet it is extremely important when trouble shooting pumping systems. The second reason is the different units used lead to confusion – they must be clearly defined and understood.

APPENDIX AIV

Equations Of Compressible Flow, Derivations And Applications

AIV-1: PURPOSE – PROVIDING CHAPTER IV EQUATION DETAILS

This appendix covers in much greater detail some of the equations of compressible flow given in Chapter IV. Relationships are described in detail for the following:

- the use of thermodynamic variables, in particular the enthalpy;
- adiabatic flow with irreversibilities, basic equations. (Equations that do not involve the concept of the Mach number will be fully developed. They will be applied to flow of fluid in straight conduit with irreversibilities. This group of equations is the most important within the context of this book.);
- the Peter Paige[xxiv] equation (This equation is an excellent tool to help understand flashing flow and the choking phenomenon. It will be developed and covered in some detail.);
- adiabatic flow with irreversibilities involving equations using the Mach number (Equations that involve the concept of the Mach number will be developed. They will be applied to flow of fluid in straight pipe with irreversibilities. This development is simply for completeness to give information to those who are interested and who may need it in their work. Most computations performed in the context of this book do not require the Mach number.);
- the concept of choked flow. This is critical to an understanding of compressible flow. The use of the Redlich-Kwong[xxviii] equation in order to obtain greater accuracy than one can obtain from the ideal gas equation will be described in detail.

APPENDIX AIV | Equations of Compressible Flow

AIV-2: USING THERMODYNAMIC VARIABLES – IN PARTICULAR, ENTHALPY

Enthalpy is a concept. It is a variable that is composed of the internal energy plus the pressure-volume energy. In other words, $h = u + Pv$, or $H = U + PV$ in total units. It stems from the study of open (flowing) systems. The variable, enthalpy, appears frequently in flow formulae.

Like other thermodynamic variables, enthalpy is not an absolute number, since it is obtained by integration. Enthalpy is measured as the difference between two quantities whose absolute values are unknown. Therefore, one of the quantities is fixed in an arbitrary reference state. The arbitrariness of the state makes no difference to the result, so long as we all agree on the reference state.

The reference state for a pure (single component) substance is chosen for convenience. It differs with different substances. For instance, the reference quantity for the internal energy of pure water is usually set at zero for liquid water at 0.01 degrees Celsius and 0.6113 kPa (the triple point is the reference state) for the steam tables, but *this choice is not the only possibility*. Since the enthalpy concept is derived from the internal energy, its reference point for water is also 0.01°C and 0.6113 kPa. Its reference value is not zero, however, since it has the value of the product Pv for the liquid.

Caveat — Reference states for thermodynamic properties

When comparing enthalpies or any other thermodynamic property from different sources, it is important to check the reference state. Otherwise, the numbers can be misleading.

The choice of different reference states for different substances does not present a problem for pure substances. For mixtures, it is more convenient to have a common reference state mainly for purposes of automatic computation. Prausnitz, Anderson and Grens use the ideal gas state at 300°K as the reference state for the enthalpy of mixtures.

Liquid enthalpies are found by integration from this ideal gas state to the liquid state. Vapor enthalpies are found by integration from the same reference state to the temperature of interest using the ideal gas specific heats and then using a departure function from the hypothetical ideal gas state to the real state.

Vapor enthalpies are important in fluid flow computation; liquid enthalpies are less important.

AIV-3: ADIABATIC AND IRREVERSIBLE FLOW IN UNIFORM CONDUITS – BASIC EQUATIONS

Bernoulli equation including 'lost work'

The Bernoulli[i] equation (including the lost work term, Equation IV-22) can be transformed by multiplying all terms by density squared. We can constrain it by assuming horizontal conduit and no compressor or turbine in the piping. This eliminates the potential energy term and the work term. With most changes in elevation in an industrial plant and with most gas or vapor densities, this is acceptable (but the assumption must be checked). We will use the assumption in order to reduce the complexity of the mathematics.

$$\frac{dP}{\rho} + \frac{UdU}{g_c} + \frac{U^2 f_M dL}{2g_c(4r_H)} = 0 \quad \textbf{(AIV-1)}$$

$$\rho dP + \frac{\rho^2 UdU}{g_c} + \frac{\rho^2 U^2 f_M dL}{2g_c(4r_H)} = 0$$

Mass velocity (flux)

Then we introduce the concept of mass velocity (mass flow per unit area or mass flux). Even with compressible fluids, the mass velocity along a constant cross section conduit is a constant in steady-state flow as can be seen by examining the units (lb_m/ft^2-s).

The following development only applies to constant cross section conduits.

$$G \equiv \rho U \quad \textbf{(AIV-2)}$$

$$U = G\rho^{-1}$$

$$dU = -G\rho^{-2} d\rho$$

$$UdU = -G^2 \rho^{-3} d\rho$$

APPENDIX AIV | **Equations of Compressible Flow**

Transformed Bernoulli equation

The above relationship is useful when dealing with compressible fluids, as it simplifies integration. At steady-state flow, the mass flux is constant throughout the length of a constant section conduit, even when the velocity changes, because it is the mass flow rate divided by the cross sectional area. This should be borne in mind for use with rectangular or other shaped ducts. We will concentrate on conduits of circular section. Therefore, $4r_H$ in equation AIV-1 will be replaced by the diameter, D. The velocity, U, can be eliminated from the transformed Bernoulli[i] equation, as follows

$$\rho dP - \frac{G^2 d\rho}{g_c \rho} + \frac{f_M G^2 dL}{2 g_c D} = 0 \tag{AIV-3}$$

In Equation AIV-3, since the mass velocity, G, is constant at steady state in a constant section conduit, even with compressible fluids, it can be taken outside an integral. The friction factor can be assumed constant over small steps. An average value can be used by estimating the initial and final value of the step and dividing by two. The above approach reduces the number of variables from 6 (ρ, P, U, f_M, L, D) to three (ρ, P, L). In case it is felt too many assumptions are being made, it is possible to check computations against actual tests or published results

Partially integrated, transformed Bernoulli equation

The partially integrated form of the transformed Bernoulli equation is given below. The first term is left under the integral until it is decided what further constraints to impose. In particular, it must be decided which equation-of-state will give the optimal accuracy. The relationship of density to pressure over a path must be known in order to complete the integration. For now, we will use this equation to analyze flow in conduits qualitatively.

$$\int_1^2 \rho dP - \frac{G^2}{g_c} \int_1^2 \frac{d\rho}{\rho} + \frac{f_{Mav} G^2}{2 g_c D} \int_1^2 dL = 0 \tag{AIV-4}$$

$$\int_1^2 \rho dP - \frac{G^2}{g_c} \ln \frac{\rho_2}{\rho_1} + \frac{f_{Mav} G^2}{2 g_c D}(L_2 - L_1) = 0$$

The first term is derived from the driving differential pressure energy. The second term is derived from the kinetic energy term in the governing Bernoulli equation. The third term contains the irreversibilities. All terms are related at the same end points, sections 1 and 2, of the integration. The terms, although derived from specific energy, no longer have units of energy per unit mass.

Equations of Compressible Flow | APPENDIX AIV

Perhaps a more common form of the equation is obtained by multiplying and dividing the second term by two. Note the change in signs and the change in subscripts on the densities.

$$-\int_1^2 \rho dP = \frac{G^2}{2g_c}\left[\ln\left(\frac{\rho_1}{\rho_2}\right)^2 + \frac{f_{Mav}}{D}(L_2 - L_1)\right] \quad \text{(AIV-5)}$$

Equations AIV-4 and AIV-5 are equivalent. Their use requires the choice of an equation-of-state to relate density, pressure, temperature and composition and the choice of a sufficiently short length of conduit, or sufficiently turbulent flow, so an average friction factor may be estimated. These two equations are useful for computer simulations.

AIV-4: THE PETER PAIGE EQUATION, CHOKED FLOW

Peter Paige[xxiv] transformed Equation AIV-5 to solve for the length of conduit necessary to balance a given change in the other two terms. He assumed, if the length were short enough, or if the pressure difference were small enough, the average density could be used in the first term above to simplify the integration. The equation is general. Constraints such as the adiabatic one arise only when estimating densities and the friction factor.

$$\left[\frac{-\int_1^2 \rho dP}{\frac{G^2}{2g_c}} - \ln\left(\frac{\rho_1}{\rho_2}\right)^2\right]\frac{D}{f_{Mav}} = L_2 - L_1 \quad \text{(AIV-6)}$$

$$L_2 - L_1 = \frac{2g_c}{G^2}\frac{D}{f_{Mav}}\left(-\int_1^2 \rho dP\right) - \frac{D}{f_{Mav}}\ln\left(\frac{\rho_1}{\rho_2}\right)^2$$

$$L_2 - L_1 = \frac{2g_c}{G^2}\frac{D}{f_{Mav}}\rho_{av}(P_1 - P_2) - \frac{D}{f_{Mav}}\ln\left(\frac{\rho_1}{\rho_2}\right)^2$$

The driving differential pressure establishing flow is essentially the pressure difference in the first term on the right of the last equation. The values of the densities in the second term on the right are strictly related to P and T at the corresponding sections designated by the subscripts on P, L, and ρ.

The average density and the average friction factor are also averages between the two corresponding sections. Note the pressure difference is the upstream less the downstream pressure – what is normally called differential pressure.

APPENDIX AIV | Equations of Compressible Flow

Equation AIV-6 can be used to estimate the length necessary to effect a given change in pressure and density. It can also be used to estimate a choking pressure. It is not only useful for gases and vapors, but it can be used for mixed phase fluids as well. In the latter case, the densities and friction factors must be those of the mixture.

Choked flow occurs with gases, vapors and mixed flows. It is the phenomenon whereby for fixed reservoir conditions, in spite of a lower downstream pressure, the mass flow rate in a conduit remains unchanged. The pressure, temperature and density within the exit from the conduit also remain fixed, in spite of the lower downstream pressure.

Note the expression "choked flow" refers to the uncoupling of the dependence of the mass flow rate in a segment and the downstream pressure, nothing more. The pressure at the most downstream section of a segment of conduit is higher than, and is independent of, that in the next downstream segment. The choked mass flow rate will be a function of all the usual variables, but the pressure in the most downstream section of a segment is substituted for the upstream pressure of the next segment or the sink pressure. Exit losses do not play a role in the flow computations for the segment in which choking occurs.

Derivation of choked flow equations based on observation

In an experiment performed on flashing water in a pipe, Benjamin and Miller[xxxii] used 1 psi as a constant pressure drop. They performed measurements on the lengths necessary to create this pressure drop in a pipe. The length necessary to create one psi of pressure drop upstream in the pipe was 20.6 feet. Downstream, close to the exit, it was only 0.57 feet.

These facts, when observed often enough with different fluids, allow us to reason that, for fixed pressure drops proceeding down the pipe, the average fluid density falls, causing the first term on the right of the last of Equations AIV-6 to become smaller. The second term on the right becomes smaller also, but at a slower rate as we proceed downstream. The length necessary to complete the length balance decreases towards zero.

The driving energy is shared between the kinetic energy and the conversion (the "losses") to internal energy and heat energy flow. The previous statement can be verified by considering the first equation of the set AIV-1. Since the distance over which the conversion takes place tends to zero and the location of the zero point is at the end of pipe, we can reason all of

Equations of Compressible Flow — APPENDIX AIV

the driving energy is converted to kinetic energy and none is left for conversion to internal energy (losses in mechanical energy) at the end of the pipe. Note the first law of thermodynamics equates energy change in a body to the sum of work energy flows and heat energy flows.

Joule clearly established the equivalence of mechanical energy and thermal energy. Mechanical energy is considered to be the only form of "useful" energy, and mechanical energy converted to thermal energy is considered to be a "loss".

The Bernoulli[i] equation with irreversibilities in differential form, Equation AIV-1, can be transformed to the following equation:

$$dP + \rho d\left(\frac{U^2}{2g_c}\right) + \frac{\rho f_M}{D}\left(\frac{U^2}{2g_c}\right)dL = 0 \quad \text{(AIV-7)}$$

At steady-state mass flow, compressible fluid still accelerates as it flows down a conduit. If we arbitrarily fix the differential pressure increment at some low value, say 1 psi or 1 kPa. We can examine the above equation for each segment of the conduit that has the same pressure drop.

Starting at the upstream section, it will be noticed the densities decrease progressively as we proceed down the conduit. The friction factor increases a little, but not much.

Equation AIV-7 can be integrated over a one unit pressure drop using average values of density, friction factor and velocity over the length of the increment in the last term to give equation AIV-8.

$$1.0 + \rho_{av}\frac{\left[U_2^2 - U_1^2\right]}{2g_c} + \frac{\rho_{av} f_{Mav}}{D}\frac{U_{av}^2}{2g_c}(L_2 - L_1) = 0 \quad \text{(AIV-8)}$$

The first term in Equation AIV-8 has units of pressure, as do all the other groups of terms. The difference between the squares of the outlet and inlet velocities of a segment is much smaller than the square of the average velocity in the length over which the pressure drop takes place. To maintain the validity of the equation as we proceed down the conduit, the difference in length has to be reduced because the average velocity squared is increasing at a greater rate than the difference of the squares. This conclusion is confirmed by observations (Benjamin-Miller[xxxii]).

APPENDIX AIV | Equations of Compressible Flow

The difference in length, associated with a fixed pressure drop necessary to effect a change in kinetic energy and to overcome permanent conversion of mechanical energy, decreases as we move down the conduit in the direction of flow. If the upstream pressure is sufficiently high, or the downstream pressure sufficiently low, and the conduit sufficiently long, the length increment can be reduced to approximately zero. The equation will still balance. Physically, we cannot go beyond this state – we cannot create negative length.

The choking phenomenon, which is the condition we are describing, will always take place at the end of the conduit. There is less resistance when the fluid suddenly leaves the conduit. The acceleration is greatest at this point and the conversion of mechanical to thermal energy is greatest for a minimal length increment.

Criterion for choked flow

A mathematical switch for the incremental method of establishing choked pressure

Mathematically, we could keep on integrating equation AIV-7 with negative differences in lengths. Physically, we cannot. This fact gives a nice switch for computer and calculator computations. We can approximate the point at which flow chokes by setting dL in equation AIV-7 equal to zero.

(AIV-9)

$$dP + \rho d\left(\frac{U^2}{2g_c}\right) = 0$$

$$dP + \rho \frac{U dU}{g_c} = 0$$

$$dP + \rho \frac{G}{\rho} \frac{dU}{g_c} = 0$$

$$\frac{dP}{dU} = -\frac{G}{g_c}$$

At the end of a constant cross section line, the change in pressure with change in velocity is equal to a constant, minus G/g_c when flow is choked. This conclusion stems from the observation that the length increment for a fixed pressure drop tends to zero at the end of the line or at a sudden expansion when flow is choked.

If we can write a general equation for the derivative in AIV-9 and equate it to the term on the right, we can solve for the choking (so-called critical) conditions. This "critical pressure" is the pressure immediately inside

Equations of Compressible Flow | APPENDIX AIV

the conduit exit when the length increment to cause an infinitely small pressure drop tends to zero. It is not the thermodynamic "critical pressure". Specific volume and velocity at this point are also of interest.

Note we now have a criterion for choked flow derived from thermodynamic considerations and observations of experimental facts. The criterion relates changes in two variables, pressure and velocity, to a constant mass flux. This criterion applies to all fluids under all conditions, not only the adiabatic one.

First law with constraints (adiabatic flow, horizontal conduit, no work)

Starting with the first law energy balance for open (flow) systems (true for all fluids, reversible or not) and assuming horizontal conduit (no elevation change), an adiabatic process, and no fluid work, we have equations AIV-10. The units of each term are thermal (Btu/lb$_m$) in the customary U.S. system.

(AIV-10)
$$\delta q + \delta w = dh + \frac{dU^2}{2g_c J}$$

$$0 = dh + \frac{dU^2}{2g_c J}$$

The dh term represents differential enthalpy. We can integrate the second differential equation of AIV-10 to obtain the following:

(AIV-11)
$$h_2 - h_1 + \frac{U_2^2}{2g_c J} - \frac{U_1^2}{2g_c J} = 0$$

$$h_2 + \frac{U_2^2}{2g_c J} = h_1 + \frac{U_1^2}{2g_c J} \equiv h_s$$

$$h + \frac{U^2}{2g_c J} = h_s$$

Adiabatic 'stagnation enthalpy'

The sum of the enthalpy and the kinetic energy is the same anywhere in the conduit under the adiabatic constraint. The kinetic energy terms and the enthalpy terms have identical units when the J factor is included. The sum of the flowing enthalpy and the kinetic energy is given the name "stagnation enthalpy". Stagnation enthalpy is the enthalpy that would be measured or estimated if the fluid were brought to rest adiabatically. At steady state in adiabatic flow, it is constant.

APPENDIX AIV | Equations of Compressible Flow

If there is no heat transfer across the conduit walls, the stagnation enthalpy is constant. We can use this fact to derive terminal conditions from known starting conditions. In particular, if the system is close to adiabatic (good insulation), the reservoir (source) enthalpy at the point where the velocity is close to zero can be used.

To simplify the development, we will use the ideal gas approximation with a constant heat capacity ratio, γ. In the set AIV-12, we define enthalpy, h, in Btu/lb_m in terms of the ratio of specific heats, γ, pressure, P, and specific volume, v. Since the term containing the ratios of specific heats is constant, we call it B to reduce the number of symbols. We then substitute for enthalpy in the stagnation enthalpy equation. Lastly, we convert to mass velocity and a differential equation is derived.

The differential equation gives the change in pressure with specific volume of an ideal gas under conditions of constant stagnation enthalpy.

The ratio of specific heats, γ, of an ideal gas is reasonably constant over normal adiabatic changes in conduits. For instance, the specific heat of steam (as an ideal gas) varies by 0.8% from 368°F to 317°F, a difference in temperature of 51°F. This corresponds to an adiabatic pressure drop from 170 psia to 39 psia or 131 psi in the example that will follow. For the sake of simplifying the differentiation, we can consider the ratio a constant. In what follows, the subscript, s, refers to the source or the stagnation condition; the subscript, o, refers to the reference temperature (which drops out of the equation).

From Equation AIV-12, the flowing temperature, T, will be lower than the stagnation temperature, T_s, because of kinetic energy's contribution to total energy. The gas constant, R', has units of $Btu/lb_m\text{-R}$ in the customary U.S. version of the ideal gas equation and is so used here. The dimensional conversion factor, J $ft\text{-}lb_f/Btu$, has been included and can be made equal to one, dimensionless, if SI units are used.

Also, if the gas constant must be replaced by the difference between the specific heats at constant pressure and at constant volume, it must be converted from mechanical to thermal units, because the specific heats are normally in thermal units in the customary U.S. system.

Equations of Compressible Flow — APPENDIX AIV

AIV-5: CHOKED FLOW USING THE IDEAL GAS EQUATION

Constrained first law including stagnation enthalpy (adiabatic model)

For an ideal gas flowing adiabatically, the constrained first law equation including the stagnation enthalpy is as shown as one of the last three equations of the set AIV-12.

Initially at least, we are interested in average velocity, U, at a section of conduit rather than specific volume, v, so we will effect a change in variables by using the mass flux relationship. The assumption of steady state makes the mass flux constant in a straight conduit.

The equation set AIV-12 makes use of "thermal" energy terms (Btu/lb$_m$) and mechanical energy units, ft-lb$_f$/lb$_m$, hence the conversion factor, J.

The last equation of the set AIV-13 is a first law equation for ideal gases. It simply states that the negative of the pressure change with velocity anywhere along a conduit with constant section is inversely proportional to velocity squared at a fixed mass flux. It uses mechanical energy units.

$$h + \frac{U^2}{2g_c J} = h_s$$

$$c_p(T - T_o) + \frac{U^2}{2g_c J} = c_p(T_s - T_o)$$

$$c_p T + \frac{U^2}{2g_c J} = c_p T_s$$

$$c_p T = c_p \frac{Pv}{R'J} = \frac{c_p}{(c_P - c_V)J} Pv = \frac{\gamma}{\gamma - 1} \frac{Pv}{J}$$

$$B \equiv \frac{\gamma}{\gamma - 1}$$

$$c_p T = B \frac{Pv}{J}$$

$$B \frac{Pv}{J} + \frac{U^2}{2g_c J} = c_p T_s$$

$$\frac{Pv}{J} + \frac{G^2 v^2}{2g_c JB} = \frac{c_p T_s}{B}$$

$$\frac{P}{J} + \frac{G^2 v}{2g_c JB} = \frac{c_p T_s}{B} v^{-1}$$

$$\frac{dP}{J} + \frac{G^2}{2g_c JB} dv = -\frac{c_p T_s}{B} \frac{dv}{v^2}$$

(AIV-12)

Note the first equation of the set AIV-12 imposes an adiabatic constraint on the development.

APPENDIX AIV | Equations of Compressible Flow

(AIV-13)
$$G = U\rho = \frac{U}{v}$$
$$v = \frac{U}{G}$$
$$dv = \frac{1}{G}dU$$
$$dP + \frac{G}{2g_cB}dU = -\frac{c_pJT_sGdU}{BU^2}$$
$$\frac{dP}{dU} + \frac{G}{2g_cB} = -\frac{c_pJT_sG}{BU^2}$$
$$-\frac{dP}{dU} = \frac{G}{2g_cB} + \frac{c_pJT_sG}{BU^2} = G\left[\frac{1}{2g_cB} + \frac{c_pJT_s}{BU^2}\right]$$

The fourth line of AIV-13 is derived by making substitutions for dv and v in the last line of AIV-12.

Computing ideal gas choked quantities (adiabatic conduit)

Equation AIV-9 can be used to make AIV-13 specific to conditions of an ideal gas at the end of the line when flow is choked.

(AIV-14)
$$-\frac{dP}{dU}\bigg|_{choked} = \frac{G*}{g_c} = G*\left[\frac{1}{2g_cB} + \frac{c_pJT_s}{BU^2}\right]_{choked}$$
$$1 = \frac{1}{2B} + \frac{g_cc_pJT_s}{BU*^2}$$
$$\frac{g_cc_pJT_s}{U*^2} = B - \frac{1}{2} = \frac{\gamma}{\gamma-1} - \frac{1}{2} = \frac{\gamma+1}{2(\gamma-1)}$$
$$U*^2 = \frac{2(\gamma-1)}{\gamma+1}g_cc_pJT_s$$
$$U* = \sqrt{\frac{2(\gamma-1)}{\gamma+1}g_cc_pJT_s}$$
$$G*v* = \sqrt{\frac{2(\gamma-1)}{\gamma+1}g_cc_pJT_s}$$
$$v* = \frac{1}{G*}\sqrt{\frac{2(\gamma-1)}{\gamma+1}g_cc_pJT_s}$$

Equations of Compressible Flow | APPENDIX AIV

The mechanical equivalent of heat, J, is only required when using customary U.S. units. The expression for U* in the equation set of AIV-14 gives the velocity at the conduit exit under choked conditions in an adiabatic conduit for an ideal gas. The expression for v* gives the specific volume under the same conditions for a given mass flux as a function of the stagnation temperature. We now need the critical (choked) pressures. Note the mass flux at the choked condition is imposed by the upstream fluid properties and conduit sizes and configuration. The asterisk/star (*) as a superscript has no real significance other than correspondence to the choked condition.

Choked pressures, ideal gas (adiabatic conduit)

Choked variables are the values of those variables within the exit from a conduit when the mass flow rate becomes constant in spite of lower downstream pressure. Choked variables are independent of the downstream variables.

The use of the ideal gas equation-of-state leads to relatively simple expressions for variables at the choked condition. Be warned that, although simpler to use than the equations for real gases, these expressions may be grossly in error for gases that are not ideal and for regimes that are not adiabatic.

We start with the third from last of the equations of the group AIV-12 and immediately transform it and differentiate it.

(AIV-15)

$$Pv + \frac{G^2 v^2}{2 g_c B} = \frac{c_p J T_s}{B}$$

$$Pdv + vdP + \frac{G^2 v}{g_c B} dv = 0$$

$$dv = \frac{dU}{G}$$

$$\frac{P}{G} dU + vdP + \frac{Gv}{g_c B} dU = 0$$

$$-\frac{dP}{dU} = \frac{P}{Gv} + \frac{G}{g_c B}$$

$$-\frac{dP}{dU}\bigg|_{choked} = \frac{G^*}{g_c} = \left[\frac{P}{Gv} + \frac{G}{g_c B}\right]_{choked}$$

$$1 = \frac{g_c P_{choked}}{G^{*2} v^*} + \frac{1}{B}$$

$$P_{choked} = \left[1 - \frac{1}{B}\right]\frac{G^{*2} v^*}{g_c} = \frac{G^{*2} v^*}{\gamma g_c}$$

APPENDIX AIV | Equations of Compressible Flow

Equations AIV-14 gave an expression for the choked specific volume, v*. This can be substituted into the last equation of the set AIV-15.

$$P^* = \frac{G^*}{\gamma g_c}\sqrt{\frac{2(\gamma-1)}{\gamma+1}g_c c_p J T_s} = G^*\sqrt{\frac{2(\gamma-1)}{\gamma^2(\gamma+1)}\frac{c_p J T_s}{g_c}} \quad \text{(AIV-16)}$$

A check of the units of Equation AIV-16 will show the choked pressure has units of pounds-force per foot squared.

Equation AIV-16 gives the choked flow pressure of an ideal gas as a function of the choked mass flux (the mass flux at the same point that choked pressure is established) and the stagnation enthalpy (specific heat times the absolute temperature at the source). The mass flux is usually obtained from the required design conditions, for instance, when it is necessary to relieve an amount of fluid under emergency conditions. If the design conditions do not impose the mass flow, it must be estimated, usually by iteration, after careful logical analysis of the situation. We will give the criteria shortly. The stagnation enthalpy can be estimated from the quiescent conditions in the source vessel – where the velocity is close to zero.

These equations are useful for quick estimates, for an understanding of gas behavior, or as the starting point in an iteration. If the estimated choked flow pressure for the assumed or estimated mass flux is less than the sink pressure, the flow is not choked under these assumed or estimated conditions and the downstream pressure for further flow computation is the sink pressure. If it is greater, flow is choked and the downstream pressure (within the conduit exit) to be used is the estimated choked pressure. If the flow is choked, the conduit exit loss coefficient is not used in conduit pressure drop computations. If the flow is not choked the exit loss coefficient must be included.

Proof that equation AIV-16 for choked pressure corresponds to the traditional equation for sonic velocity in an ideal gas undergoing an adiabatic process

The velocity of sound in an ideal gas is given traditionally as the square root of the product of the ratio of specific heats, the dimensional constant, the gas constant and the absolute temperature. This is true, but how do we obtain the temperature that would permit us to estimate the sonic velocity if the flow is choked? The above development gives one method.

Starting from the concept of stagnation enthalpy in AIV-11, we can develop the concept of stagnation temperature. This is the temperature an

Equations of Compressible Flow | APPENDIX AIV

ideal gas would reach if brought to rest adiabatically from its flowing condition. It is a temperature that can be approached in carefully designed experiments using a stationary thermal sensor in the flowing stream. The stagnation temperature is also equal to the reservoir temperature in a carefully insulated system.

For an ideal gas, the enthalpy is a function only of temperature. We can substitute the functional relationship into equation AIV-11 as follows:

(AIV-17)

$$h_1 + \frac{U_1^2}{2g_c J} = h_s$$

$$c_p T_1 + \frac{U_1^2}{2g_c J} = c_p T_s$$

$$T_s = T_1 + \frac{U_1^2}{2g_c c_p J} = T_2 + \frac{U_2^2}{2g_c c_p J}$$

The mechanical equivalent of heat, J, equals 778.16 ft-lb$_f$/Btu and has been included to remind the users of customary U.S. units to convert thermal energy units (Btu) to mechanical energy units (ft-lb$_f$). This constant is one in SI units. The numerical subscripts refer to (1) a fixed upstream section and (2) a fixed downstream section. The duplication of the term containing the average velocity squared is permitted because the stagnation temperature of an ideal gas is a constant in an adiabatic process, just as is the stagnation enthalpy.

The combination of terms involving velocity squared can be verified to have units of temperature. The kinetic energy corrections are assumed equal to one.

Traditional form of equation for speed of sound

The temperature form of the equation for acoustic velocity in terms of the Mach number is

(AIV-18)

$$U_a = \sqrt{g_c \gamma R T}$$

$$N_{Ma} = \frac{U}{U_a} = \frac{U}{\sqrt{g_c \gamma R T}}$$

$$N_{Ma}^2 = \frac{U^2}{g_c \gamma R T}$$

$$U^2 = g_c \gamma R T N_{Ma}^2$$

APPENDIX AIV | **Equations of Compressible Flow**

Equation AIV-18 uses a gas "constant" with mechanical energy units (ft-lb$_f$/lb$_m$). We can substitute the velocity, U, from AIV-18 into AIV-17.

(AIV-19)
$$T_1 + \frac{g_c \gamma R T_1 N_{1Ma}^2}{2g_c J c_P} = T_2 + \frac{g_c \gamma R T_2 N_{2Ma}^2}{2g_c J c_P}$$

$$\gamma = \frac{c_P}{c_v} = \frac{c_P}{c_P - R/J}$$

$$\gamma c_P - \gamma \frac{R}{J} = c_P$$

$$c_P(\gamma - 1) = \gamma \frac{R}{J}$$

$$\frac{J c_P}{R} = \frac{\gamma}{\gamma - 1}$$

$$T_1 + \frac{(\gamma-1) T_1 N_{Ma1}^2}{2} = T_2 + \frac{(\gamma-1) T_2 N_{Ma2}^2}{2}$$

$$T_1 \left[1 + \frac{(\gamma-1) N_{Ma1}^2}{2} \right] = T_2 \left[1 + \frac{(\gamma-1) N_{Ma2}^2}{2} \right]$$

$$\frac{T_1}{T_2} = \frac{\left[1 + \frac{(\gamma-1) N_{Ma2}^2}{2} \right]}{\left[1 + \frac{(\gamma-1) N_{Ma1}^2}{2} \right]}$$

The equation developed in AIV-19 gives the relationship for an ideal gas between the temperatures at two sections in a straight length of adiabatic conduit and the Mach numbers at the same two sections. For an ideal gas, we can arbitrarily pick point one as the source where the velocity, therefore the Mach number, is zero and we can pick point two as the choked flow point where the Mach number is one,

(AIV-20)
$$\frac{T_{source}}{T_{choked}} = \frac{\left[1 + \frac{(\gamma-1)}{2} \right]}{[1]} = \frac{\gamma + 1}{2}$$

Equation AIV-20 shows the relationship between the temperature in the vessel, the source, and the temperature at the choke point, the "critical" temperature, is a very simple function of the ratio of specific heats for an ideal gas undergoing an adiabatic expansion.

Link to other thermodynamic variables

Equation AIV-20 shows, if we can assume ideal gas behavior and an adiabatic process in an ideal, choked nozzle, we can estimate the temperature at the choke point. This may sound like a lot of ifs, but the second part, at least, is not so far fetched. An adiabatic process in an ideal, choked nozzle is relatively easy to approximate. Ideal gas behavior is more difficult to ensure.

Given the temperature at the choke point and the temperature at the source, we can relate other thermodynamic variables such as pressure and density to the temperature ratio during the same adiabatic process. For a detailed development of what follows, see any text on fundamental thermodynamics.

Essentially, any fluid undergoing an ideal, adiabatic process, as idealized by an insulated flow nozzle at the entrance to a pipe (or by a PSV), has its flowing properties related by $T_2/T_1 = [\rho_2/\rho_1]^{\gamma-1} = [P_2/P_1]^{(\gamma-1)/\gamma}$. As we already have fixed conditions in the reservoir, and we can estimate the choked temperature by AIV-20, we can estimate the choked density and pressure from the above property relationship. The choked average velocity, U, at the section is given by the fifth equation of the set AIV-14. The choked mass flux, G^* $lb_m/s\text{-}ft^2$, may be obtained from the product $U^*\rho^*$.

The stagnation temperature, T_s, from equation AIV-20, is substituted into equation AIV-16 and the mechanical energy gas constant, R, equals R´J.

(AIV-21)

$$P^* = G^* \sqrt{\frac{2(\gamma-1)}{\gamma^2(\gamma+1)} T^* \frac{\gamma+1}{2} \frac{c_p J}{g_c}} = G^* \sqrt{\frac{(\gamma-1)}{\gamma^2} T^* \frac{c_p J}{g_c}}$$

$$P^* = G^* \sqrt{\frac{(c_p/c_v - 1)}{(c_p/c_v)^2} T^* \frac{c_p J}{g_c}} = G^* \sqrt{\frac{(c_p/c_v - 1)c_v^2}{c_p} \frac{T^* J}{g_c}}$$

$$P^* = G^* \sqrt{\frac{(c_p - c_v)c_v}{c_p} \frac{T^* J}{g_c}} = G^* \sqrt{\frac{R'}{\gamma} \frac{J T^*}{g_c}} = G^* \sqrt{\frac{R T^*}{\gamma g_c}}$$

$$P^* = \frac{R T^*}{v^*} = \frac{U^*}{v^*} \sqrt{\frac{R\, T^*}{\gamma\, g_c}}$$

$$U^* = R T^* \sqrt{\frac{\gamma g_c}{R T^*}} = \sqrt{\gamma g_c R T^*}$$

APPENDIX AIV | Equations of Compressible Flow

The last equation in the group AIV-21 is the conventional one for sonic velocity in an ideal gas. It was derived by equating "critical" (sonic) pressure estimated from the ideal gas "law" with choked pressure estimated from the choked flow criterion, minus dP/dU equals G^*/g_c.

The "mechanical equivalent of heat", J, is a conversion factor required only when the "constant" R is in "thermal" energy units of Btu/lb_m-R.

The equations for properties of choked flow were derived without the use of the conventional equation for sonic velocity. The fact they reduce to the conventional equation for sonic velocity of an ideal gas helps establish their validity. They are also more useful than the canned equation, since they permit computations based on known upstream conditions. The conventional sonic velocity equation for an ideal gas simply tells you what the velocity is *if you know the corresponding temperature*.

Conclusions to be drawn for the adiabatic, choked flow condition

The following concepts apply quantitatively to ideal gases and qualitatively to real gases.

1. From AIV-14, the choked velocity of an ideal gas flowing adiabatically is a function only of conditions in the reservoir (source) – fundamentally of the enthalpy, practically of the temperature.

2. From AIV-17, the temperature at the choke point is also a function only of the conditions in the reservoir, but, in order to estimate temperature, the velocity must first be estimated. If there are multiple choke points, due to enlargements, the velocity and the temperature will be the same at each choke point from the velocity-stagnation temperature relationship of AIV-17.

3. The variables P^* (AIV-16) and v^* (AIV-14) are also functions of the reservoir conditions, but, in addition, the mass flux first must be established. The choked pressure, P^*, is proportional to G and the choked mass volume, v^*, is inversely proportional to G.

4. If flow is choked across an inlet restriction to the conduit (a PSV, for example), the mass flux at the throat is a function only of the reservoir conditions and the effective throat area. The flow might choke again at the exit from the conduit, but mass flow rate remains the same – it is governed only by the first choke point. In this latter case, a new value of G is estimated by dividing the constant mass flow rate by the area of the cross section. This value of G is used to estimate an upstream pressure to make

Equations of Compressible Flow | APPENDIX AIV

sure this pressure is below either the critical pressure for a balanced PSV or below the manufacturer's recommended limit for an unbalanced one.

5. To have multiple choke points, it is necessary to have additional conduits with increased cross sections. The smaller value of G in each additional segment means the specific volume at the choke point of each segment will be greater (AIV-14) and that the pressure will be less (AIV-16). Although G is different at each section and at the throat of any inlet nozzle, the mass flow rate is constant throughout all sections, if there are no branches.

6. If flow is choked only at a downstream point in a constant diameter pipe, then G is a function of the choked pressure, the reservoir pressure, the conduit L/D and relative roughness and the variable fluid density and viscosity. The exit loss is not considered.

7. If flow is not choked at all, then G is a function of the fluid, the reservoir conditions, the sink conditions, the conduit cross sectional area and L/D, the relative roughness and the exit loss.

8. The enthalpy at the source is a function only of the temperature at the source for an ideal gas.

The importance of the choking phenomenon should be evident from the above discussion.

Example AIV-1: Multiple choke points with ideal gases, adiabatic models

Figure AIV-1 shows a pressure relief valve with three horizontal lengths of different diameter conduit. If the conduit is insulated, or if it is short enough, the adiabatic model can be used. If the gas is hotter than the surroundings, the adiabatic model gives conservative results for safely sizing relief headers. If the gas is colder, the results are not conservative and heat transfer must be taken into consideration.

The example in Figure AIV-1 was chosen to demonstrate the choking phenomenon and for purposes of comparison with a real gas model. It is hypothetical. The use of an ideal gas model permits an analytic solution to the problem as opposed to an iterative one.

Note that, with an ideal gas flowing adiabatically with no heat energy flow to the environment, the temperatures at the choke points will be identical by Equation AIV-20. This temperature will be less than the source temperature for gamma greater than one. If the deceleration in the sink is adiabatic and isentropic, the source temperature will be recovered.

APPENDIX AIV | Equations of Compressible Flow

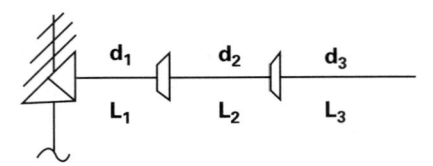

Figure AIV-1. Multiple choke points

Example AIV-1: (Continued from previous page)

With multiple choke points, one choke point (in this case it must be the PSV) will set the mass flow rate. This rate will be constant throughout the conduit. The mass flux will be different in each conduit segment, L_i. When sizing or checking relief header piping, one first sizes the relief valve to relieve the estimated mass flow rate based on the worst case for the vessel being protected – heat input, power loss, blocked discharge, etc. Then the smallest, commercially available relief valve that will pass this mass flow rate is chosen. A new mass flow is estimated based on the actual orifice size and the accumulated pressure in the vessel. The allowable accumulated pressure depends upon the vessel code. The allowable accumulated pressure will be a multiple of the design pressure such as 1.1 or 1.25, depending on the safety design basis. The new, greater, mass flow rate is the one that must be used to verify that the backpressure at the valve is not greater than that allowable for the type of valve, balanced or unbalanced.

In a situation such as the one depicted in Figure AIV-1, we cannot just assume flow will be choked only at the relief valve. The backpressure at the relief valve must be checked, because, if it is too great, it will prevent the valve from passing the required flow rate. We will show that, in the example, flow can choke at four locations: the PSV nozzle, the two expander inlets and the exit to the sink. Because flow chokes at the inlet to an expansion, the effective length of the downstream conduit is greater for pressure drop computations than the actual length by the equivalent length of the inlet swage.

The governing mass flux is G_I. This is the mass flow rate required of the PSV divided by the effective area of the PSV nozzle. The other mass fluxes are related to G_I by the ratios of the internal nozzle to internal conduit areas or the squares of the internal nozzle to internal pipe diameters.

Equations of Compressible Flow — APPENDIX AIV

Example AIV-1: (Continued from previous page)

$$\dot{m} = G_I A_I = G_1 A_1 = G_2 A_2 = G_3 A_3$$

$$G_1 = G_I \frac{A_I}{A_1} = G_I \left(\frac{d_I}{d_1}\right)^2$$

$$G_2 = G_I \left(\frac{d_I}{d_2}\right)^2$$

$$G_3 = G_I \left(\frac{d_I}{d_3}\right)^2$$

From Equation AIV-16 we can estimate the choked pressure at each expansion using the ideal gas assumption,

$$P_I^* = G_I \sqrt{\frac{2(\gamma-1)}{\gamma^2(\gamma+1)} \frac{c_p J T_s}{g_c}}$$

$$\frac{P_1^*}{P_I^*} = \frac{G_1}{G_I}$$

$$P_1^* = P_I^* \frac{G_1}{G_I}$$

$$P_2^* = P_I^* \frac{G_2}{G_I}$$

$$P_3^* = P_I^* \frac{G_3}{G_I}$$

Since the ratios of the mass fluxes are smaller sequentially, subsequent downstream choked pressures are lower. The choked pressure at the nozzle, P^*_I, is the first choked pressure. Subscripts 1 to 3 give the subsequent ones.

If the estimated backpressure at Segment 1 exceeds the maximum allowable backpressure for the particular PSV, then the pipe diameter must be increased for one or more segments of conduit. It is assumed the noise associated with "sonic" velocity can be tolerated for the length of time relief flow occurs.

APPENDIX AIV | Equations of Compressible Flow

Example AIV-1: (Continued from previous page)

The temperature of the fluid at the choke point can be estimated from Equation AIV-20. It is the same for all choke points for an ideal gas. It is proportional to the stagnation temperature and to a function of gamma, the ratio of specific heats.

$$T_I^* = \frac{2}{\gamma+1} T_s = T_1^* = T_2^* = T_3^*$$

The temperature also can be estimated for the nozzle by the following relationship. Both formulae give the same results for ideal gases. Note the presence of the exponent signifies an adiabatic process. See most elementary thermodynamic texts for a derivation.

$$T_I^* = \left[\frac{P_I^*}{P_s}\right]^{1-1/\gamma} T_s$$

The above equation is applicable only to the first choke point, the nozzle. Subsequent choke points have lower pressures but the same temperature. The densities at the choke points will differ because the choked pressures change.

Caveat — Integration with constraints

When performing an integration under an isothermal or an adiabatic constraint, it is well to remember the difference in the mechanisms. In the first, energy is exchanged to maintain isothermality. In the second, the energy remains with the mass flowing so the equilibrium properties are different due to a higher temperature in the adiabatic case.

Data for the example (ideal gas, pressure relief)

PSV: 1/2-inch ideal nozzle.
Pipe: Schedule 40, ϕ_1 = 3/4 inch, ϕ_2 = 1 inch, ϕ_3 = 1 1/2 inch, L_1 = 8 feet, L_2 = 1 foot, L_3 = 8 feet of equivalent length (lengths will differ due to entrance effects).
Source fluid: Ideal air at 30 atmospheres, 1000 R, ratio of specific heats is 1.4.

The mass flow through a choked nozzle is a function of the reservoir conditions and of the throat area. This may be obtained by multiplying the mass flux by the effective area of the nozzle throat. The mass flux is found from the development following AIV-20.

Equations of Compressible Flow — APPENDIX AIV

Example AIV-1: (Continued from previous page)

Choked pressure

The choked pressure for the first choke point, the PSV, the asterisk pressure, can be estimated directly for ideal gases from the equation that follows. The derivation will not be given because it is lengthy and it involves considerations that go beyond the intended scope of this book. Equation AIV-16, which was derived fully, could also have been used. The functional relationship is among the reservoir (source) pressure, the choked pressure and the ratio of specific heats only for ideal gases (ideal air, in this case).

$$r_c = \frac{P_I^*}{P_s} = \left[\frac{2}{1.4+1}\right]^{\frac{1}{1-1/1.4}} = 0.528$$

$$P_I^* = 0.528(30) = 15.840 \quad atm$$

The nozzle pressure is 15.84 atmospheres, 232.8 psia or 33,530 psfa.

Choked temperature for an ideal gas

The temperature at the choke point can be estimated from the following equation. The derivation of this equation will not be given for the same reasons cited above. This temperature is a function of the critical pressure ratio and the reservoir (source, subscript s) temperature. It can also be estimated by using equation AIV-17 that was derived fully.

$$T_I^* = \frac{P_I^{*1-1/\gamma} T_s}{P_s^{1-1/\gamma}} = \left[\frac{P_I^*}{P_s}\right]^{1-1/\gamma} T_s = \left[\frac{15.84}{30}\right]^{1-1/\gamma} 1000 = 833.33 \quad R$$

Choked densities

In order to estimate the densities at the choke points, we must first obtain the density of the gas in the reservoir from the ideal gas relationship.

$$Pv = \frac{P}{\rho} = RT$$

$$\rho_s = \frac{P_s}{RT_s} = \frac{30(14.7)(144)29}{1546(1000)} = 1.191 \quad lb_m/ft^3$$

APPENDIX AIV | Equations of Compressible Flow

Example AIV-1: (Continued from previous page)

The density at the first choke point is obtained from the following equation.

$$\frac{\rho}{\rho_s} = \left(\frac{P}{P_s}\right)^{1/\gamma}$$

$$\rho_I^* = \rho_s \left(\frac{P_I^*}{P_s}\right)^{1/\gamma} = 1.191 \left(\frac{15.84}{30}\right)^{1/1.4} = 0.755 \quad lb_m/ft^3$$

Choked velocity (speed)

The speed of sound in an ideal gas is given by the well-known "sonic" velocity equation as a function of the temperature and the ratio of specific heats only. It can also be found from equation AIV-21 by realising RT^* equals P^*/ρ^*.

$$U^* = \sqrt{g_c \gamma P^*/\rho^*} = \sqrt{32.17(1.4)15.84(144)14.7/0.755} = 1414.3 \quad ft/s$$

Choked mass flux

The mass flux, G_I^*, at the first choked flow point can be found from its definition. It must be remembered this mass flux is not constant throughout the nozzle and mass flux is not mass flow (which is constant at steady state).

$$G_I^* = U_I^* \rho_I^* = 1414.5(0.755) = 1067.8 \quad lb_m/s\text{-}ft^2$$

The internal diameters of the Sch. 40 pipes are 0.824, 1.049 and 1.610 inches. The mass fluxes downstream are 393.1, 242.5 and 103.0 lb_m/s-ft² respectively from the equation on page AIV-18.

Equations of Compressible Flow — APPENDIX AIV

Example AIV-1: (continued from previous page)

Choked mass flow

The mass flow rate is constant throughout the different diameters of conduit. The mass flux equals the mass flow divided by the area.

$$\dot{m} = G_i^* A_i = 1067.8 \left(\frac{\pi (0.5/12)^2}{4} \right) = 1.456 \quad lb_m/s$$

Choked pressures

The pressures at the downstream sections when flow is choked at each point may be estimated from AIV-16. The specific heat of air at constant pressure, c_p, is taken as 0.24 Btu/lb_m-F and gamma, the ratio of specific heats, is taken as 1.4.

$$P_I^* = G_I^* \sqrt{\frac{2(\gamma-1)}{\gamma^2(\gamma+1)} \frac{c_p J T_s}{g_c}}$$

$$\frac{G_i}{G_I} = \frac{A_I}{A_i} = \left(\frac{d_I}{d_i} \right)^2$$

$$P_1^* = 1068.2 \left(\frac{0.5}{0.824} \right)^2 \left[\frac{2(1.4-1)0.24(778.16)1000}{1.4^2(1.4+1)32.174} \right]^{1/2}$$

$$P_1^* = 12358.5\, psfa = 85.8\, psia = 5.84\, atm$$

$$P_2^* = P_1^* \left(\frac{d_1}{d_2} \right)^2 = 5.84 \left(\frac{0.824}{1.049} \right)^2 = 3.6 \quad atm \quad (7625.3\ psfa)$$

$$P_3^* = P_2^* \left(\frac{d_2}{d_3} \right)^2 = 3.6 \left(\frac{1.049}{1.610} \right)^2 = 1.53 \quad atm \quad (3237.1\ psfa)$$

APPENDIX AIV | Equations of Compressible Flow

Reynolds numbers, for pressure drops in conduit

The Reynolds[v] number for the ideal gas case will depend on the mass flux at the choke points and will vary with varying temperature upstream. In what follows, we ignore that variation for the sake of brevity.

To estimate the Reynolds number at the choke point, the viscosity at 833.33 R is required. Sutherland's general formula for viscosity is given by Crane[xiii] for several gases.

$$\mu_{cP} = \mu_{cP_o} \left(\frac{0.555 T_o + C}{0.555 T + C} \right) \left(\frac{T}{T_o} \right)^{3/2}$$

For air the reference temperature can be taken as 459.67 R. The corresponding reference viscosity is 0.016 centipoise. Crane gives the value of the constant, C, as 120. Using these values in the above formula gives a viscosity at 833.33 R of 0.0252 centipoise. The temperature and the viscosity at the choke points are constant for an ideal gas.

Crane gives a formula for the Reynolds number (which can be checked by the formulae given in this book) as follows:

$$N_{Re} = 123.9 \frac{dG}{\mu_{cP}}$$

Using the two preceding formulae, we obtain the following Reynolds numbers corresponding to the conduit sections at the choke points:

$$N_{Re,1} = 1.597 E6$$
$$N_{Re,2} = 1.254 E6$$
$$N_{Re,3} = 0.817 E6$$

From the Churchill-Usagi[xli] equation, Equation AII-10, the corresponding Moody friction factors are:

$$f_{M1} = 0.0241$$
$$f_{M2} = 0.0225$$
$$f_{M3} = 0.0206$$

Equations of Compressible Flow | APPENDIX AIV

Limitations of the Peter Paige equation: Two unknowns, second equation needed

The Peter Paige[xxiv] equation, Equation AIV-6, can be used for gases as well as for mixed flow. It is a force balance, second law (of thermodynamics). The equation is not limited to ideal gases. It can be rearranged as a functional relationship equal to zero.

$$F(P,\rho) = 0 = \frac{g_c}{G^2} \frac{D}{f_{Mav}} (\rho_1 + \rho_2)(P_1 - P_2) - \frac{D}{f_{Mav}} \ln\left(\frac{\rho_1}{\rho_2}\right)^2 - (L_2 - L_1)$$

Starting at the end of the conduit, fixed length increments can be chosen. The downstream choked conditions are known, if it exists. We have two unknowns in each length increment, the upstream pressure and the upstream density, P_1 and ρ_1. The mass flux can be fixed from the relief requirements.

If we start at the beginning of the conduit, the upstream conditions must be known and the unknowns are subscripted with a 2.

A second independent equation is needed in order to solve for the unknowns. The last equation of the set AIV-12 gives an independent first law relationship among the same variables for adiabatic flow. It can be integrated and the specific volumes can be transformed to densities. This equation is limited to ideal gases.

$$\int_1^2 dP + \frac{G^2}{2g_c B} \int_1^2 dv = -\frac{c_p JT_s}{B} \int_1^2 \frac{dv}{v^2}$$

$$P_2 - P_1 + \frac{G^2}{2g_c B}\left(\frac{1}{\rho_2} - \frac{1}{\rho_1}\right) = \frac{c_p JT_s}{B}(\rho_2 - \rho_1)$$

$$E(P,\rho) = 0 = (P_2 - P_1) + \frac{G^2}{2g_c B}\left(\frac{1}{\rho_2} - \frac{1}{\rho_1}\right) - \frac{c_p JT_s}{B}(\rho_2 - \rho_1)$$

The functional relationships, $E(P_1,\rho_1)$ and $F(P_1,\rho_1)$ are derived respectively from a first law energy balance and a second law force balance. The first law equation is independent of considerations of losses (irreversibilities), the second law equation includes them.

Newton-Raphson iteration, an algorithm from Balzhiser, Samuels and Eliassen

A very useful little book no longer in print, *Chemical Engineering Thermodynamics*, by Balzhiser, Samuels and Eliassen (BS&E)[xlv] gave an algorithm for the solution of simultaneous equations of two variables when the partial derivatives are available.

APPENDIX AIV | Equations of Compressible Flow

As always when there are two unknowns, two independent equations must be solved, simultaneously. The two equations can be arranged as two functional relationships, each equal to zero. The functions can be represented as two curves in three dimensional space as in Figure AIV-2.

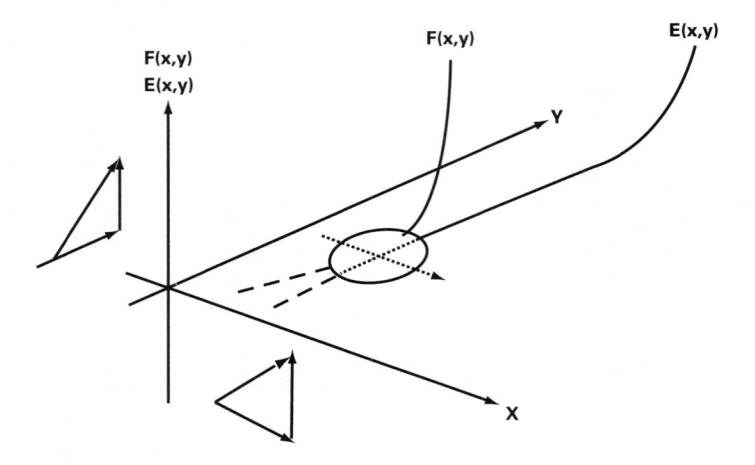

Figure AIV-2. Simultaneous equations — functions of two variables

It is to be noted that setting the value of the function to zero only applies to a single point. Each pair of values of the independent variables will result in an associated value of the function. The expression "each pair" refers to the values at the same cross section of conduit. The single point at which the two curves cross the horizontal plane gives the simultaneous values of the two independent variables. The two partial derivatives of each of the two functions (four partial derivatives) give approximations to the change in distance along their respective axes. The approximations become more and more exact as the curves approach the zero transition. The BS&E algorithm gives the zero transition point.

Equations of Compressible Flow | APPENDIX AIV

The two vector triangles in Figure AIV-2 represent the partial derivatives along their respective planes.

$$E(P,\rho) = 0$$

$$F(P,\rho) = 0$$

$$\left(\frac{\partial E(P,\rho)}{\partial \rho}\right)_P = x$$

$$\left(\frac{\partial E(P,\rho)}{\partial P}\right)_\rho = y$$

$$\left(\frac{\partial F(P,\rho)}{\partial \rho}\right)_P = p$$

$$\left(\frac{\partial F(P,\rho)}{\partial P}\right)_\rho = q$$

$$Det = xq - yp$$

$$\Delta P = \frac{E(P,\rho)p - F(P,\rho)x}{Det}$$

$$\Delta \rho = \frac{F(P,\rho)y - E(P,\rho)q}{Det}$$

The two functions equal zero when the values of P and ρ are correct simultaneously. The new values of P and ρ for the next iteration are P + ΔP and ρ + Δρ. The algorithm is remarkably robust for well-behaved, smooth functions. We will use it for the solution of the example problem.

Note the partial derivatives will have different values depending on whether the upstream or downstream values are held constant.

The term, Det, means "determinant".

Partial derivatives of the equations E(P,ρ) and F(P,ρ)

The terms x, y, p and q of the above set are required in order to implement the algorithm. Those terms used for an ideal gas when the upstream quantities change are given below. It should be apparent, and it will become more apparent, solutions to complex problems involving multi-component mixtures only can be obtained on a computer.

APPENDIX AIV | Equations of Compressible Flow

$$x = \frac{\partial E(P,\rho)}{\partial \rho_1} = \frac{G^2}{2g_c B \rho_1^2} + \frac{c_P JT_s}{B}$$

$$y = \frac{\partial E(P,\rho)}{\partial P_1} = -1$$

$$p = \frac{\partial F(P,\rho)}{\partial \rho_1} = \frac{2D}{f_{Mav}}\left[\frac{g_c}{2G^2}(P_1 - P_2) - \frac{1}{\rho_1}\right]$$

$$q = \frac{\partial F(P,\rho)}{\partial P_1} = \frac{g_c D}{G^2 f_{Mav}}(\rho_2 + \rho_1)$$

Tabulated results of simulation of Example AIV-1 (ideal gas, adiabatic model)

Following are the tabulated results of the simulation of Example AIV-1. The choked properties are those of the downstream sections: P2(1,1), P2(2,1), P2(3,1), T2(1,1), T2(2,1), T2(3,1), U2(1,1), U2(2,1), U2(3,1), and RHO2(1,1) RHO2(2,1) RHO2(3,1). The numbering system used is that the first letter or letter series signifies the variable; the first number indicates upstream, 1, or downstream, 2; the first number in brackets gives the conduit segment and the second one gives the increment within the segment.

As has already been pointed out, for an ideal gas, the velocities and temperatures, U2(i,1) and T2(i,1) at each choke point are identical to one another, the pressures, the densities and the temperatures fall as we proceed downstream. The velocities increase.

One phenomenon not clear in the results of the simulation is that the sink temperature, on the basis of an ideal gas undergoing an adiabatic process, is the same as the source temperature, in spite of the lower, intermediate temperature at the choke points.

Each conduit segment was divided arbitrarily into twenty lengths (counting upstream) for the simulation. A comparison will show the upstream pressures of each segment are lower than the corresponding upstream choked pressures of the next, upstream, segment; that is P1(1,20)<PI (32336<33530), P1(2,20)<P2(1,1) (11857<12258), P1(3,20),<P2(2,1) (7198<7625) and Patm<P2(3,1) (2116<3237).

The simulation shows how the variables behave throughout the header.

Equations of Compressible Flow | APPENDIX AIV

Simulation of Example AIV-1

Pipe segment counting upstream from choke point

J ->	1	2	3	4	5	6	7	8	9	10	11	12	13	14	15	16	17	18	19	20
Upstream pressures, PSFA																				
P1(1,J)	16610	18418	19816	21001	22052	23005	23886	24709	25484	26220	26922	27595	28241	28865	29468	30053	30620	31173	31711	32236
P1(2,J)	8455	8803	9071	9298	9498	9679	9845	10001	10147	10285	10417	10543	10664	10780	10893	11002	11107	11210	11310	11407
P1(3,J)	4077	4432	4707	4939	5145	5332	5504	5665	5816	5960	6097	6228	6354	6476	6593	6707	6818	6925	7030	7133
Downstream pressures, PSFA																				
P2(1,J)	12358	16610	18418	19816	21001	22052	23005	23886	24709	25484	26220	26922	27595	28241	28865	29468	30053	30620	31173	31711
P2(2,J)	7625	8455	8803	9071	9298	9498	9679	9845	10001	10147	10285	10417	10543	10664	10780	10893	11002	11107	11210	11310
P2(3,J)	3237	4077	4432	4707	4939	5145	5332	5504	5665	5816	5960	6097	6228	6354	6476	6593	6707	6818	6925	7030
Downstream temperatures, R																				
T2(1,J)	834	894	911	921	928	934	939	943	946	949	952	954	956	958	960	961	962	964	965	966
T2(2,J)	834	857	865	871	876	880	884	887	890	892	895	897	899	901	902	904	906	907	908	910
T2(3,J)	834	882	897	907	914	920	924	929	932	935	938	940	942	945	946	948	950	951	953	954
Downstream velocities, FPS																				
U2(1,J)	1415	1129	1037	975	927	889	856	828	804	781	762	743	727	712	697	684	672	660	649	639
U2(2,J)	1415	1312	1272	1243	1220	1200	1182	1166	1151	1138	1126	1114	1103	1093	1083	1074	1065	1057	1049	1041
U3(3,J)	1415	1189	1112	1059	1017	982	953	927	904	884	865	848	832	817	803	790	778	767	756	746
Downstream densities, LBm/FT3																				
RHO2(1,J)	0.278	0.348	0.379	0.403	0.424	0.443	0.459	0.475	0.490	0.503	0.516	0.529	0.541	0.553	0.564	0.575	0.585	0.596	0.606	0.615
RHO2(2,J)	0.172	0.185	0.191	0.195	0.199	0.202	0.205	0.208	0.211	0.213	0.216	0.218	0.220	0.222	0.224	0.226	0.228	0.230	0.231	0.233
RHO2(3,J)	0.073	0.087	0.093	0.097	0.101	0.105	0.108	0.111	0.114	0.117	0.119	0.122	0.124	0.126	0.128	0.130	0.132	0.134	0.136	0.138

APPENDIX AIV | Equations of Compressible Flow

AIV-6: ADIABATIC CHOKED FLOW; P, v, T RELATIONSHIPS USING THE REDLICH-KWONG EQUATION

Example AIV-2: Multiple choke points with a real gas, incremental solution

We will give an example based on a real gas. The gas chosen is air and the choice was made in order to permit a comparison of two different methods of estimation of a similar problem. The piping in example, AIV-2, has the same configuration and diameters as that given in Figure AIV-1 and the conditions in the reservoir (source) and the mass flow rate through the PSV are identical. The major difference is air, although almost an ideal gas, is treated as consisting of three components: nitrogen, oxygen and argon, in order to test the R-K EOS for mixtures against the previous results for air as an ideal gas. The trace amounts of water vapor and carbon dioxide have been lumped into nitrogen.

This section is concerned with developing relationships for choked flow using the Redlich-Kwong[xxviii] equation-of-state. We are no longer dealing with an ideal gas. The original R-K EOS is one of the simpler ones but it gives surprising accuracy even for mixtures. The derivations necessary for its implementation are far from simple. These derivations will be avoided by using the Peter Paige[xxiv] equation and the fact the incremental length to balance this equation tends to zero at the end of a straight segment when the flow are choked. The derivations of the analytical choked flow equations, however, are given in the paper, *Choked Flow* by *R. Mulley* (Available from ISA).

The Peter Paige equation was given in Section AIV-4 as Equation AIV-6. The use of the Bernoulli[i] equation, from which the Peter Paige equation was derived, to create a mathematical switch was discussed in the same section. The mathematical switch gives the signal for incipient choked flow and avoids the necessity of some onerous mathematical derivations.

The isenthalpic flow relationship will be used to estimate changes from the starting point, the source. It is,

$$h_T = h_s = h + \frac{U^2}{2g_c J} = h + \frac{G^2 v^2}{2g_c J} \tag{AIV-22}$$

The enthalpy of an adiabatic process with zero flow is symbolized by h_T or h_s (or h_{ST}), the total or stagnation enthalpy (mass or molar basis). It is equal to the enthalpy at the source in the case of an adiabatic network.

Example AIV-2: (Continued from previous page)

It also is equal to the enthalpy at a downstream section plus the kinetic energy at the same section. The kinetic energy can be expressed either as a function of the average velocity across a section or as a function of the mass flux and specific volume. The customary U.S. units of enthalpy are Btu/lb_m and J is 778.16 ft-lb_f/Btu. The SI units resolve to N·m/kg or J/kg and the coefficients J and g_c equal one and are dimensionless. The coefficient, J, is also equal to one and it is dimensionless when the units of enthalpy are foot-pounds-force per pound-mass.

The enthalpy at the source is estimated from the departure function in the next section.

Enthalpy departure function

The enthalpy departure function estimates the change in enthalpy from the ideal to the real gas state at the system temperature. It gives a means of using an ideal gas relationship to estimate the enthalpy change from the ideal gas state at a base temperature to the system temperature and then it adds the departure function to obtain the real enthalpy. The departure function may be established in terms of any of the equations-of-state.

The reader is reminded once more that enthalpy is not absolute. Its value will depend not only on the units chosen, but also on the temperature chosen as a base.

The original R-K equation (Reid, Prausnitz, and Poling[xxix], *Fourth Edition*) gives the enthalpy of non–ideal gas mixtures as follows in equation AIV-23.

$$\underline{h} = \int_{T_0}^{T} c^o_{Pmix} dT + (\text{departure function}) \quad J/g-\text{mole} \quad \textbf{(AIV-23)}$$

$$\underline{h} = (T-T_0)\sum_i y_i A_i + \frac{(T^2 - T_0^2)}{2}\sum_i y_i B_i$$

$$+ \frac{(T^3 - T_0^3)}{3}\sum_i y_i C_i + \frac{(T^4 - T_0^4)}{4}\sum_i y_i D_i$$

$$+ \left[P\underline{v} - \underline{R}T - \frac{3a}{2b}\frac{1}{T^{0.5}} \ln \frac{v+b}{v} \right] \frac{8.3144}{83.144} \quad J/g-\text{mole}$$

The first four terms on the right of the second equation of the set AIV-23 give the ideal gas mixture molar enthalpy in J/g-mole. The last three terms give the departure of the enthalpy from the ideal gas state of the

APPENDIX AIV | Equations of Compressible Flow

Example AIV-2: (Continued from previous page)

mixture. The four coefficients given in Reid, Prausnitz and Poling are the coefficients of the isobaric heat capacity for ideal gases. The units of the heat capacity coefficients give enthalpy in J/g-mole. The units of all the terms of the departure function are bar·cm^3/g-mole. So, pressure is in bars and molar volume is in cm^3/g-mole. Multiplying the departure function by the ratio of the two appropriate "universal" molar gas constants converts the departure function to the same units as the ideal gas function, joules/g-mole. The units of the conversion factor (the ratio) are (J/g-mole-K)/(bar-cm^3/g-mol-K) or joules/bar-cm^3.

The author apologizes for harping on about terminology, but he has found that the habit of otherwise intelligent people of playing fast and loose with logic and common sense serves as an impediment to intellectual progress.

Note in the above book by Reid, Prausnitz and Poling, the term, $T^{0.5}$, in the denominator of the R-K equation is included in the coefficient, a. This leads to errors if one is not fully aware of the fact. In this book we will separate the system temperature from the term, a. The coefficient of attraction, a, of the original R-K equation is $0.427\,48\,\underline{R}^2 T_c^{2.5}/P_c$ and the coefficient of repulsion, b, is $0.08664\,\underline{R}\,T_c/P_c$. Neither coefficient is a function of the system temperature, T. The reference temperature, T_0, is usually 298.2°K (25°C, the usual temperature in laboratories). The units of R underbar, T_c and P_c are bar-cm^3/g-mole-K, K and bars. Be aware of the sudden switch to a molar basis (g-moles). On a mass basis (grams), h = $\underline{h}$/(MW).

The coefficients of the terms, a and b, are pure numbers, but the other components of the terms possess units. In the set AIV-23, a and b are found using T in degrees K and P in bars with R underbar in bar-cm^3/g-mole-K.

Equation AIV-23 gives a way of estimating a real enthalpy at a fixed temperature starting from ideal conditions at a reference temperature. Note enthalpy, in this equation, is a function of P, T, $\underline{v}$ and composition.

The numerical value of the enthalpy depends upon the units chosen and on the reference state. In this case, the reference state is an ideal gas at some stated temperature, subscripted with a zero. The reference temperature can be any convenient temperature. Enthalpy is not "absolute".

The units chosen for the accompanying simulations were customary U.S. units.

Example AIV-2: (Continued from previous page)

All terms in equation AIV-23 are on a per unit g-mole basis. The coefficients of the ideal gas heat capacity terms, A_i, B_i, C_i and D_i are those of individual components of a mixture. They can be found for a large number of components in the Property Data Bank of Reid, Prausnitz and Poling. The moles referred to are g-moles for the mixture. The R-K coefficients, a and b, refer to the mixture as do $\underline{v}$ and $\underline{R}$. The reader is reminded to check each term for consistency of units.

Equivalent terms for equation AIV-23

On the last line of AIV-23, the first two terms, $P\underline{v} - \underline{R}T$, the difference between the real molar $P\underline{v}$ energy and the ideal energy, can be replaced by their equivalent difference as expressed in AIV-24.

$$\frac{b\underline{R}T}{\underline{v}-b} - \frac{a}{\left(\underline{v}+b\right)T^{0.5}} \quad \text{(AIV-24)}$$

The use of one or the other expression depends upon whether or not one wishes pressure to appear explicitly. Equation AIV-24 can be proved by substituting the Redlich-Kwong[xxviii] equation into $P\underline{v}$-$\underline{R}T$ and simplifying the result. By substituting equation AIV-24 into the set AIV-23, the enthalpy of the source for an adiabatic system can be found as a function of T and $\underline{v}$. The velocity or the mass flux, in this case, is equal to zero.

A more accurate heat capacity correlation (if needed)

A more accurate equation than that used in AIV-23 for the ideal gas heat capacity between the strict limits of 200°K and 600°K (the normal industrial temperature range) is,

$$\underline{c}_p^o = D_{1i} + \frac{D_{2i}}{T} + D_{3i}T + D_{4i}\ln T \quad \text{(AIV-25)}$$

The coefficients for this equation can be found in Prausnitz, Anderson and Grens. The units of specific heat in this case are J/g-mole-K.

To convert enthalpy in J/g-mole to Btu/lb-mole, multiply $\underline{h}$ by I = 0.43021.

Equation AIV-25 is given only for completeness. It was not used in the simulations associated with the problems given in this book.

APPENDIX AIV | Equations of Compressible Flow

Example AIV-2: (Continued from previous page)

Minimum number of simultaneous equations required

For a fluid of known or fixed composition, a thermodynamic relationship is one among three variables: P, T and $\underline{v}$. From Willard Gibbs' phase rule, we only need to specify any two of the thermodynamic variables of a single phase, single or fixed component system, before all other variables become fixed. We need two independent equations for simultaneous solution. For uniform section segments, the first independent equation comes from the first law of thermodynamics for flowing systems. This law defines the energy relationships of a fluid. The second independent equation comes from the second law, which describes the direction in which first law changes tend naturally. In our case, we will use the adiabatic constraint to simplify the mathematics and to conform more closely to industrial reality.

Energy relationship among variables, E(T,v), for adiabatic flow (first law)

The first of the two equations needed is derived by substituting equation AIV-23, the R-K description of enthalpy, into AIV-22, the isenthalpic (adiabatic) flow relationship. It is transformed to a functional relationship for later use in a Newton-Raphson iteration, E(T,v).

All terms have to be on a consistent basis. We choose a mass basis (lb_m or kg). The reader would be wise to check all the conversions before using the algorithm. The energy units will be Btu's or joules. We must check equation AIV-26 to see all the algebraic summations are Btu/lb_m or joules/kg.

The last term of equation AIV-26 has units of Btu/lb_m in the customary U.S. system. The conversion factor, J, is necessary in the denominator to change mechanical energy units to thermal units. In the SI system, two factors, g_c and J, in the denominator can be made equal to one, dimensionless, to give joules/kg.

We have already established the units of the terms in the large parentheses as joules per gram-mole due to the source of the data used. In the customary U.S. system, we have to convert both the energy and the mass terms. Using the factor, I, in the numerator converts joules to British thermal units. The factor (MW) (mixture molecular mass, g/g-mole) in the denominator converts molar units to mass units (grams). The factor L in the denominator converts grams to lb_m. In the SI system, K equal to 1,000 g/kg is necessary in the numerator. Other terms that must be consistent are: the molar gas constant in joules/(g-mole-K); the temperature is in degrees Kelvin; and the molar volume is in cm^3/g-mole.

Example AIV-2: (Continued from previous page)

$$E(T,v) = 0 = \qquad \text{(AIV-26)}$$

$$\frac{IK}{(MW)L}\left(\begin{array}{c}-\underline{h}_T + (T-T_0)\sum_i y_i A_i + \dfrac{(T^2 - T_0^2)}{2}\sum_i y_i B_i \\ + \dfrac{(T^3 - T_0^3)}{3}\sum_i y_i C_i + \dfrac{(T^4 - T_0^4)}{4}\sum_i y_i D_i \\ +\left[\dfrac{b\underline{R}T}{\underline{v}-b} - \dfrac{a}{(\underline{v}+b)T^{0.5}} - \dfrac{3a}{2b}\dfrac{1}{T^{0.5}}\ln\dfrac{\underline{v}+b}{\underline{v}}\right]\dfrac{8.3144}{83.144}\end{array}\right)$$

$$+\dfrac{G^2 v^2}{2 g_c J}$$

U.S. $I = 0.43021\ Btu/J$, $g_c = 32.17\ ft - lb_m / lb_f - s^2$,
$J = 778.16\ ft - lb_f / Btu$, $K = 1.0$, $L = 0.0026792\ lb_m / g$

SI $I = 1.0$ $J = 1.0$, $K = 1000.0\ g/kg$, $L = 1.0$

Equation AIV-26 is a functional relationship for adiabatic flow down a conduit. It relates v and T at a section to mass flux, G, and to the total enthalpy (stagnation enthalpy) or the enthalpy at the source. The equation is written in terms of enthalpy, not irreversibilities. Because it does not account for irreversibilities, it cannot be used directly to solve for losses in mechanically useful energy. The accuracy is as good as the accuracy of the Redlich-Kwong[xxviii] equation.

Note both molar and mass units are used in Equation AIV-26. The overall molecular mass (mixture) has to be included to give consistent units. The equation gives the values of variables at a section.

The advantage of using the mass flux, G, is it simplifies the derivations of compressible flow. A molar basis is helpful, in the general case, when dealing with mixtures. Most of the available data will be on this basis.

The functional relationship of AIV-26 is a good example of the difficulties engineers face in the real world where data comes from different sources. It is well to analyze each combination of algebraically additive terms. The term $G^2v^2/(2G_c J)$ has units of Btu/lb_m in customary U.S. units and joules/kg in SI units (after manipulation, with both g_c and J equal to unity with no units).

APPENDIX AIV | Equations of Compressible Flow

Example AIV-2: (Continued from previous page)

Force-balance relationship among variables, F(T,v), (second law)

The other equation used in the solution of the simultaneous equations will be the Peter Paige[xxiv] equation. The Peter Paige equation is a second law equation. An analytical method for finding choked pressure for real gases using the Redlich-Kwong[xxviii] equation is given in the paper *Choked Flow* by R. Mulley (Available from ISA).

Viscosity computations to establish loss of mechanical energy

Viscosity is a naturally arising resistance that liquid and gaseous fluids offer to flow. It is one of the most difficult to grasp concepts in science and engineering probably due to the way in which it is taught. We have already pointed out the logical inconsistency of the concept of "fluid friction". How can the behavior of molecules that never touch be described by the macroscopic phenomenon of friction? Nevertheless, eminent scientists and engineers still try to find ways of using the term "fluid friction" to describe energy relationships in flow phenomena.

The author prefers to consider the phenomenon to be due to the compression of gases and liquids during flow and to the subsequent local increase in temperature. When the temperature differential dissipates by energy transfer, the result is a loss in available mechanical energy. As was clearly established by Sadi Carnot[ii], transfer of energy under a temperature gradient without recovery of mechanical energy is an irreversible loss of mechanical energy.

Viscosity is nothing more than an experimentally established ratio between the shear stress (force per unit area in the direction of the force) applied to a layer of fluid and the velocity gradient (change in velocity per unit distance perpendicular to the force). This ratio, which is called viscosity, is measured practically by various means and, therefore, receives various names and units, all equally confusing. Viscosity is a dynamic quantity as opposed to a static one such as pressure, temperature or specific volume. It is sometimes called a non-equilibrium quantity, but the term "equilibrium" is applied to dynamic equilibrium as well as to static equilibrium so viscosity could be described as an equilibrium property under certain circumstances.

Viscosity values are a means to an end in terms of irreversibilities associated with fluid flow. They must be correlated with such quantities as the Reynolds[v] number in order to estimated the mechanical energy losses that occur with flow. In addition, viscosities of pure components are normally

Example AIV-2: (Continued from previous page)

available as functions of temperature, but mixture viscosities are not available. Mixture viscosities must be estimated by the use of various algorithms.

Algorithm that uses Chapman-Enskog theory

The Chapman-Enskog theory gives a means of estimating viscosities for low pressure (undefined) multicomponent gas mixtures. The book by Reid, Prausnitz and Sherwood[xlvi], The *Properties of Gases and Liquids*, gives various means of estimating mixture viscosities. We will use a method from this book (*RPS*). It is based on estimating the pure component viscosities in micropoise and then using a weighting method to estimate the mixture viscosity. When compared to experimental data, the error is usually less than 2% if one omits data on mixtures of wildly varying molecular masses.

The pure component viscosity in micropoise is first estimated using one of the three qualifications: (1), non polar, (2), polar, non hydrogen bonding and (3), polar, hydrogen bonding, then the weighting equation is applied:

$$\eta_m = \sum_{i=1}^{n} \frac{y_i \eta_i}{\sum_{j=1}^{n} y_j \phi_{ij}}$$

The term, phi, is found by Wilke's approximation:

$$\phi_{ij} = \frac{[1 + (\eta_i / \eta_j)^{1/2} (M_j / M_i)^{1/4}]^2}{[8(1 + M_i / M_j)]^{1/2}}$$

Phi is found for each binary pair by interchanging the subscripts. The theory requires the gas be of low density and the molecules be so far apart only binary interactions take place.

For a binary system with subscripts 1 and 2, the first equation becomes,

$$\eta_m = \frac{y_1 \eta_1}{y_1 + y_2 \phi_{12}} + \frac{y_2 \eta_2}{y_2 + y_1 \phi_{21}}$$

The above equation can be manipulated to show the value of phi with identical subscripts (11, 22, 33, etc.) is one.

There are more complicated and more accurate approximations for viscosity given in the subsequent revision of the above book, but it was felt the above set of equations was adequate for the task at hand.

APPENDIX AIV | Equations of Compressible Flow

Analysis of simulation of example AIV-2

The first six sets of row information in the table giving the results of the simulation were established for the purpose of checking the input data to the Fortran simulation of example AIV-2. The remaining sets of row information were intended to allow comparison with the tabulation of the results of the simulation of Example AIV-1, the ideal gas simulation. The choked pressures of example AIV-2 should be compared with the first column of downstream pressures of AIV-1. The choked temperatures, volumes and average velocities should be compared with the corresponding columns.

Analysis of simulation of example AIV-2

The real case choked pressures vary from minus 1.1% to plus 1.6% of the ideal gas pressures. The choked temperatures are about 4% higher and are not constant. The choked velocities are about 4% higher and are not constant. These differences are to be expected and are in the right direction.

The results of the simulation in example AIV-2 should be verified to see the upstream pressures of each segment are lower than the choked pressures upstream. The sink pressure should also be lower than the choked pressure of segment 3. The sink pressure is 2,117 psfa and the choked pressure of segment 3 is 3,288 psfa. The pressure upstream in segment 3 is 3,337 psfa and the choked pressure in segment 2 is 7,711 psfa. The pressure upstream in segment 2 is 7,731 psfa and the choked pressure in segment 1 is 12,280. The pressure upstream in segment 1 is 25,000 psfa and the choked pressure in the PSV nozzle is about 33,530 psfa. This latter pressure was not computed for the real case. It was taken from the ideal nozzle calculation. The difference in pressure was considered sufficient to justify this shortcut.

Equations of Compressible Flow — APPENDIX AIV

Results of simulation of example AIV-2, customary U.S. units					
WMOL	**R**	**PCM**	**TCM**	**VCM**	**SUMZJ**
28.96	53.39	76950.0	237.0	.04757	.2896
ARKM	**BRKM**				
1.377E+04	1.436E-02				
CG(I)	**CG(1)**	**CG(2)**	**CG(3)**		
1068.0	393.2	242.6	103.0		
CP A	**CP B**	**CP C**	**CP D**		
.7440E+01	-.1800E-02	.1975E-05	-.4784E-09	N2	
.6713E+01	-.4883E-06	.1287E-05	-.4362E-09	O2	
.4969E+0	-.4261E-05	.3809E-08	0.000E+01	AR	
SUMYA	**SUMYB**	**SUMYC**	**SUMYD**		
.7265E+01	-.1406E-02	.1813E-05	-.4651E-09		
ENTH0	**PSRCE**	**TSRCE**	**VSRCE**		
.8752E+05	.6350E+05	.1000E+04	.8472E+00		
	PSINK	**TSINK**	**VSINK**		
	2117.0	998.3	25.18		
CHOKED P	**CHOKED T**	**CHOKED V**	**CHOKED U**		
.1228E+05	.8636E+03	.3759E+01	.1478E+04	SEG 1	
.7711E+04	.8686E+03	.6018E+01	.1460E+04	SEG 2	
.3288E+04	.8698E+03	.1413E+02	.1455E+04	SEG 3	
P_{UP} SEG 1	P_{UP} SEG 2	P_{UP} SEG 3			
.2500E+05	.7731E+04	.3337E+04			
V_{UP} SEG 1	V_{UP} SEG 2	VUP SEG 3			
.1827E+00	.1059E+00	.4501E-01			
Legend					
WMOL	mixture molar mass, lb_m/lb_m-mole				
R	mixture gas constant, $ft\text{-}lb_f/lb_m\text{-}R$				
PCM	critical pressure, mixture, psfa				
TCM	critical temperature, mixture, R				
VCM	critical volume, mixture, ft^3/lb_m				
SUMZJ	sums of critical compressibility times mole fraction				
ARKM	Redlich-Kwong coefficient, a, mixture				
BRKM	Redlich-Kwong coefficient, b, mixture				
CG(i)	mass velocity in throat or segment, $lb_m/s\text{-}ft^2$				
CP A,B,C,D	component heat capacity coefficients, cp in $ft\text{-}lb_f/lb_m\text{-}R$				
SUMYi	mole fraction weighted sum of heat capacity coefficients				
ENTH0	source or stagnation enthalpy, $ft\text{-}lb_f/lbm$				
PSINK	sink pressure, psfa				
TSINK	sink temperature, R				
VSINK	specific volume gas in sink, ft^3/lb_m				
P_{UP} SEG	upstream pressure in segment 1, 2 or 3, psfa				

APPENDIX AIV | **Equations of Compressible Flow**

Analysis of simulation of example AIV-2

The simulation of AIV-2 was much more complicated than AIV-1 in that it considered three components of a non-ideal (real) mixture. The enthalpy computations and the friction factor (from viscosity and Reynolds[v] number) computations were more detailed. Also, the choked conditions were estimated from a constraint on the Peter Paige[xxiv] equation, not from a theoretical equation based on an ideal gas. The use of the R-K EOS makes the second simulation more useful generally, especially for gas mixtures.

The reader is invited to make his own analysis and to compare the results with experimental data.

Example AIV-3: Real gas, analytic solution

The author finally resolved the problem of finding an analytic solution to the choked flow phenomenon. The solution is described in the paper *Choked Flow* by Raymond Mulley and the Fortran code is found in the paper *Simulations* by the same author (Available from ISA). The tabulated results are given below. If the tabulations of examples AIV-2 and AIV-3 are compared, it will be seen they are very close.

The analytic solution is inherently more accurate. It will be demonstrated in Appendix AV for a very complex problem.

Equations of Compressible Flow | APPENDIX AIV

Real Gas Example AIV-3
Analytic Method
Results of simulation of analytic method, real gas, configuration of Figure AIV-1

WMOL	R	PCM	TCM	VCM	SUMZJ
28.96	53.39	76950.0	237.0	.04757	.2896
ARKM	**BRKM**				
1.377E+04	1.436E-02				
CG(I)	**CG(1)**	**CG(2)**	**CG(3)**		
1068.0	393.2	242.6	103.0		
CP A	**CP B**	**CP C**	**CP D**		
.7440E+01	-.1800E-02	.1975E-05	-.4784E-09	N2	
.6713E+01	-.4883E-06	.1287E-05	-.4362E-09	O2	
.4969E+0	-.4261E-05	.3809E-08	0.000E+01	AR	
SUMYA	**SUMYB**	**SUMYC**	**SUMYD**		
.7265E+01	-.1406E-02	.1813E-05	-.4651E-09		
ENTH0	**PSRCE**	**TSRCE**	**VSRCE**		
.8752E+05	.6350E+05	.1000E+04	.8472E+00		
	PSINK	**TSINK**	**VSINK**		
	2117.0	998.3	25.18		
CHOKED P	**CHOKED T**	**CHOKED V**	**CHOKED U**		
1.24E+04	8.37E+02	3.60E+00	1.42E+03	SEG1	
7.66E+03	8.37E+02	5.83E+00	1.42E+03	SEG2	
3.25E+03	8.37E+02	1.37E+01	.1414E+04	SEG3	
P_{UP} SEG 1	**P_{UP} SEG 2**	**P_{UP} SEG 3**			
2.51E+004	7.68E+003	3.30E+003			
V_{UP} SEG 1	**V_{UP} SEG 2**	**V_{UP} SEG 3**			
1.83E-01	1.06E-01	4.50E-02			

Legend

WMOL	mixture molar mass, lb_m/lb_m-mole
R	mixture gas constant, ft-lb_f/lb_m-R
PCM	critical pressure, mixture, psfa
TCM	critical temperature, mixture, R
VCM	critical volume, mixture, ft^3/lb_m
SUMZJ	sums of critical compressibility times mole fraction
ARKM	Redlich-Kwong coefficient, a, mixture
BRKM	Redlich-Kwong coefficient, b, mixture
CG(i)	mass velocity in throat or segment, lb_m/s-ft^2
CP A,B,C,D	component heat capacity coefficients, cp in ft-lb_f/lb_m-R
SUMYi	mole fraction weighted sum of heat capacity coefficients
ENTH0	source or stagnation enthalpy, ft-lb_f/lb_m
PSINK	sink pressure, psfa
TSINK	sink temperature, R
VSINK	specific volume gas in sink, ft^3/lb_m
P_{UP} SEG	upstream pressure in segment 1, 2 or 3, psfa

| APPENDIX AIV | **Equations of Compressible Flow** |

AIV-7: SUMMARY OF APPENDIX AIV

Appendix AIV has given a good deal of theoretical and practical information on the basic equations of compressible flow.

The very useful and very practical Peter Paige[xxiv] equations were developed and the Benjamin and Miller[xxxii] experiment on the choking effects of flashing water was referenced. It was shown the Peter Paige equation is equivalent to, and more general than, the commonly used equation to describe sonic flow.

An extensive set of equations involving the Redlich-Kwong[xxviii] equation-of-state was developed. An algorithm for computer implementation of the R-K equation was given.

Finally, a real gas example using the analytical method was given. A complete development of the real gas equations will be found in the paper *Choked Flow* by R. Mulley and in the paper *Simulations* by the same author (Available from ISA).

A Glossary of terms used in this appendix will be found at the end of this book.

APPENDIX AV

Compressible Fluid Flow – Complex Systems

AV-1: SCOPE — ESTIMATING COMPLICATED PRESSURE DROPS AND FLOWS

Appendix AV lays the groundwork for detailed estimations of pressure drops and flows of some very complicated compressible fluid systems. By "complicated compressible fluid systems" is meant vent headers with flows of multicomponent mixtures from multiple, different sources. An example of a computer simulation of an adiabatic vapor relief problem will be given. The scope of the appendix will be confined to the problems of the industrial plant.

The system chosen for demonstration is a complex safety relief vent header. Some of the main concepts have already been discussed in Chapter IV and in Appendix AIV. We will put these concepts into the context of the network shown in Figure AV-1. This example is taken from a real case. The piping network existed and had to be analyzed. The example was chosen to demonstrate real-life problems. The flow rates were estimated for the worst-case scenario. The input data was changed a little to emphasize certain phenomena and to protect the original client. The results from a run of the simulation will be given with suitable commentary.

AV-2: DESCRIBING THE PIPING NETWORK

Figure AV-1 describes the piping network. The plant layout and modifications to it generally fix the overall piping configuration of a vent header system. The design configuration generally is changed due to additions and deletions, as vessels are put to different use or added or eliminated over the course of time. Figure AV-1 is based on a real chemical plant. In this case, it was necessary to confirm or infirm the safe operation of the system.

APPENDIX V | **Compressible Fluid Flow — Complex Systems**

It was stated in Chapter V that a computer simulation requires the physical network be described by a simple coding procedure. Any combination of possible flows can then be imposed on the system to study the effects on the relief devices of changed piping configurations, changed diameters, or different combinations of flows. Many design disciplines benefit from the output data from such a program. The program can give information of interest to process engineers, piping design personnel and control systems engineers (and many others) such as velocities, densities, temperatures and pressures at every fitting. This information can be used in piping support studies and in pipe stress studies, as well as for a general analysis of how the system will perform.

A sketch is meant for a human, not a computer. The sketch is a record of what exists and what the basis of the simulation was. It is used during the walk-down of the system for checking purposes. It must be up-to-date, even if it exists only in marked-up form.

The isometric sketch was described simply in Chapter V. Figure AV-1 is a more complex version of such a sketch. All relief devices in a system should be shown on an isometric sketch.

The original model allowed the possibility of three conduit feeders (a connection to a single segment with a cross) to each segment. We have not changed this option even though the author has not observed its use. Most junctions resolve themselves to simple pipe tees (two possible feeders to any particular segment).

Once the piping network has been described by an isometric sketch, it can be coded into a computer. If output data is not needed for individual bends, the fittings of constant diameter pipe can be included as an effective length in the data for the segment. The author prefers not to use effective lengths, but to describe each segment individually so that data for each fitting and segment may be generated.

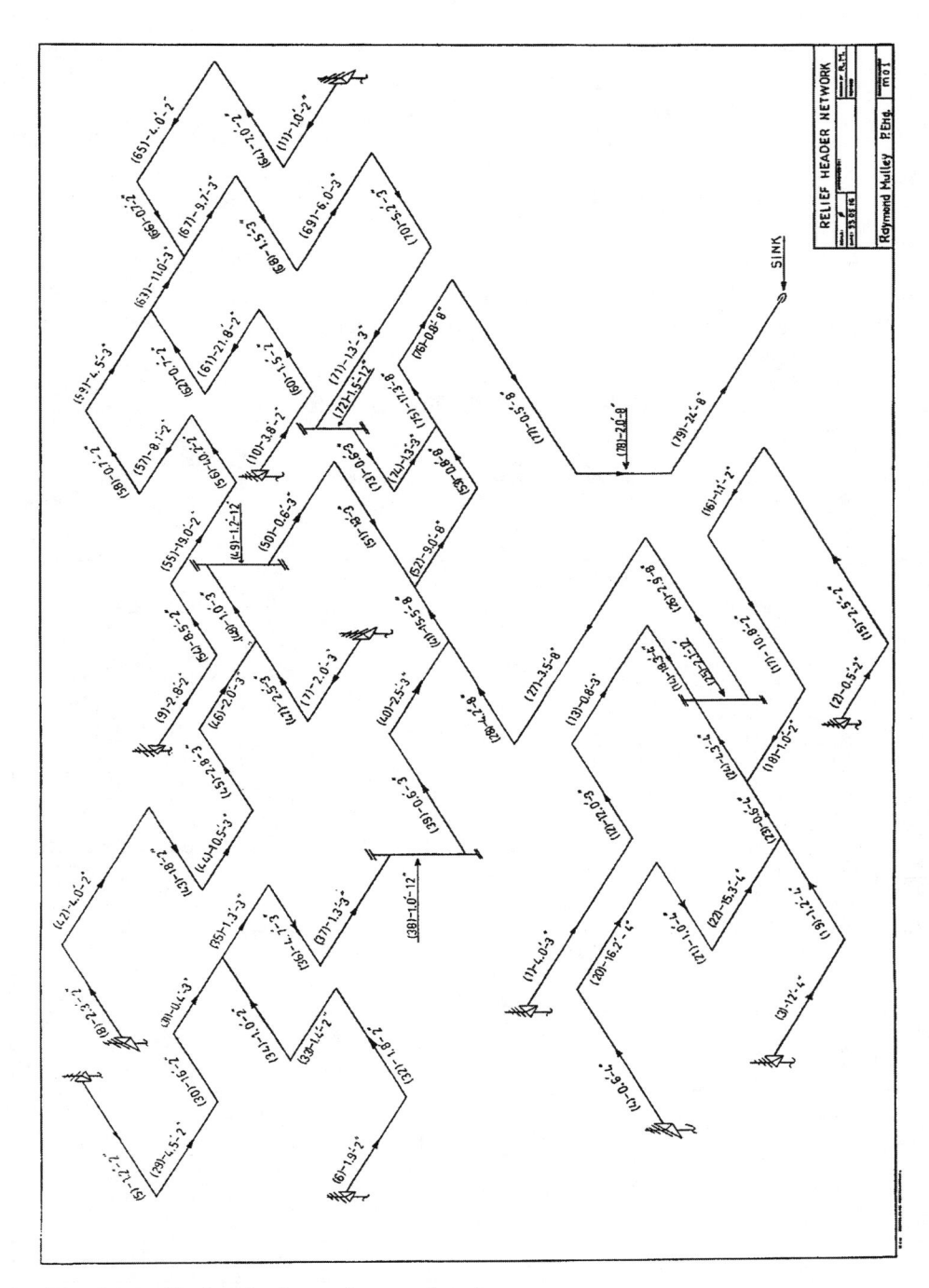

Compressible Fluid Flow — Complex Systems | APPENDIX V

Figure AV-1. Isometric sketch of complex vent header

APPENDIX V | Compressible Fluid Flow — Complex Systems

Examination of the isometric sketch of Figure AV-1 reveals four vertical 12-inch diameter segments: numbers 25, 38, 49 and 72. These segments are treated as any other segment but with zero length. The segments are essentially modeled as small vessels, with the K factor for the inlet fitting attributed to the upstream segment and only that for outlet fitting attributed to the nominal segment. The K factors are transformed to delta factors internally using a method that adapts the work of Benedict et al[xxxv].

The column with the heading Transition type identifies each of the eight types of transitions of the paper *Changes to Adiabatic Vent Header Simulation*. These transitions are listed after the piping configuration table.

Pipe Configuration for Vent Header									
SEGMENTS: 79 || SOURCES: 11								
SEG #	Length, ft	I.D. inches	Roughness, ft	Nominal terminal K	Transition type	FEEDERS			End Connection
						I	J	K	
1	4.0	3.068	1.50E-04	0.252	2	0	0	0	L3x3
2	0.5	2.067	1.50E-04	0.266	2	0	0	0	L2x2
3	12.0	4.026	1.50E-04	0.238	2	0	0	0	L4x4
4	0.6	4.026	1.50E-04	0.238	2	0	0	0	L4x4
5	1.2	2.067	1.50E-04	0.266	2	0	0	0	L2x2
6	1.9	2.067	1.50E-04	0.266	2	0	0	0	L2x2
7	2.0	3.068	1.50E-04	0.252	2	0	0	0	L3x3
8	2.3	2.067	1.50E-04	0.266	2	0	0	0	L2x2
9	2.8	2.067	1.50E-04	0.266	2	0	0	0	L2x2
10	3.8	2.067	1.50E-04	0.266	2	0	0	0	L2x2
11	1.0	2.067	1.50E-04	0.266	2	0	0	0	L2x2
12	12.0	3.068	1.50E-04	0.252	2	1	0	0	L3x3
13	0.8	3.068	1.50E-04	0.252	8	12	0	0	L3x4
14	18.3	4.026	1.50E-04	0.790	3	13	0	0	E4x12
15	2.5	2.067	1.50E-04	0.266	2	2	0	0	L2x2
16	1.1	2.067	1.50E-04	0.266	2	15	0	0	L2x2
17	10.8	2.067	1.50E-04	0.266	2	16	0	0	L2x2
18	1.0	2.067	1.50E-04	0.266	6	17	0	0	B-T4x2x4
19	1.2	4.026	1.50E-04	0.0	4	3	0	0	R-T4x4x4
20	16.2	4.026	1.50E-04	0.238	2	4	0	0	L4x4
21	1.0	4.026	1.50E-04	0.238	2	20	0	0	L4x4
22	15.3	4.026	1.50E-04	0.238	6	21	0	0	B-T4x4x4
23	0.6	4.026	1.50E-04	0.0	4	19	22	0	R-T4x2x4
24	4.3	4.026	1.50E-04	0.196	3	18	23	0	E4X12
25	2.3	12	1.50E-04	0.219	7	14	24	0	Contract.
26	2.9	7.981	1.50E-04	0.196	2	25	0	0	L8x8
27	3.5	7.981	1.50E-04	0.196	2	26	0	0	L8x8

Pipe Configuration for Vent Header

	SEGMENTS: 79		SOURCES: 11			FEEDERS			
SEG #	Length, ft	I.D. inches	Roughness, ft	Nominal terminal K	Transition type	I	J	K	End Connection
28	4.2	7.981	1.50E-04	0.0	4	27	0	0	R-T**8**x3x8
29	4.5	2.067	1.50E-04	0.266	2	5	0	0	L2x2
30	16	2.067	1.50E-04	0.266	2	29	0	0	L2x3
31	0.4	3.068	1.50E-04	0.0	4	30	0	0	R-T**3**x2x3
32	1.8	2.067	1.50E-04	0.266	2	6	0	0	L2x2
33	1.4	2.067	1.50E-04	0.266	2	32	0	0	L2x2
34	1.0	2.067	1.50E-04	0.266	6	33	0	0	B-T3x**2**x3
35	1.3	3.068	1.50E-04	0.252	2	31	34	0	L3x3
36	4.7	3.068	1.50E-04	0.252	2	35	0	0	L3x3
37	1.3	3.068	1.50E-04	0.879	3	36	0	0	E3x12
38	1.0	12	1.50E-04	0.469	7	37	0	0	Contract.
39	0.6	3.068	1.50E-04	0.252	2	38	0	0	L3x3
40	2.5	3.068	1.50E-04	0.252	6	39	0	0	B-T**8**x**3**x8
41	15.5	7.981	1.50E-04	0.266	5	28	40	0	O-T**8**x3x8
42	4.0	2.067	1.50E-04	0.266	2	8	0	0	L2x2
43	18.0	2.067	1.50E-04	0.266	8	42	0	0	L2x3
44	10.5	3.068	1.50E-04	0.252	2	43	0	0	L3x3
45	2.8	3.068	1.50E-04	0.252	2	44	0	0	L3x3
46	2.0	3.068	1.50E-04	0.252	6	45	0	0	B-T3x**3**x3
47	2.5	3.068	1.50E-04	0.0	4	7	0	0	R-T**3**x3x3
48	1.0	3.068	1.50E-04	0.879	3	46	47	0	E3x12
49	1.2	12	1.50E-04	0.459	7	48	0	0	Contract.
50	0.6	3.068	1.50E-04	0.252	2	49	0	0	L3x3
51	1.3	3.068	1.50E-04	0.252	5	50	0	0	O-T**3**x8x8
52	9.0	7.981	1.50E-04	0.196	2	51	41	0	L8x8
53	0.8	7.981	1.50E-04	0.0	4	52	0	0	R-T**8**x3x8
54	8.5	2.067	1.50E-04	0.266	2	9	0	0	L2x2
55	19.0	2.067	1.50E-04	0.266	2	54	0	0	L2x2
56	40.2	2.067	1.50E-04	0.266	2	55	0	0	L2x2
57	8.1	2.067	1.50E-04	0.266	2	56	0	0	L2x2
58	0.7	2.067	1.50E-04	0.266	8	57	0	0	L2x3
59	4.5	3.068	1.50E-04	0.0	4	58	0	0	R-T**3**x2x3
60	1.5	2.067	1.50E-04	0.266	2	10	0	0	L2x2
61	21.8	2.067	1.50E-04	0.266	2	60	0	0	L2x2
62	0.7	2.067	1.50E-04	0.266	6	61	0	0	B-T3x**2**x3
63	11	3.068	1.50E-04	0.0	4	59	62	0	R-T**3**x2x3
64	7.0	2.067	1.50E-04	0.266	2	11	0	0	L2x2
65	4.0	2.067	1.50E-04	0.266	2	64	0	0	L2x2
66	0.7	2.067	1.50E-04	0.266	6	65	0	0	B-T3x**2**x3
67	9.7	3.068	1.50E-04	0.252	2	66	63	0	L3x3
68	1.5	3.068	1.50E-04	0.252	2	67	0	0	L3x3

APPENDIX V | Compressible Fluid Flow — Complex Systems

| \multicolumn{9}{c|}{Pipe Configuration for Vent Header} |

	SEGMENTS: 79		SOURCES: 11			\multicolumn{3}{c	}{FEEDERS}	
SEG #	Length, ft	I.D. inches	Roughness, ft	Nominal terminal K	Transition type	I	J	K	End Connection
69	6.0	3.068	1.50E-04	0.252	2	68	0	0	L3x3
70	5.2	3.068	1.50E-04	0.252	2	69	0	0	L3x3
71	1.3	3.068	1.50E-04	0.879	3	70	0	0	E3x12
72	1.5	12	1.50E-04	0.469	7	71	0	0	Contract.
73	0.6	3.068	1.50E-04	0.252	2	72	0	0	L3x3
74	1.3	3.068	1.50E-04	0.252	6	73	0	0	B-T8x3x8
75	17.3	7.981	1.50E-04	0.196	2	74	53	0	L8x8
76	0.8	7.981	1.50E-04	0.196	2	75	0	0	L8x8
77	0.5	7.981	1.50E-04	0.196	2	76	0	0	L8x8
78	2.0	7.981	1.50E-04	0.196	2	77	0	0	L8x8
79	24.0	7.981	1.50E-04	1.0	1	78	0	0	L8x∞

The transition types are:

1. the exit to atmospheric pressure or to the lowest sink pressure;
2. bends;
3. pipe enlargements other than the exit to the sink;
4. tees with one inlet contiguous with the run;
5. tees with two opposed inlets at 90 degrees to the run;
6. tees with one branched inlet;
7. contractions in the exits from knockout pots;
8. combination bends and enlargements that were combined and called transition 8 in the program.

AV-3: DESCRIBING THE FLOW REGIME

Recommendations made in Chapter V will be repeated. It is not necessary to size all vent headers for simultaneous relief of all safety devices. It is recommended to divide the system into groups of devices that will relieve together under similar conditions such as fire exposure (by fire zone), cooling water failure, power failure, etc., and then to investigate the groups for the worst case backpressure at each source. The worst case backpressure can then be compared to the maximum allowable backpressure, which is a function of the type of relief device – rupture disk, conventional PSV or balanced bellows PSV. In addition, the safety device will have been chosen based on engineering judgment and the maximum allowable backpressure may be lower than that required by the vessel code.

Compressible Fluid Flow — Complex Systems | APPENDIX V

If it is necessary to provide data for piping support or stress analysis, segments can be described as being terminated by bends and fittings where the data are needed. This technique is favored by the author and was followed in the example case.

It is necessary to decide on a model of fluid flow: isothermal, adiabatic vapor or gas, or adiabatic flashing flow. The most general model will involve multicomponent mixtures. The model chosen for demonstration is the multicomponent, adiabatic vapor model.

AV-4: COMPONENT INPUT DATA, ELEVEN SOURCES

	Properties of Components at Sources								
	Psink, Psia	Patm, Psia	# comps.						
	14.7	14.7	3						
Comp. #	MW	Pc, Psia	Tc, R	Vc, ft^3/lb$_m$	Zc	Coeff. A	Coeff. B	Coeff. C	Coeff. D
1	3.20E+01	1.17E+03	9.23E+02	5.90E-02	2.24E-01	1.23E+02	2.29E-01	4.63E-05	-2.84E-08
2	1.80E+01	3.20E+03	1.17E+03	4.98E-02	2.29E-01	3.23E+02	1.10E-02	3.36E-05	-6.36E-09
3	9.31E+01	7.70E+02	1.26E+03	4.64E-02	2.47E-01	-8.08E+01	7.08E-01	-3.16E-04	5.59E-08
	COMPONENTS								
	METHANOL		WATER		ANILINE				
	VISCOSITY CODE								
	3		3		3				

	Data on Sources						
Source	Max * permissible backpressure. Psig	P acc. Rated by Code, Psig	T acc. Rated by Code, F	P. Rating Rated by Code, Psig	Component flow rates at sources, lbm/h		
					MeOh	H$_2$O	Analine
1	60	60	308	50	9	6000	40
2	60	60	308	50	6	6000	9
3	30	30	274	25	6	889	6
4	30	60	439	50	6	3640.1	8041.9
5	60	60	310	50	6	6000	93.5
6	18	18	257	15	0.6	975	9.2
7	20	60	326	50	0	2973.2	447.8
8	60	60	308	50	0	13.3	4.6
9	60	60	307	50	0	3008	0
10	60	60	308	50	0	18.4	7.4
11	60	60	308	50	1.9	1899	19

* *The maximum permissible backpressure is arrived at by engineering judgment. It is always equal to or lower than backpressure permitted by any vessel code. It sometimes is lower due to known equipment connections that are not rated according to the vessel code. A system is only as good as its weakest member.*

APPENDIX V | Compressible Fluid Flow — Complex Systems

AV-5: PLAN OF ATTACK

This simulation was originally written using customary U.S. units. The units have not been changed but it is not difficult to do so. The iteration variables were chosen to be temperature and specific volume. For fixed composition mixtures and for pure fluids, any two thermodynamic variables may be chosen. Even though composition in the network varies due to mixing, once mixed the composition remains constant until the next mixing point.

Each simulation project must have its plan of attack. Our plan is to:
1. establish the mass flow rate (normally the mass flow at accumulated pressure) for each safety device based on the worst-case failure of each vessel;
2. sum the mass flow rates of feeders to each successive downstream segment;
3. convert mass flow rates to mass velocities for each segment;
4. for each segment, convert mass flow to mole flow, mole fractions per segment and total mole flow;
5. establish the molar heat capacity of each mixture in each segment by using the ideal gas heat capacity coefficients, A, B, C and D, for each component;
6. establish the real gas enthalpies using the departure functions from the R-K equation;
7. establish the pseudocritical properties for each mixture. These are, at segment "n":

    $(Z_{cm})_n = \Sigma_i y_i (Z_c)_i$, = critical mixture compressibility

    The subscript, m, refers to a property of the mixture, c, refers to a critical property and, i, to the component.

    $(v_{cm})_n = \Sigma_i y_i (v_c)_i$, ft^3/lb$_m$-mol, critical mixture molar volume in segment n

    $(M_m)_n = \Sigma_i y_i (M_c)_i$, lb$_m$/lb$_m$-mol, mixture molar mass in segment n

    $(T_{cm})_n = \Sigma_i y_i (T_c)_i$, R, critical mixture temperature in segment n

    $(P_{cm})_n = \Sigma_i (y_i (P_{ci}))_n$, lb$_f$/ft^2 critical mixture pressure in segment n

    $(R_m)_n = 1545/(M_m)_n$, mixture gas "constant", lb$_f$-ft/lb$_m$-R in segment

8. establish the constants, a and b, of the Redlich-Kwong[xxviii] equation-of-state for each new pipe segment;
9. estimate the enthalpy at each source and sum the feeder enthalpies for the subsequent enthalpies in each pipe segment;

10. identify the choke points (they may be multiple) based on the analytic method;
11. estimate the downstream temperature and specific volume of the fluid in each pipe segment from pressure and stagnation enthalpy using Newton-Raphson iteration on simultaneous equations for T and v (Start from sink pressure plus the pressure difference across the exit or from the choked pressure.);
12. estimate the upstream pressure, temperature and specific volume of the fluid in each pipe segment using the Peter Paige[xxiv] equation;
13. pass the estimated data to the upstream pipe segment or use choked properties and repeat until a safety device discharge is reached;
14. compare the estimated discharge pressures with the maximum permissible discharge pressures and make any necessary changes, and;
15. iterate to a satisfactory solution.

'No one said it was going to be easy! Fortunately, the estimations and the data manipulation can be programmed to be executed automatically.

AV-6: IRREVERSIBILITIES DUE TO FORM (AND MIXING) EFFECTS

In simulations of safety vent headers, permanent changes in mechanical energy across a fitting or some other piping transition may have to be estimated for incompressible fluids (liquids), compressible fluids (vapors and gases) and for mixed phase fluids (flashing fluids). Permanent changes in mechanical energy, in general, fall into two categories: those due to the wall effect of conduits ("skin friction") and those due to an obstruction or a change in fluid path ("form friction"). Form effects are usually more substantial than skin effects, and yet there is much less formal analysis of the former than of the latter.

As has been pointed out throughout this book, there is no such thing as fluid friction. It is a misnomer for an irreversible change in kinetic energy to thermal energy. This is why the author prefers the term "irreversibilities" to "friction" although it is impossible to avoid the term "friction" consistently since it is so ingrained in the literature.

There have been reams of information published on wall effect "friction" factors. The K factors used to estimate mechanical energy conversions to thermal energy due to form effects have been treated more summarily.

APPENDIX V | Compressible Fluid Flow — Complex Systems

Approach to treatment of 'form friction'

There are three different practical approaches to estimating form effects:

1. the use of conventional K factors;
2. the use of Miller's K factors;
3. the Benedict et al[xxxv] approach.

The first two approaches bear the same name, K factor, but the factor is only identical in the case of incompressible fluids (liquids). Both approaches ratio permanent changes in mechanical energy across a transition to kinetic energy at some section.

Approach number 1, conventional K factors, is the most widely used and most widely documented. It is the most inaccurate for compressible fluids, however, and requires judicially chosen safety factors. We use it for incompressible fluids.

The Miller approach is more accurate than the traditional one, if good data is available, but it is not so well documented.

The Benedict et al methods abandon use of the ratios of permanent changes in mechanical energy to kinetic energy at a section in favor of a more general, empirical relationship related to entropy increase across a transition.

The Benedict et al methods have great potential, but they are not well documented. The author believes the Benedict et al methods to be potentially the most accurate of all the methods, but this remains to be proven. The original simulation used conventional K factors exclusively. The simulation described in this appendix uses a method based on Benedict et al.

Data reduction

If we study the numerous fitting and piping transitions of the isometric sketch of Figure AV-1, we see one sudden enlargement at the sink and five more at the entrances to the four vertical segments. Each sudden enlargement, including those at the branch inlets to pipe tees and those at changes in pipe diameter must be examined for the possibility of choked flow. The four vertical segments also have sudden contractions at the inlets to the downstream segments.

Compressible Fluid Flow — Complex Systems | APPENDIX V

The many elbows present represent single input-single output devices. A few have different, larger diameters at their exits. Many are of different size.

The tees represent mixing devices. They may have different configurations, sizes and connections.

Categories of causes of permanent changes in mechanical energy

We can reduce the number of categories of transitions and conduit for analysis as follows:

Non-mixing categories
1. sudden expansion;
2. pure bend;
3. enlargement;

Mixing categories
4. contiguous inlet tee;
5. opposed inlet tee;
6. branched inlet tee;

Non-mixing categories
7. contraction;
8. bend plus expander.

Estimation methods

We use our own adaptation of the methods of Benedict et al[xxxv]. In postulating this adaptation, we were not too concerned with absolute accuracy. Establishing the computer simulation was the primary concern. It is expected someone else will verify the method against reality and will signal any errors. Note the computer simulation has been used successfully in simpler form (incremental method of finding the choked pressure and original K factors), so we think this approach is reasonable.

The categories of conduit and transitions require different estimation methods. For the straight conduit, we use conventional techniques and "friction" factors. For sudden enlargements and contractions, the Benedict et al methods are used. For elbows and tees, we use methods that were modified from those of Benedict et al. The nominal K factors referred to in the printouts are simply to give a starting point for corrections.

APPENDIX V | Compressible Fluid Flow — Complex Systems

In what follows, remember the simulation calculates in the direction of the flow for mass balances but in the opposite direction for pressure drops.

For elbows, bends and sudden enlargements, we use a method based on the first sketch from *Optimal-Systems*[xlvii] from the paper by Benedict et al[xxxv]. This method estimates the ratio of stagnation pressures across a fitting (downstream to upstream) when the upstream ratio of flowing to stagnation pressures is known as seen by equation (1). The method demands iteration on the unknown stagnation pressure ratio and the unknown upstream pressure to stagnation pressure ratio. The input is the pseudo K value for translation to a value of the slope, δ, internal to the program.

$$\frac{P_{ST2}}{P_{ST1}} = \frac{P_{F1}}{P_{ST1}}\delta + 1.0 - \delta \qquad (1)$$

For sudden contractions and combining tees, we use equation (2), which can be solved directly since downstream conditions are known. This method is based on the second sketch from *Optimal-Systems*[xlvii].

$$\frac{P_{ST2}}{P_{ST1}} = \frac{P_{F2}}{P_{ST2}}\delta + 1.0 - \delta \qquad (2)$$

The first ratio in equations [1] and [2] is known as α; the second ratio, R, is either R_1, the upstream flowing pressure to stagnation pressure ratio, or R_2, the downstream flowing pressure to stagnation pressure ratio. When performing tests, the projection method of eliminating recovery effects should be adhered to.

AV-7: MANIFOLD FLOW

Manifold flow is defined as splitting a single stream into two or more streams or combining two or more streams into a single one. The combining fittings of safety relief vent headers fall under the general designation of manifold flow.

The Bernoulli[i] equation still applies in manifold flow. If a velocity reduction across a fitting produces a negative change in kinetic energy whose absolute value is greater than the "friction losses" (permanent changes in mechanical energy), the downstream static pressure will increase. This is a fairly common, though often totally unexpected result.

The losses of mechanically useful energy are often expressed in terms of "velocity heads" meaning a multiplying factor times the kinetic energy at

Compressible Fluid Flow — Complex Systems | APPENDIX V

a specific location. The location is associated with a K value and the user must know whether it is a downstream or an upstream location.

The K factors or the "loss" coefficients are usually defined as the change in specific mechanical energy across a path through a fitting divided by the specific kinetic energy of the flow before it was divided or after it was combined. Miller[xvi] defines the coefficients somewhat differently and the coefficients of the two methods are not strictly interchangeable for compressible fluids although they are the same for incompressible fluids. Benedict et al[xxxv] do not use the K factors at all.

The numbering system used for piping networks in this book identifies the connections of combining or dividing tees, so the major flow is always number 3. This makes it easier to perform correlations on experimental data and the data can be reduced. D.S. Miller has developed charts of his K factors versus the fractional volumetric flow.

Caveat — Verifying definitions

Given the propensity of otherwise intelligent people to use the same name for different things or to use different names for the same thing, it is wise always to check definitions carefully.

The correlations of Benedict et al can replace the K factor correlations for conversion of mechanical to thermal energy in compressible flow when greater accuracy is needed. The accuracy obtained will depend on that of the correlation. These correlations are used to predict loss in mechanical energy when a fluid traverses a fitting.

Estimating losses of mechanical energy across a pipe tee, three methods

The problem of estimating mechanical energy conversion to thermal energy across a pipe tee is attacked by either one of three methods. In each case, appropriate safety factors must be included to minimize error in the wrong direction. The three methods are:

1. the traditional K factor method used for compressible and incompressible fluids;
2. Miller's K factor method for compressible fluids. The Figure V-4 is used as the basis from experimental data when Miller's K factors are used. This sketch shows the loss coefficients plotted against the ratio of volumetric flow through the branch to volumetric flow in the combined leg. Two curves are shown: one for the run and one for the branch. These curves are established experimentally;

3. a method based on the Benedict et al[xxxv] correlations for compressible fluids.

At the beginning of an estimation, only the downstream conditions are known. Of the upstream conditions, only the source properties and the mass flows in each leg are known unless one or more flows are choked. These facts suggest that iteration is necessary.

For the traditional K factor method, the iteration is started by assuming the ratio of volumetric flows is equal to that of the mass flows. With this initial ratio, the trial loss coefficients can be found, and the trial losses can be estimated from the known downstream conditions. The trial losses are estimated from $h_{ik} = K_{ik} U^2_{ik}/2g_c$ with i referring to upstream and k downstream. The subscripts ik mean use the one that is appropriate.

The sum of the trial losses, the downstream kinetic energy and the downstream static energy equals the trial sum of the upstream kinetic and static energies. The unknowns can be transformed into functions of T and v using the Redlich-Kwong[xxviii] equation and $U=Gv$. The total energy equation also is given as $E(T_1,v_1)$. We now have two equations for Newton-Raphson iteration on trial values of upstream temperature and specific volume. The next trial of the volumetric flow ratio can be estimated from the mass flow rates and the specific volumes. The algorithm used in our original program made use of an external "false position" loop to achieve convergence. Once converged on temperature and specific volume, pressure can be found from the R-K equation.

Note for both K factor methods, losses across the exit to a sink are based on the upstream velocity (kinetic energy). The velocity in the sink is essentially zero.

AV-8: VISCOSITY CONSIDERATIONS

Viscosity is a property of a fluid used to estimate its resistance to flow. Viscosity is particularly important in estimating pressure drop through uniform section pipe segments. It is not usually a factor in estimating changes in mechanical energy through fittings.

The treatment of the subject of viscosity in most texts is entirely unsatisfactory. The use of concepts such as "fluid friction" flies in the face of common sense. Even a high school student knows molecules do not rub

Compressible Fluid Flow — Complex Systems | APPENDIX V

together. Friction is a macroscopic phenomenon. The concept is applied well to solid materials such as brake pads and drums. It falls flat on its face when applied to molecules that have force fields that repel other molecules when they come too close.

The problem seems to be that a satisfactory set of equations that describe viscosity theoretically is extremely difficult to formulate. However, viscosity can be measured experimentally quite readily. Rather than abandon, or at least postpone, theoretical considerations in favor of practical ones, most authors seem to compound the difficulty by using confusing terminology and by ignoring some fundamental thermodynamic considerations.

Friction factor

It is not possible to avoid using the term, friction, in all cases – especially since it is incorporated into textbooks and much reference material, such as Crane[xiii]. The friction factor is a case in point.

The friction factor helps give a measure of the resistance to flow of fluids that extends the utility of the concept of viscosity. It really only applies to conduits of constant internal section. Viscosity is one of the properties influencing the magnitude of the friction factor. When considering changes in mechanical energy through fittings or expansions and contractions, the concept of "form friction" is more useful and the K factors come into play. Friction factors are well established and have been thoroughly correlated for liquids, at least, with the various flowing quantities. The K factors are less well established. The work of Benedict et al[xxxv] is promising as a more accurate method that replaces the use of K factors.

Caveat — Using correct K factors

When using the various coefficients, the reader is reminded to use them only with their applicable equations. Two definitions for K factors, the traditional one and that of Miller, should serve as an example of differences in the governing equations.

Viscosity plays a role in friction factor estimations. It is a fluid property that serves to give a measure of the resistance to motion of the fluid. Viscosity arises out of the forces, attractive and repulsive, that exist between molecules. When fluid is put into bulk motion by an external force, the intermolecular forces move with the molecular movement. Forces moving through a distance in the direction of the forces constitute the definition of work. Work is done and the mechanical work energy of compression is partially converted to thermal energy.

| APPENDIX V | Compressible Fluid Flow — Complex Systems |

The difference in mechanical work energy along a pipe or across a fitting constitutes a loss of available mechanical energy and is treated as lost energy even though first law considerations teach that total energy is conserved. One has to remember the word "losses" applies only to mechanical energy, not to mechanical plus thermal energy (the total energy remains constant).

Three indices for viscosity

In the case of uniform cross section pipe, estimations for the friction factor require an index of the type of molecular interaction that will occur. There are three major types of interaction that have been modeled: that between hard spheres with no attractive or repulsive forces; that between nonspherical molecules that can form electrical dipoles due to charge redistribution; and, that between molecules with isolated positive hydrogen atoms that can attract or bond with the negative portions of similar or different molecules.

Our simulation relies on the work of Reid, Prausnitz and Sherwood[xlvi] in the use of these indices. The methods used are corresponding states methods. The assumptions for mixtures of gases are the molecules are sufficiently separated and their velocities are so great only two-body interactions need be considered. The bodies can be similar or different molecules.

The first model used for gases was the ideal gas model. The ideal gas model concerns itself with spherical molecules that have no intermolecular forces except at the time of closest approach. This model subsequently was modified to take into consideration polar molecules that definitely were not spherical and that could form dipoles due to weak van der Waals forces. An additional modification considered those molecules that form stronger bonds between some of their positive hydrogen atoms and a negative kernel of a similar or a different type of molecule.

Three formulae for two-body interactions resulted for:
    nonpolar gases;
    polar gases, nonhydrogen bonding;
    polar gases, hydrogen bonding.

The first category, nonpolar gases, includes the vast majority of pure gases. Examples are: acetylene, benzene, isobutane, n-butane, carbon dioxide, carbon disulphide, carbon tetrachloride, chlorine, cyclohexane, ethane, ethylene, methane, n-pentane, propane, propylene, sulphur dioxide and toluene.

The second category, non-hydrogen bonding polar gases, includes dimethyl ether and chloromethane.

The third category, hydrogen bonding polar gases, includes water and most alcohols, ethers and phenols. The main manifestation of hydrogen bonding is a higher boiling temperature than that of close molecular weight hydrocarbons. This category includes water, methanol, dimethyl ether, ethanol, n-propanol, i-propanol, n-butanol, i-butanol, n-pentanol, n-hexanol, n-heptanol and acetic acid. Note the positive hydrogen has to bond to an electronegative component either in a similar or in a different molecule.

The simulation makes use of a simple index, M = 1, 2 or 3, that is applied to each individual component of a gas mixture to instruct the program on the formula to apply.

Gas viscosity

The theory of gas viscosity depends heavily on that of incompressible fluids. The following assumes laminar flow.

Gas molecules in a container move randomly, recoil from one another, transfer momentum, and recoil from the walls of the container. The changes in momentum at the walls are measured as pressure.

If a fluid is put into motion by movement of a solid wall oriented parallel to the motion, the molecules will have a directed motion superposed on their random motion. This directed motion is fluid flow. The directed motion has a velocity that is the same as that of the moving element close to the element, and that falls off perpendicular to the element. The directed flow can be separated into layers. The random movement of molecules still takes place between layers. This random movement now has a directed component that imparts negative or positive momentum to the receiving layer of fluid.

The faster moving layer will receive negative momentum; it will be slowed down. The slower moving layer will receive positive momentum; it will have its velocity increased in the direction of flow. Even in the ideal case, the velocities will tend to equalize.

Practically, an external force is required to produce flow. It is known that a force produces an acceleration. Therefore, if an external force produces a steady-state velocity without acceleration, other forces, equal and opposite to the external force, must arise in the fluid in order to maintain the steady state. These forces are called viscous forces.

APPENDIX V | Compressible Fluid Flow — Complex Systems

Usual description of viscosity

The usual description of viscosity starts by showing two parallel plates, one of which is in motion. The fluid under investigation is between the plates. Practical viscometers are either formed of capillary tubes, or of a rotating drum around a stationary cylinder, or vice versa. If the cylinders and drums were sufficiently large, they could approximate the two parallel plates.

The analysis usually starts with a moving plate of known area to which a force is applied causing the motion. From experiments with real viscometers, the force is proportional to the area of the plate in motion and to the velocity gradient. The velocity gradient is the rate of change of the velocity in the direction of motion with the change in distance in the perpendicular direction. It is important to distinguish the directions and to remember the property, viscosity, is measured under laminar flow conditions. Further correlations are required to make it useful for turbulent flow.

The equations describing the above are:

$$F_{ext} \propto S\frac{dv}{dr}$$
$$F_{ext} = \eta S\frac{dv}{dr}$$
$$F_{int} = -\eta S\frac{dv}{dr}$$
(AV-1)

An equal and opposite internal force is necessary to prevent acceleration from the steady state. Velocity, v, is perpendicular to area, S.

Flow through tubes

J. S. Poiseuille[xv] (1844) established the basic practical equations for laminar, incompressible flow through tubes. At each velocity or flow rate, external force is necessary to overcome the internal force or resistance to flow. To maintain steady-state flow, this external force must be applied constantly.

For flow to occur in tubes or pipes, there must be a pressure difference. In a uniform diameter pipe, the force in the direction of flow is the differential pressure times the cross sectional area of the pipe. To prevent acceleration, an internal force must arise and must be equal to and opposite from the external force.

$$F_{ext} = (P_1 - P_2)\pi r^2 = F_{int} = -\eta S\frac{dv}{dr}$$
(AV-2)

Compressible Fluid Flow — Complex Systems | APPENDIX V

Viscosity is the name given to the coefficient of proportionality, eta, of the above relationship.

The units of viscosity may be obtained from the set AV-3.

$$[F] \equiv \left[kg \frac{dv}{dt} \right] = \left[\eta S \frac{dv}{dr} \right]$$

$$[\eta] = \left[\frac{kg}{m-s} \right] \quad \text{or} \quad \left[\frac{lb_m}{ft-s} \right]$$

(AV-3)

Functional relationship between point velocity and radius of pipe

From equation AV-2, the relationship between point velocity and radius requires an indefinite integral. The stationary plate is a cylindrical reference area in this case.

$$dv = \frac{-(P_1 - P_2)\pi r^2 dr}{\eta S}$$

(AV-4)

$$dv = \frac{-(P_1 - P_2)\pi r^2 dr}{\eta 2\pi r l} = \frac{(P_2 - P_1) r dr}{2\eta l}$$

$$v = \int dv + c = \int \frac{(P_2 - P_1) r dr}{2\eta l} + c$$

$$v = \frac{(P_2 - P_1) r^2}{4\eta l} + c$$

When v = 0, r = R.

$$v = \frac{(P_2 - P_1) r^2}{4\eta l} + c$$

(AV-5)

$$0 = \frac{(P_2 - P_1) R^2}{4\eta l} + c$$

$$c = -\frac{(P_2 - P_1) R^2}{4\eta l}$$

So that

$$v = \frac{(P_2 - P_1) r^2}{4\eta l} - \frac{(P_2 - P_1) R^2}{4\eta l}$$

(AV-6)

$$v = \frac{(P_1 - P_2)}{4\eta l} (R^2 - r^2)$$

Flow of Industrial Fluids—Theory and Equations

APPENDIX V | Compressible Fluid Flow — Complex Systems

Total volumetric flow, incompressible fluids

The point velocity in steady state, incompressible, laminar flow is proportional to the product of the differential pressure and the difference of the squares of the pipe radius and that of the point velocity. It is inversely proportional to the product of viscosity and flow length.

To obtain the total volumetric flow, we must integrate the point velocity across the normal cylindrical cross sections of the pipe. The term $2\pi r dr$ is an annulus of area to be multiplied by the point velocity, v, in order to give volumetric flow through the annulus.

$$q = \frac{dV}{dt} = \int_0^R v(2\pi r)dr = \int_0^R \frac{(P_1 - P_2)}{2\eta l}(R^2 - r^2)(\pi r)dr \quad \text{(AV-7)}$$

$$q = \frac{\pi(P_1 - P_2)}{2\eta l}\int_0^R (R^2 - r^2)rdr$$

$$q = \frac{\pi(P_1 - P_2)}{2\eta l}\left(\frac{R^2 r^2}{2} - \frac{r^4}{4}\right)\Big|_0^R = \frac{\pi(P_1 - P_2)}{2\eta l}\left(\frac{R^4}{2} - \frac{R^4}{4}\right)$$

$$q = \frac{\pi(P_1 - P_2)}{2\eta l}\frac{R^4}{4} = \frac{\pi(P_1 - P_2)R^4}{8\eta l}$$

The last equation in the above development is Poiseuille's[xv] equation for the laminar flow of incompressible fluids through uniform pipes or tubes. It shows laminar flow is proportional to differential pressure and the fourth power of the radius of the conduit. It is inversely proportional to the viscosity. The equation permits the experimental determination of the viscosity of incompressible fluids from measurable quantities

Total volumetric flow, compressible fluids, standardized condition

To extend the use of the Poiseuille[xv] equation to compressible fluids, it is usual to assume ideal gas behavior. The subscript zero refers to the standardized condition. The terms V_{av} and V_0 represent average and standard total volumes.

$$P_{av} = \frac{P_1 + P_2}{2} \tag{AV-8}$$

$$\frac{P_{av}}{P_0} = \frac{P_1 + P_2}{2P_0}$$

$$\frac{P_{av} v_{av}}{P_0 v_0} = \frac{RT}{RT} = 1$$

$$\frac{V_0}{V_{av}} = \frac{v_0}{v_{av}}$$

$$\frac{V_0}{V_{av}} = \frac{P_{av}}{P_0} = \frac{P_1 + P_2}{2P_0}$$

$$q_0 = \frac{\pi (P_1 - P_2) R^4}{8\eta l} \frac{P_1 + P_2}{2P_0}$$

The last equation of the above development permits the experimental measurement of the viscosity of compressible ideal gases using standardized quantities.

'Elementary' kinetic theory

Transport properties are properties used in the description of mass, momentum and energy fluxes. Flux is the transport across a unit area under a density differential. The density is mass per unit volume in the case of mass flux; it is momentum density (nmv_y) in the case of moment flux and it is energy density ($c_v nT$) in the case of energy flux. The three associated transport coefficients are D, for mass, η, for viscosity and λ for energy. We will only be concerned with viscosity at this point.
Depending on the model used, a value can be estimated for the viscosity coefficient. For the noninteracting, rigid-sphere model, the coefficient is estimated from AV-9.

$$\eta = 26.69 \frac{(MT)^{1/2}}{\sigma^2} \tag{AV-9}$$

APPENDIX V | Compressible Fluid Flow — Complex Systems

The viscosity coefficient, eta, has units of micropoise; M is the molecular mass (weight) in grams per g-mole; T is the temperature in degrees K, and sigma is the hard-sphere diameter in angstrom units.

The theoretical treatment of the effect of intermolecular forces is usually handled by Chapman-Enskog equations. These equations involve an intermolecular potential function and are extremely difficult to grasp. Practical equations usually use corresponding states methodology. Thodos and co-workers developed the equations (slightly modified by Reid, Prausnitz and Sherwood[xlvi]) that follow:

Nonpolar gases

$$\eta = (4.610 T_r^{0.618} - 2.04 e^{-0.449 T_r} + 1.94 e^{-4.058 T_r} + 0.1)/\xi \qquad \text{(AV-10)}$$

Non-hydrogen-bonding polar gases at reduced temperatures below 2.5

$$\eta = ((1.9 T_r - 0.29)^{4/5} z_c^{-2/3})/\xi \qquad \text{(AV-11)}$$

Hydrogen-bonding polar gases at reduced temperatures below 2.0

$$\eta = ((0.755 T_r - 0.055) z_c^{-5/4})/\xi \qquad \text{(AV-12)}$$

In the above equations, M is molecular mass in grams/g-mole, P_c is the critical pressure in atmospheres, T_r is the reduced temperature, T/T_c, with temperatures in K and eta is in micropoise. Epsilon is given by AV-13.

$$\xi = T_c^{1/6} M^{-1/2} P_c^{-2/3} \qquad \text{(AV-13)}$$

The Thodos equations should not be used with hydrogen, helium or with diatomic halogens. Reid, Prausnitz and Sherwood[xlvi] suggest using the Golubev equations in this case, AV-14. The Golubev equations do not work for gases that associate significantly.

$$\eta = \frac{3.5 M^{1/2} P_c^{2/3}}{T_c^{1/6}} T_r^{0.965}, \quad T_r < 1 \qquad \text{(AV-14)}$$

$$\eta = \frac{3.5 M^{1/2} P_c^{2/3}}{T_c^{1/6}} T_r^{0.71 + 0.29/T_r}, \quad T_r > 1$$

Average error

Reid, Prausnitz and Sherwood[xlvi] gave average percentage errors compared with experimental values of 18 nonpolar gases using the Thodos and co-workers method as 2.2%. For eight polar gases, the error was 3.1%. There may be more accurate methods but, given the ease of programming the above equations and the acceptable accuracy for industrial purposes, the Thodos method will be used in this work.

Viscosities of low-pressure gas mixtures

The experimental viscosities of gas mixtures are shown in Figure AV-2, taken from Reid, Prausnitz and Sherwood. There is sometimes a maximum as shown by the ammonia-hydrogen binary mixture and to a lesser extent by the ethylene-ammonia binary mixture. Minimums are not encountered. An approximation to the rigorous solution is,

$$\eta_{mix} = \sum_{i=1}^{n} \frac{y_i \eta_i}{\sum_{j=1}^{n} y_j \phi_{ij}} \quad \textbf{(AV-15)}$$

Wilke approximated the binary coefficient, phi, by the equations shown in AV-16.

$$\phi_{ij} = \frac{\left[1 + (\eta_i / \eta_j)^{1/2} (M_j / M_i)^{1/4}\right]^2}{\left[8(1 + M_i / M_j)\right]^{1/2}} \quad \textbf{(AV-16)}$$

$$\phi_{ji} = \frac{\eta_j}{\eta_i} \frac{M_i}{M_j} \phi_{ij}$$

The error varies with the mixture, but is usually less than 2.0%.

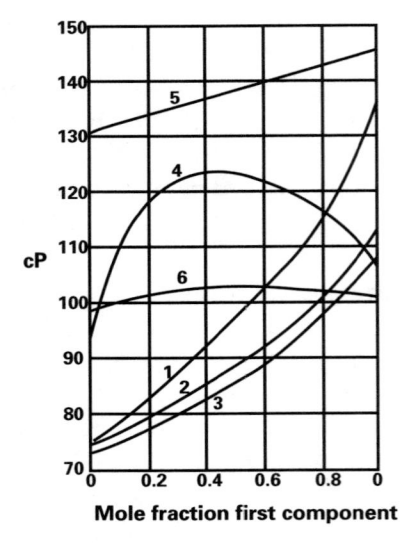

APPENDIX V | Compressible Fluid Flow — Complex Systems

Figure AV-2. Gas mixture viscosities

No	System	Temperature, K
1	Hydrogen sulfide-ethyl ether	331
2	Hydrogen sulfide-ammonia	331
3	Methane-n-butane	293
4	Ammonia-hydrogen	306
5	Ammonia-methylamine	423
6	Ethylene-ammonia	293

Source: R. Reid and R. Prausnitz, Properties of Gases and Liquids, 1987, reprinted by permission McGraw-Hill Companies.

The expansion of equation AV-16 is

$$\eta_m = \frac{y_1 \eta_1}{y_1 \phi_{11} + y_2 \phi_{12} + y_3 \phi_{13}}$$ (AV-17)

$$+ \frac{y_2 \eta_2}{y_1 \phi_{21} + y_2 \phi_{22} + y_3 \phi_{23}}$$

$$+ \frac{y_3 \eta_3}{y_1 \phi_{31} + y_2 \phi_{32} + y_3 \phi_{33}} \quad micropoise$$

$$\mu = \eta_m / 10000 \quad centipoise$$

The Reynolds[v] number, N_{Re}, is given in AV-18.

$$N_{Re} = \frac{6.31 W_n}{d\mu} \qquad \text{(AV-18)}$$

$W_n = lb_m/h$

$d = id$ inches

$\mu =$ viscosity coefficient, centipoise

The friction factor, f_M, can be estimated with the Churchill-Usagi[xli] equations.

Fixed length increments can be chosen such as five feet and the upstream pressure can be found by using the Peter Paige equation for the mechanical energy balance with losses and the total energy equation to solve by Newton-Raphson iteration for T and v. The Redlich-Kwong[xxviii] equation will permit a solution for pressure once the upstream values of temperature and specific volume are known.

AV-9: SIMULATION RESULTS (ANALYTIC METHOD)

	Simulation of Complex Piping Network– Multicomponent Vapors									
SEG #	FLOW lb_m/h	P1 psig	T1 F	v1 ft^3/lb_m	U1 ft/s	P2 psig	T2 F	v2 ft^3/lb_m	U2 ft/s	VMACH ratio
1	6049.0	17.9	300.9	13.6	446.6	17.8	292.0	13.7	452.7	0.270
2	6015.0	37.2	304.3	8.6	616.3	36.9	288.1	8.5	619.8	0.369
3	901.0	19.7	272.3	12.4	35.1	19.7	272.3	12.4	35.2	0.021
4	11688.0	25.5	434.6	5.8	212.1	25.5	432.6	5.8	212.3	0.175
5	6099.5	38.0	306.4	8.4	609.8	38.0	290.3	8.3	618.1	0.369
6	984.8	29.7	259.6	9.4	110.4	29.7	259.0	9.4	110.4	0.068
7	3421.0	12.9	318.3	14.8	274.7	12.9	314.9	14.8	275.4	0.171
8	17.9	17.3	300.6	11.0	2.4	17.3	300.6	11.0	2.4	0.002
9	3008.0	32.0	302.4	9.6	342.9	31.9	297.1	9.6	346.0	0.206
10	25.8	22.4	301.5	9.2	2.8	22.4	301.5	9.2	2.8	0.002
11	1919.9	22.6	301.7	11.9	272.1	22.5	298.5	11.9	272.6	0.163
12	6049.0	17.0	300.7	14.0	459.0	16.7	290.6	14.5	480.0	0.287
13	6049.0	15.2	300.4	14.9	487.6	15.1	290.1	14.7	489.1	0.292
14	6049.0	17.4	300.8	13.9	263.5	17.3	297.6	14.0	267.4	0.160
15	6015.0	35.4	304.0	8.9	638.5	35.2	285.6	9.0	659.0	0.392
16	6015.0	32.2	303.4	9.5	682.0	32.2	283.3	9.4	692.8	0.412
17	6015.0	29.7	303.0	10.0	720.2	29.7	268.9	12.0	897.7	0.535
18	6015.0	18.4	301.0	13.5	969.3	18.3	260.9	13.2	1002.8	0.598
19	901.0	19.7	272.3	12.4	35.2	19.7	272.3	12.4	35.2	0.021
20	11688.0	25.3	434.6	5.8	213.4	25.0	432.4	5.9	216.9	0.179
21	11688.0	24.4	434.4	5.9	218.1	24.4	432.3	5.9	218.4	0.180
22	11688.0	24.2	434.4	6.0	219.6	23.9	432.1	6.1	223.2	0.184
23	12589.0	17.6	421.7	7.8	306.6	17.5	417.5	7.7	306.9	0.243
24	18604.0	13.3	381.0	11.5	671.5	13.0	359.9	11.6	696.7	0.487
25	24653.0	13.1	360.7	12.8	111.3	13.1	360.1	12.8	111.3	0.074
26	24653.0	12.5	360.5	13.1	257.8	12.4	357.5	13.0	258.1	0.172
27	24653.0	12.3	360.5	13.1	259.2	12.3	357.5	13.1	259.6	0.173
28	24653.0	12.2	360.5	13.2	260.6	12.1	357.4	13.2	261.1	0.174
29	6099.5	35.8	306.1	8.8	636.8	35.7	286.5	9.0	675.4	0.403
30	6099.5	31.2	305.3	9.6	700.6	30.9	262.8	13.1	1004.5	0.601
31	6099.5	22.6	303.7	11.9	392.1	22.5	297.0	11.8	392.5	0.235
32	984.8	29.6	259.6	9.4	110.5	29.6	259.0	9.4	110.6	0.068
33	984.8	29.6	259.6	9.4	110.7	29.5	259.0	9.4	110.7	0.068
34	984.8	29.5	259.5	9.4	110.8	29.5	259.0	9.4	110.9	0.068
35	7084.3	21.6	297.2	12.1	463.9	21.4	287.8	12.0	466.1	0.280
36	7084.3	20.9	297.1	12.3	473.1	20.6	287.0	12.4	481.9	0.289
37	7084.3	19.7	296.9	12.8	489.7	19.5	286.4	12.7	492.3	0.296
38	7084.3	19.8	296.9	12.8	32.0	19.8	296.9	12.8	32.0	0.019
39	7084.3	16.1	296.3	14.3	546.5	16.1	283.4	14.1	548.2	0.329

Simulation of Complex Piping Network– Multicomponent Vapors

SEG #	FLOW lb$_m$/h	P1 psig	T1 F	v1 ft^3/lb$_m$	U1 ft/s	P2 psig	T2 F	v2 ft^3/lb$_m$	U2 ft/s	VMACH ratio
40	7084.3	15.4	296.1	14.6	560.1	15.3	282.3	14.5	567.9	0.341
41	31737.3	11.4	345.6	14.4	364.9	11.1	339.4	14.4	369.3	0.239
42	17.9	17.3	300.6	11.0	2.4	17.3	300.6	11.0	2.4	0.002
43	17.9	17.3	300.6	11.0	2.4	17.3	300.6	11.0	2.4	0.002
44	17.9	17.5	300.6	11.0	1.1	17.5	300.6	11.0	1.1	0.001
45	17.9	17.5	300.6	11.0	1.1	17.5	300.6	11.0	1.1	0.001
46	17.9	17.5	300.6	11.0	1.1	17.5	300.6	11.0	1.1	0.001
47	3421.0	12.7	318.2	15.0	276.9	12.6	314.8	14.9	277.8	0.172
48	3438.9	12.5	318.1	15.0	280.0	12.5	314.6	15.0	280.3	0.174
49	3438.9	12.6	318.1	15.0	18.3	12.6	318.1	15.0	18.3	0.011
50	3438.9	11.7	317.9	15.6	289.5	11.6	314.2	15.5	289.7	0.180
51	3438.9	11.5	317.9	15.7	291.4	11.4	314.1	15.6	292.0	0.181
52	35176.2	6.7	342.1	17.7	498.3	6.4	330.8	17.7	504.9	0.326
53	35176.2	6.1	342.0	18.2	513.3	6.1	330.3	18.0	513.9	0.332
54	3008.0	31.2	302.2	9.7	348.9	30.9	296.4	9.9	359.4	0.214
55	3008.0	29.5	301.9	10.1	362.6	29.4	294.8	10.8	390.8	0.233
56	3008.0	25.9	301.3	11.0	395.0	25.6	289.5	13.5	491.5	0.293
57	3008.0	17.4	299.8	13.9	500.1	17.2	287.3	14.6	532.0	0.317
58	3008.0	14.8	299.3	15.1	543.0	14.7	286.5	15.0	546.3	0.326
59	3008.0	17.1	299.7	14.1	229.0	17.0	297.4	14.1	229.9	0.137
60	25.8	22.4	301.5	9.2	2.8	22.4	301.5	9.2	2.8	0.002
61	25.8	22.4	301.5	9.2	2.8	22.4	301.5	9.2	2.8	0.002
62	25.8	22.4	301.5	9.2	2.8	22.4	301.5	9.2	2.8	0.002
63	3033.8	16.9	299.7	14.1	231.9	16.6	297.2	14.2	234.2	0.140
64	1919.9	22.4	301.7	12.0	274.0	22.2	298.2	12.1	278.1	0.166
65	1919.9	21.6	301.5	12.2	279.5	21.3	298.0	12.3	282.0	0.168
66	1919.9	21.1	301.5	12.4	283.5	21.1	297.9	12.3	283.9	0.170
67	4953.7	15.5	299.8	14.8	395.6	15.4	292.6	15.0	405.8	0.242
68	4953.7	14.4	299.7	15.3	410.3	14.3	292.3	15.2	412.0	0.246
69	4953.7	14.0	299.6	15.5	416.7	13.8	291.7	15.7	424.1	0.253
70	4953.7	13.1	299.4	16.0	429.2	13.0	291.1	16.1	436.2	0.261
71	4953.7	12.3	299.3	16.5	441.8	12.2	290.7	16.4	443.7	0.265
72	4953.7	12.4	299.3	16.5	28.8	12.4	299.3	16.5	28.8	0.017
73	4953.7	10.2	298.9	17.9	480.6	10.1	288.9	17.7	481.8	0.288
74	4953.7	9.7	298.8	18.3	489.4	9.6	288.4	18.1	492.0	0.294
75	40129.9	5.0	336.3	19.7	633.2	4.9	317.2	20.1	660.9	0.422
76	40129.9	3.6	336.0	21.2	681.4	3.6	315.9	20.7	682.9	0.436
77	40129.9	3.0	335.9	22.0	705.8	2.9	314.4	21.4	706.9	0.451
78	40129.9	2.3	335.8	22.8	732.9	2.2	312.5	22.3	737.8	0.471
79	40129.9	1.5	335.7	23.9	768.2	1.4	305.1	25.4	848.3	0.542

APPENDIX V | Compressible Fluid Flow — Complex Systems

AV-10: SUMMARY OF APPENDIX AV

This appendix has tried to outline the most important aspects of simulating complex piping systems relieving gaseous mixtures. The method chosen was to describe a common industrial situation involving a complex piping network and complex mixtures of components. The system was a vent header, common to most industrial plants and refineries.

The importance of the following was emphasized:
- defining the piping network by isometric sketches;
- defining the flow regimes by grouping devices that may relieve simultaneously;
- having an organized plan of attack;
- establishing the choke points;
- understanding manifold flow and when to make use of worst-case data;
- understanding the K factors used and the importance of picking the correct kinetic energy term;
- the differences between the three methods of estimating permanent mechanical energy changes across a transition:
  1. Conventional K factors
  2. Miller's K factors
  3. Benedict et al[xxxv] correlations
- using engineering judgment in influencing design to avoid unknowns;
- the significance of the various drawings and data sheets;
- personally verifying data by a final walk-down of the system using an isometric sketch as a basis for recording "as-built" information.

The results of a Fortran simulation of multicomponent gas flow in a very complex piping network were given. The simulation was based on most of the ideas presented in this book.

APPENDIX B

Endnotes

i   Bernoulli, Daniel, 1700-1782. Swiss Mathematician and Physicist. Developed the ideal Bernoulli equation that was later modified to include irreversibilities. Historical reference.

ii   Carnot, Sadi. French Engineer. 'Father of Thermodynamics'. The father of thermodynamics, Sadi Carnot, had a father, General Hypolite Carnot. The grandfather of thermodynamics not only was an army general under Napoleon, he was an important politician (member of the Directorate) during the French revolution and he was a hydraulician who wrote extensively on such things as shock losses in turbomachinery. Fox, Robert., "Réflexions sur la puissance motrice du feu", Vrin, Paris. Historical reference.

iii   Coriolis, Gustave-Gaspard, Circa 1835. French engineer-scientist who described the force named after him that is seen as a force perpendicular to an object's motion. Historical reference.

iv   Kelvin, Lord, William Thompson, 1824-1907. English scientist who, along with Clausius, German, was active in promoting Carnot's ideas. Responsible for the absolute temperature scale named after him. Historical reference.

v   Reynolds, Osbourne, 1842-1912. English physicist and mechanical engineer. Historical reference.

vi   Venturi, Giovanni Battista, Circa 1800. Italian physicist and hydraulician. Historical reference.

vii   Darcy, Henry Philibert Gaspard, 1803-1858. French hydraulician. Known for the equation of the same name that relates irreversibilities to fluid viscosity, density and velocity. The equation is sometimes called the Darcy-Weisbach equation. Historical reference.

viii   Moody, Louis F., Circa 1944. Developed the 'friction factor' charts used extensively in this book.

APPENDIX B | Endnotes

ix   Kegel (Flow Measurement, ISA)

x   Perry, Chemical Engineers' Handbook, 50 th edition, McGraw-Hill. General reference.

xi   Ostwald (1853-1932)and de Waele, Ostwald was born in Latvia but was professor in Leipsig and is claimed by Germany. Nobel prize for catalysis in 1909. Taught Arrhenius, van't Hoff and Nernst among others. The name Ostwald-de Waele is given to a power law function describing apparent viscosity. Historical reference.

xii   Bernhardt, Processing of Polymeric Materials, Reinhold, N.Y.,1959. Source of further information.

xiii   Crane, Technical Paper No. 410. Seventeenth Printing, 1978 – but see latest printing. General reference.

xiv   Fanning, Circa 1880. Responsible for the 'friction factor' named after him. Historical reference.

xv   Hagen-Poiseuille. G. H. L. Hagen, German hydraulic engineer and Louis Marie Poiseuille, French physician and physicist , Circa 1844. Historical reference.

xvi   Miller, D.S. Internal Flow Systems, 2nd Edition, BHRA Cranfield, UK, 1990.

xvii   Driskell, L.R., ISA Handbook of Control Valves, Table of Representative Valve Factors

xviii   Simpson, L.L., Chem. Eng., July 17, 1968, Head loss (mechanical energy) formula for square edged, concentric orifice plates, Equation II-16

xix   Miller, Richard.W., Empirical formula for permanent loss across an orifice curve-fitted to the ASME head loss curve (ASME Research Report on Fluid Meters). Source of further information: Flow Measurement Engineering Handbook.

xx   Hero, Greek, Second century BC, Developed a fire pump with two cylinders and a hand-operated rocking beam to supply motive force. Historical reference.

xxi   Agricola, German, 16th century. Described the extensive use of pumps in the mining industry for mine dewatering purposes. Historical reference.

xxii   Papin, Denis, French inventor, Circa 1689. The invention of the centrifugal pump is generally credited to him. Its use had spread around the world by the mid 1800's. Today, the centrifugal pump is an omnipresent part of our existence. Historical reference.

xxiii   Schutte and Koerting, Manufacturers of ejectors, amongst other things.

xxiv   Paige, Peter, Engineer responsible for the equation of the same name. (Chemical Engineering, Aug 14, 1967)

xxv   Richter, S.H., Size Relief Systems for Two Phase Flow, Hydocarbon Processing, July 1978.

xxvi   Laplace, Simon Laplace, 1749-1827. French scientist. Corrected Newton's hypothesis for isothermal flow in computing the speed of sound in air to an adiabatic hypothesis (1816). Historical reference.

xxvii   van der Waals, Dutch physicist, circa 1870, one of the first to try to improve the ideal gas model with a model that involved two parameters. Historical reference.

xxviii Redlich-Kwong equation, original, circa 1950, is probably still the most famous EOS.

xxix   Reid, Prausnitz and Poling, The Properties of Gases and Liquids, McGraw-Hill, Fourth Edition, 1987. Successor to an equally useful and prestigeous book by the same name but authored by Reid, Prausnitz and Sherwood. General reference for physical-chemical data.

xxx   Prausnitz, Anderson and Grens, Computer Calculations for Multicomponent Vapor-Liquid and Liquid-Liquid Equilibria. Prentice-Hall, 1980, out of print but full of good ideas. Developed a solid database for its use in normal plant situations.

xxxi   McCabe and Smith, Unit Operations of Chemical Engineering, McGraw-Hill, 1967. Gerneral reference.

xxxii   Benjamin and Miller, The Flow of a Flashing Mixture of Water and Steam through Pipes, ASME Transactions, October 1942. Cited by Peter Paige who used the Benjamin and Miller Data to arrive at some of his conclusions. Historical reference.

xxxiii Duckler, A.E. et al, "Frictional pressure drop in two-phase flow", AIChE J., 10 44-51 (1964)

xxxiv   DeGance and Atherton, Published useful information on two phase flow systems. Ch. Eng, March 23, 1970

xxxv   Benedict et al (Benedict, R.P., N.A. Carlucci and S.D. Swetz, Flow Losses in Abrupt Enlargements and Contractions, Journal of Power Engineering, January 1966). Cited as a source of data and ideas by Optimal-Systems.

xxxvi   ASME steam tables, General reference.

xxxvii Blasius (student of Prandtl), circa 1911. First to establish that the friction factor followed a functional relationship in Reynolds[v] number below a Reynolds[v] number of 100 000 in hydraulically smooth pipes. Equation AI-30. Historical reference.

xxxviii Prandtl, circa 1911, Prandtl developed another equation for the relationship between a friction factor (a wall friction factor, superscripted fs) and the pipe Reynolds[v] number. Historical reference.

xxxix   Nikuradse, circa 1900, correlated Darcy's[vii] data, He carefully glued the sand on the internal walls of pipes. Using this method, he was able to vary the relative roughnesses of pipe walls between 1/500 and 1/15, about 33:1. Historical reference.

APPENDIX B | Endnotes

xl Colebrook, circa 1911, empirically arranged the Prandtl equation and one developed by von Kármán to produce a formula that gave remarkably good results in correlating published data. Historical reference.

xli Churchill-Usagi. Stuart W. Churchill of the University of Pennsylvania presented a convenient formula of estimating the friction factor from the Reynolds[v] number and the relative surface roughness (Chem. Eng. Nov. 7, 1977)

xlii Weber and Meissner, Thermodynamics for Chemical Engineers, 1957

xliii Benedict, R.P., Fundamentals of Pipe Flow, Wiley 1980.

xliv Hydraulics Institute. Source of reputable data

xlv Chemical Engineering Thermodynamics, Balzhiser, Samuels and Eliassen (BS&E). No longer in print.

xlvi Reid, Prausnitz and Sherwood, The Properties of Gases and Liquids, gives various means of estimating mixture viscosities. See successor book by Reid, Prausnitz and Poling.

xlvii Optimal-Systems, Website – http://optimal-systems.demon.co.uk. Commercial organisation, source of good data.

xlviii Golubev equations for estimations of mixture viscosity. Used by Reid, Prausnitz and Sherwood.

APPENDIX C

Table of Principal Symbols and Glossary of Principal Terms and Units

Note that most of the symbolism, including the subscripts and superscripts, used in this book is found in the general literature. A small amount is the author's invention. The author only included in these tables that which he felt important. If the reader wants further enlightenment, it can usually be found in the narrative surrounding the equation in which the symbol was first used in the main text.

Caveat —Always check the units of each term of each equation at the start of a computation. Each term should have the same units as every other term of the same equation. The author has attempted to be consistent and to define each term carefully, but he has been constrained by common useage in some cases.

Term	Example Eq./Note	Description	Customary U.S. Units	SI Units (or submultiples)
A	I-2	Cross sectional area, conduit	feet^2	m^2
A	AII-10	Calculated term used in Churchill-Usagi equation	-	-
a	IV-9	'Acoustic' velocity	feet/s	m/s
a	IV-17A	R-K repulsion coefficient	(ft-lbf/lbm)R$^{0.5}$	(J/kg)K$^{0.5}$
a	Fig. AIII-5	Pump impeller inlet blade angle	degrees	degrees
α	II-36	Correction factor (value of ratio of a point velocity to the average velocity across a section)	dimensionless	dimensionless
B	A1V-12	Ratio $\gamma/(\gamma-1)$	dimensionless	dimensionless
B	AII-10	Calculated term used in Churchill-Usagi equation	-	-
b	IV-17A	R-K attraction coefficient	ft^3/lb$_m$	m^3/kg
β	II-12	Value of ratio of a small diameter to a larger one	dimensionless	dimensionless
β	AIII-14	Pump impeller outlet blade angle	degrees	degrees
c	AV-4	Constant point velocity	ft^3/s	NA
c_p	IV-12	Isobaric 'heat' capacity, mass	Btu/lb$_m$-R	J/kg-K
c_p	not used	Isobaric 'heat' capacity, molar	Btu/lb$_{mole}$-R	J/kg-mole-K
c_v	IV-13	Isometric 'heat' capacity	Btu/lb$_m$-R	J/kg-K
c_v	not used	Isometric 'heat' capacity, molar	Btu/lb$_{mole}$-R	J/kg-mole-K
C_v	AII-1	Valve coefficient – proportionality between flow rate of water at 60°F (U.S. gpm) and $(\Delta P)^{1/2}$(psi)$^{1/2}$	U.S.gpm/psi$^{1/2}$	NA

APPENDIX C — Table of Principal Symbols and Glossary of Principal Terms and Units

Term	Example Eq./Note	Description	Customary U.S. Units	SI Units (or submultiples)
d	I-1	Diameter	inches	NA
d	I-3	Differential operator	dimensionless	dimensionless
η	AIII-18	Pump efficiency, energy transmitted to fluid/ energy transmitted to shaft	dimensionless	dimensionless
η	AV-9	Viscosity coefficient, eta	Poise/10^6	Poise/10^6
δ	I-3	Path dependent operator, differential or small but finite	dimensionless	dimensionless
δ	1 of AV	Gradient	dimensionless	dimensionless
D	I-1	Diameter or length dimension	feet	meters
D	AII-10	Pipe ID	feet or inches	meters of mm
Δ	I-3	Finite change between limits	dimensionless	dimensionless
e	I-23	Total energy per unit mass at a section	ft-lb_f/lb_m	J/kg
ε	AII-10	Absolute roughness	feet or inches	meters or mm
E dot	AIII-31	Total mechanical transferred per unit time	ft-lb_f/s	J/s
ξ	AV-13	Epsilon, a calculated coefficient in the Thodos et al relationships	-	-
f_M	I-17	Moody friction factor (subscript F refers to Fanning)	dimensionless	dimensionless
f	AII-10	Friction factor in Churchill-Usagi equation	dimensionless	dimensionless
F	I-6	Force	lb_f	N
F	AI-57	Mechanical energy converted to irreversibilities	ft-lb_f/lb_m	J/kg
γ	I-24	'Weight' density – synthetic unit substituted for mass density	lb_f/ft^3	N/m^3
γ	IV-13	Ratio of 'heat capacities', c_p/c_v	dimensionless	dimensionless
g	I-3	Acceleration of gravity	feet/s^2	m/s^2
g_c	I-3	Dimensional constant	32.17 lb_m-ft/lb_f-s^2	1, none
G	I-2	Mass velocity (flux)	lb_m/s-ft^2	kg/s-m^2
h	I-3	Enthalpy (flow energy) (mass basis)	Btu/lb_m	J/kg
$\underline{h}$		Enthalpy (flow energy) (molar basis)	Btu/lb-mole	J/kg-mole
h_f, (h_L)	I-4	Mechanical energy converted to thermal energy ('losses')	ft-lb_f/lb_m	J/kg
h_s, h_T	AIV-22	Stagnation enthalpy	Btu/lb_m	J/kg
H	I-23	'Head' – synthetic unit substituted for energy	feet	meters
J	IV-2	Conversion factor mech/therm	776.16 lb_f-ft/Btu	1, none
K	I-21	'Loss' coefficient of an obstruction (fraction of kinetic energy converted to thermal energy).	dimensionless	dimensionless
L	AIV-4	Length or position	ft	m
μ	1-1	'Absolute' viscosity	lb_m/s-ft	Kg/s-m
μ_{cP}	1-1	viscosity (centipoise)	0.01g/cm-s	0.01g/cm-s
μ_m	I-6	Viscosity (no name, U.S. units)	lb_m/ft-s	kg/m-s
m	I-2	Mass (a dot signifies per second)	lb_m	kg
N_{Re}	I-1	Reynolds number (a further subscript, D, refers to pipe ID – o refers to orifice bore)	dimensionless	dimensionless

396   Flow of Industrial Fluids—Theory and Equations

Table of Principal Symbols and Glossary of Principal Terms and Units   APPENDIX C

Term	Example Eq./Note	Description	Customary U.S. Units	SI Units (or submultiples)
N_{Ma}	IV-16	Mach number	dimensionless	dimensionless
ϕ	AV-15, 16	Phi, calculated coefficients in the Thodos relationships	dimensionless	dimensionless
P	I-4	Pressure (absolute)	lb_f/ft^2	N/m^2 (Pa)
P	AII-1	Differential pressure	$lb_f/inch^2$	NA
P	AIII-21	Power (rate of transfer of energy)	foot-lb_f/s or hp	J/s or cv
π	II-19	Constant equal to 3.1416…..	dimensionless	NA
q	I-3	Mechanical energy per unit mass that has flowed under a temperature difference	ft-lb_f/lb_m	J/kg
q	III-1	Liquid volumetric flow rate	ft^3/s	NA
q	AIII-1	Flow rate in U.S.gpm of water at 60°F	gpm	NA
Q	III-1	Liquid volumetric flow rate	ft^3/h	NA
Q	III-2	Flow rate in U.S. gpm of water at 60°F	gpm	NA
Q dot	IV-2	Energy flow rate, thermal units	Btu/s	J/s
r_H	IV-23	Hydraulic radius	ft	m
R	IV-10	Gas constant, mass	ft-lb_f/lb_m-R	N-m/kg-K
R_o	IV-10	'Universal' gas constant, molar	ft-lb_f/lb_{mole}-R	N-m/kg-mole-K
$\underline{R}$	AIV-24	Gas constant, molar	ft-lb_f/lb_{mole}-R	N-m/kg-mole-K
R_D	AI-59	Resisting force developed by parallel layers of fluid	lb_f	N
ρ	I-1	Density (mass)	lb_m/ft^3	kg/m^3
s	AI-49	Entropy per unit mass (energy/mass-temperature)	ft-lb_f/lb_m-R	N-m/kg-K
τ	I-7	Stress (force per unit area)	lb_f/ft^2	N/m^2
t	I-8	Variable time	s	s
T	IV-10	Temperature, absolute	R, (459.67 + F)	K, (273.15+C)
T_s	IV-20	Stagnation temperature	R, (459.67 + F)	K, (273.15+C)
u	AI-57	Internal energy	ft-lb_f/lb_m	N-m/kg
U	I-1	Velocity (average across section)	ft/s	m/s
v	AV-4	Point velocity (as opposed to average velocity, U)	ft/s	m/s
v	I-2	Mass volume	ft^3/lb_m	m^3/kg
$\underline{v}$	AIV-26	Molar volume	ft^3/lb-mole	m^3/kg-mole
V	I-6	Linear velocity	ft/s	m/s
ω	AIII-6	Angular velocity	1/t	1/t
w	I-3	Net work energy per unit mass	ft-lb_f/lb_m	J/kg
W	I-1	Mass flow rate	lb_m/h	NA
W dot	IV-2	Work energy flow rate, thermal units	Btu/s	J/s
x	I-8	Variable distance	feet	meters
X	I-3	Vertical dimension	feet	meters
Z	I-6	Distance	feet	meters

APPENDIX C | Table of Principal Symbols and Glossary of Principal Terms and Units

Major subscripts and superscripts used in equations

Subscript or superscript	Example equ./note	Meaning
*	AIV-15	Qualifier to indicate value at choked section
Re, ReD	I-1, I-18	Reynolds number, number base on diameter D
cP	I-1	Centipoise
1, 2, (i)	I-2	Section specifier
n, f	I-3, I-4	Net or fluid (energy transferred)
c	I-3	Qualifier to indicate a constant or 'critical'
c, e	II-15	Qualifier to indicate compression or expansion
m	I-6	Qualifier to indicate mass based quantity
o	II-16	Qualifier to indicate orifice
0	I-9	Qualifier to indicate minimum
' (prime)	IV-28	Qualifier to indicate an ideal, perfect gas.
v	I-10	Qualifier used on shear stress of pseudoplastics
n	I-10	Ostwald and de Waele index
f	I-12	Qualifier to indicate force based quantity
fs	IV-5	Qualifier to indicate 'skin friction' (double misnomer)
F	I-19	Fanning
H	IV-8	Qualifier to indicate hydraulic
I	Example AIV-1	Qualifier to indicate quantity at initial choke point
L	II-1	Qualifier to indicate a 'loss' (conversion) in mechanical energy
M	I-18	Moody
Ma	IV-16	Mach
Mav	AIV-4	Moody, average (over integration)
P	IV-12	Qualifier to indicate constant pressure
P	I-15	Qualifier to indicate units are Poise, g/cm-s
rad	AIII-13	Radial
r	Example IV-1	Qualifier to indicate relative or divided by critical value
s	IV-9	Qualifier to indicate constant entropy
s	AIV-12	Qualifier to indicate stagnation
s	AIV-22	Qualifier to indicate source or reservoir condition
tan	AIII-3	Tangential
T	II-18	Qualifier to indicate total
v	IV-13	Qualifier to indicate constant volume
w	II-17	Qualifier to indicate wall

APPENDIX D

Table of Caveats

The word 'Caveat' has been used often in this book to warn unsuspecting readers to pay particular attention to a common intellectual trap. The word comes from the Latin 'Caveat Emptor' which roughly translated means 'Let the buyer beware!' The author feels strongly that highly intelligent people have a propensity to use words loosely, in a Mad Hatter from Alice in Wonderland fashion. It is supposed that their intent is to find a simple way of expressing complex concepts. They only succeed in confusing issues. There is no such thing as a free lunch.

Caveat Subject	Page
Reynolds numbers	9
Average or point velocities	18
Bernoulli equation versus the first law	21
Fluid classifications	29
Equivalent lengths	34
Velocity profiles	35
Hydraulic engineering practice	54
Gravity trap	56
Bernoulli balance	57
Safety factors	62
Velocity and the K factor	63
Mechanical energy 'losses' versus pressure	66
Pressure recovery	69
Source of K factor correlation	75
Proximity of disturbances	80
Viscosity units	81
Negative K values	200
Definitions of K factors	201
Steady-state assumptions	228
Hydraulic engineers' simplification	257
Velocity used with K factor, Suffixes, K factors, C_v	260
Conversion of irreversibilities to pressure drop	268
Authors' definitions of loss coefficients	268
Conversion of 'head' to pressure	283
Reference states for thermodynamic properties	320
Integration with constraints	340
Verifying definitions	375
Using correct K factors	377

APPENDIX E

Selected Bibliography

Benedict, R.P. *Fundamentals of Pipe Flow.* Wiley, 1980.

The Crane Company. *Flow of Fluids Through Valves, Fittings and Pipe, Technical Paper No. 410.* Crane, 1988 (reprint of 1942 document).

Driskell, Les. *Control Valve Selection and Sizing.* ISA, 1983.

Driskell, L.R. "Sizing Control Valves, Part I: Sizing Theory and Applications." I*SA Handbook of Control Valves*. Second Edition. Pp. 180 – 205. ISA, 1976.

Kegel, Thomas. "Chapter 22: Insertion (Sampling) Flow Measurement," *Flow Measurement*. Second Edition. D.W. Spitzer, Editor, pp. 597 – 638. ISA, 2001.

Kyle, Benjamin G. *Chemical and Process Thermodynamics.* Third edition. Prentice Hall international series in the physical and chemical engineering sciences. Prentice Hall PTR, 1999.

Miller, Donald S. *Internal Flow Systems.* Second edition. BHRA (Information Systems), 1990.

Miller, Richard W. *Flow Measurement Engineering Handbook.* Third Editon. McGraw-Hill, 1996.

Perry, Robert H., Don W. Green, and James O. Maloney, Editors. *Perry's Chemical Engineers' Handbook.* Seventh Edition. McGraw-Hill Professional, 1997.

Polling, Bruce E, Jon Prausnitz, and John P. O'Connell. *The Properties of Gases and Liquids.* Fifth Edition. McGraw-Hill, 2001.

Smith, Julian C., Peter Harriot, and Warren L. McCabe. *Unit Operations of Chemical Engineering.* Sixth edition. McGraw-Hill Science/Engineering/Math, 2000.

Hydraulic Institute — www.pumps.org
9 Sylvan Way
Parsippany NJ, 07054
Phone: (973) 267-9700

APPENDIX E | **Selected Bibliography**

Optimal Systems Limited — optimal-systems.demon.co.uk
20 - 22 Bedford Row
London
WC1R 4JS
England
Great Britain

Schutte and Koerting — www.s-k.com
2233 State Road
Bensalem, PA 19020
Phone: (215) 639-0900
Fax: (215) 639-1597

Index

*Index page numbers identified with an asterix are key citations, and as such, the reader may wish to consult these pages first.

A

Absolute pressure, 45, 54*, 87, 115, 144, 157, 165, 186

Adiabatic model, 152, 154*, 155, 186, 329, 337, 348

Adverse gradients, 266

Analytic method, 371, 388

As-built sketch, 198

Assumed (hypothetical) data, 53

Asterisk condition, 172

Atmospheric pressure, 43, 46, 54*, 92, 110, 115, 143, 294, 299, 304, 315, 368

Audience, IX

Average velocity, 8, 9, 11, 12, 13, 18*, 35, 37, 39, 48, 117*, 118, 158, 159, 161, 166, 200, 210, 254, 255, 260, 261, 264, 281, 325, 329, 333, 335, 351

Axial pumps, 106*, 108, 120

B

Balzhiser, Samuels and Eliassen, 345

Bearing losses, 288, 289

Bernoulli, XII, 1, 2, 17, 20*, 21*, 36, 37, 38, 50, 51, 53, 54, 56, 57, 64, 65, 71, 82, 89, 99, 114, 115, 141, 143, 159, 160, 183, 184, 189, 190, 200, 204, 205, 209, 210, 224, 229, 240, 241, 242, 245, 251, 252, 256, 257, 259, 265, 267, 268, 269, 276, 281, 286, 290, 291, 292, 296, 307, 308, 309, 315, 316, 317, 318, 321, 322, 374

Bingham plastic, 11*, 15, 22, 23, 24, 27

Blowers, IX, XI, 284

Brake horsepower, 293

Brief History of Pumps, 104

C

Carnot, XII, 221, 356

Casings, 120

Cavitation, XI, 87, 102, 123, 134, 137, 141, 142, 294*, 295, 297, 299, 318

Channel wall, XI

Characteristic curves, 111*, 124, 131

Chemical reactions, 17

Choked densities, 341

Choked flow, XI, 151, 163, 179, 182*, 184, 187, 188, 193, 202, 319, 323, 324, 326, 327, 329, 332, 334, 336, 342, 350, 356, 360, 362, 372, 376

Choked Flow and the Mach Number, 182

Choked mass flow, 182*, 324, 343

Choked mass flux, 332*, 335, 342

Choked pressures, 331*, 339, 340, 343, 348, 358

INDEX

Choked temperature, 335, 341

Choking, XI, 151, 153, 155, 179, 180, 183*, 186, 187, 319, 324, 326, 337, 362

Churchill-Usagi, 36, 39, 254, 257*, 258*, 271, 308, 313, 315, 344

Classification of Pumps, 105

Closed Channel, X

Compressible flow, XIII, 36, 37, 55, 58, 151*, 152, 153, 154, 155, 156, 157, 158, 161, 162, 174, 175, 192, 195, 203, 204, 241, 319, 355, 362, 375

Compressible fluids, XIII, 1

Computer simulations, 156*, 185, 188, 323

Conceptual models, 151

Conservation of energy, 17, 18

Conservation of mass, 17

Control considerations, 128

Control systems personnel, X, XII, 53*, 74, 299

Cooling water failure, 198, 368

Coriolis, XII, 15

Corrected to the pump center line, 114, 115

Corresponding states, 166*, 378, 384

Critical parameter, 170

Critical zone, 231

Cubic equations-of-state, 166

Customary U.S. System, XII

D

Darcy, 1, 21, 36, 38, 39, 209, 223, 229*, 230, 231, 232, 236, 251, 252, 286, 307

Density, 1, 8*, 9, 16, 18, 20, 21, 32, 33, 38, 39, 41, 43, 44, 45, 47, 48, 55, 58, 65, 72, 74, 77, 81, 82, 84, 88, 91, 92, 94, 95, 101, 102, 109, 110, 126, 143, 145, 146, 147, 152, 153, 154, 157, 158, 159, 160, 161, 162, 165, 166, 168, 171, 177, 185, 187, 190, 192, 201, 203, 204, 205, 213, 214, 218, 220, 229, 230, 233, 234, 235, 252, 255, 256, 257, 260, 263, 283, 288, 293, 296, 297, 302, 303, 308, 314, 315, 317, 321, 322, 323, 324, 325, 335, 337, 341, 342, 345, 357, 383

Departure function, 187, 320, 351*, 352, 370

Describing the Flow Regime, 198, 368

Describing the Piping Network, 196*, 207, 363

Diaphragm pumps, metering pumps, 125

Didactic simplifications, 2

Differential pressure, 13, 15, 45, 46, 64*, 76*, 78, 102, 109, 110, 113, 118, 120, 130, 134, 137, 186, 228, 230, 254, 265, 283, 290, 316, 322, 323, 325, 380, 382

Diffusers, 101*, 120, 266, 287

Dilatant fluids, 24, 28

Discharge characteristics, 299

Discharge throttling, 102, 137

Dispersion modeling, XI

Division of work, XII

Dravo Chem Plants, 196

Driving potential, X, 50, 65

Index page numbers identified with an asterix are key citations, and as such, the reader may wish to consult these pages first.

E

Efficiency, 56, 102, 111*, 112*, 116, 118, 119*, 121, 127*, 128, 129*, 131, 133, 223*, 274, 281, 284, 285, 286, 287, 288, 289, 290, 291, 292, 293, 294, 302, 304, 305, 306, 309, 310, 312, 316, 317, 318

Ejector, 310

Ejectors, 108, 126

Energy, IX, X, 1, 2, 3, 4, 5*, 7, 12, 16, 17, 18, 19, 20, 21, 22, 30, 33, 34, 35, 37, 39, 40, 43, 44, 45, 48, 49, 50, 51, 52, 53, 54, 55, 55-56, 57, 58, 59, 60, 61, 62, 63, 64, 65, 65-66, 67, 68, 69, 70, 70-71, 72, 73, 74, 75, 76, 77, 78, 79, 80, 81, 82, 83, 84, 85, 86, 87, 88, 89, 90, 91, 92, 93, 94, 95, 96, 97, 98, 100, 101, 102, 103, 104, 109, 110, 111, 112, 113, 114, 115, 116, 117, 118, 119, 123, 125, 127, 128, 130, 131, 132, 133, 135, 137, 139, 141, 143, 144, 145, 146, 147, 148, 153, 155, 158, 159, 160, 161, 163, 164, 165, 168, 175, 176, 179, 180, 183, 185, 191, 192, 200, 201, 202, 203, 204, 205, 207, 209, 210, 215, 220, 221, 222, 223, 224, 229, 235, 236, 239, 240, 241, 242, 243, 244, 245, 251, 253, 254, 255, 256, 257, 259, 260, 261, 264, 265, 266, 267, 268, 269, 270, 271, 274, 275, 277, 279, 280, 281, 282, 283, 284, 285, 286, 287, 288, 290, 291, 292, 293, 294, 295, 296, 297, 299, 300, 302, 304, 305, 306, 307, 309, 310, 315, 316, 317, 318, 320, 321, 322, 324, 325, 326, 327, 328, 329, 333, 334, 335, 336, 337, 340, 345, 351, 353, 354, 355, 356, 371, 372, 373, 374, 375, 376, 377, 378, 383, 387, 390

Energy transfer by centrifugal force, 101

Energy transfer by electromagnetic force, 104

Energy transfer by gravity, 101

Energy transfer by mechanical impulse, 104

Energy transfer by momentum transfer, 103

Energy transfer by volumetric displacement, 102

Engineering companies, XII

Enthalpy, 19*, 21, 155, 159, 160, 161, 164, 173, 186, 187, 191, 199, 241, 243, 270, 302, 303, 319, 320, 327, 328, 329, 332, 333, 336, 337, 350, 351, 352, 353, 354, 355, 360, 370, 371

Estimations of Irreversibilities, 80

F

Fanning, 35*, 36, 50, 235, 236, 258

Fire exposure (by fire zone), 198, 368

Flashing, XI, 151, 153, 183*, 184, 185, 187, 196, 199, 294, 295, 319, 324, 362, 369, 371

Flow control, 101, 110*, 112, 120*, 124*, 125*, 135*, 137, 149

Flow profile disturbance, 15

Flow profile influences, 15

Flow regimes, XI, 1, 6*, 7*, 49, 178, 207, 390

Fluid flow, IX

Fluid flow theory, IX

Fluid friction, 1, 2, 3, 4*, 12*, 21, 371, 376

Fluid systems, IX, 105, 318, 363*

Fluid viscosity, 1, 58

Index page numbers identified with an asterix are key citations, and as such, the reader may wish to consult these pages first.

INDEX

Force, 3, 4, 16, 20, 22, 23, 25, 26, 31, 32, 35, 37, 38, 40, 43, 45, 50, 64, 74, 81-82, 85, 86, 100, 101, 102, 104, 105, 106, 108, 110, 114, 119, 121, 122, 126, 128, 129, 143, 148, 149, 163, 185, 198, 209, 210*, 211, 212, 213, 214, 215, 216, 217, 218, 219, 220, 221, 224, 225, 226, 227, 228, 233, 237, 238, 239, 240, 245, 246, 247, 252, 256, 266, 267, 274, 275, 278, 279, 280, 282, 288, 289, 290, 292, 294, 297, 298, 300, 301, 302, 317, 332, 345, 351, 356, 377, 379, 380

Form friction, 17*, 56, 62, 371, 377

Friction factor, 10, 11, 30, 33*, 34, 35, 36, 38, 39, 50, 58, 59, 61, 62, 63, 66, 67, 68, 80, 82, 85, 86, 90, 91, 92, 98, 152, 155, 161, 181, 185, 186, 187, 188, 190, 209, 223, 229, 230, 231, 232, 235, 236, 253, 257, 258, 308, 309, 310, 313, 322, 323, 325, 360, 377, 378, 387

Friction loss, 5*, 33*, 37, 215

Fundamental Relationships, 17

G

Gas lifts, 126

Gauge pressure, 54

Geometric similarity and dissimilarity, 63, 261

Gradual contractions, 78

Gradual enlargements, 78

H

Hagen-Poiseuille, 36, 210, 230*, 245

Head loss (mechanical energy) formulae, 77

Head losses versus power losses, 284

Head tank, X, 3, 101

Hydraulic head, 43

Hydraulic practice, 44*, 46, 47, 48

Hydraulic radius, 161*, 162, 181

Hydraulic Turbines, 141*, 142, 149

Hydrostatic equilibrium, 209, 219

I

Ideal gas model, 154, 166*, 175, 192, 337, 378

Impeller, 56, 101, 102, 109*, 114, 115, 119, 120, 121, 122, 140, 273, 274, 275, 277, 280, 281, 282, 285, 287, 288, 291, 297

Impellers, 119, 149

Incompressible flow, XIII, 1*, 11, 37, 53, 55, 98, 151, 152, 153, 192, 203, 263, 380

Incompressible fluid, 1, 2

Incremental internal energy, 3*, 4, 16, 22, 34, 49, 50

Incremental method, 188*, 326, 373

Inherent characteristics, IX, 130*, 134, 148, 149

INSIGHT, IX

Installed Characteristics of Pumps, 130

Internal circulation, 288, 289

Inward projecting pipe entrance, 79

Irreversibilities, IX, 1, 5, 7, 12, 16, 17, 21, 22, 33, 34, 35, 37, 38, 42, 49, 50, 51, 56*, 57*, 58*, 62, 65, 66, 68, 69, 70, 71, 73, 74, 75, 76, 77, 78, 79, 80, 84, 86, 87, 88, 89, 90, 91, 92, 93, 95, 96, 97, 98, 112, 118, 132, 136, 137, 139, 152, 153, 158, 159, 161, 164, 175, 176, 177, 179,

Index page numbers identified with an asterix are key citations, and as such, the reader may wish to consult these pages first.

180, 181, 183, 184, 185, 186, 188, 193, 201, 209, 224, 229, 230, 236, 237, 240, 241, 242, 251, 253, 255, 256, 257, 259, 260, 262, 263, 264, 266, 267, 268, 270, 286, 289, 291, 297, 303, 304, 307, 310, 315, 319, 322, 325, 345, 355, 356, 371

Irreversibility, 4, 5, 12*, 16, 20, 33, 37, 56*, 94, 209, 215, 236, 240, 243, 244, 245, 302

Irreversible, 16*, 18, 34, 176, 232, 240, 241, 242, 243, 244, 253, 266, 300, 321, 356, 371

Isentropic, adiabatic expansion, 175, 193

Ishwar Davé, 196

Isothermal model, 154*, 155, 156, 188

J

Jet pumps, 108*, 126, 127, 128, 129, 148, 149

K

Kegel, 14

Kelvin, 163*, 223, 354

L

L.R. Driskell's Table of Representative Valve Factors, ISA Handbook of Control Valves, 75

L/D ratio, 35*, 85, 86

Laminar, 1, 6, 7, 10, 12, 13, 21, 23, 30, 33, 36, 38, 49, 50, 55, 58, 73, 210, 222, 223, 229, 230, 231, 245, 246, 248, 250, 251, 252, 258, 379, 380, 382

Language of subject matter, IX

Liquid metal pumps, 108, 129

Logical inconsistency, 1, 2, 4*, 5, 221, 356

Loss of prime, 122, 123

Loss term, hf, 21*, 35, 38

M

Mach number, 152, 162, 165, 181, 182, 184, 189, 319, 333*, 334*

Magnetic flow meters, 15

Manifold Flow, 200*, 207, 374, 390

Manifold flows (dividing and combining flows), 79

Mature technology, 148, 274

Means of energy transfer to liquids, 100

Measurement and control, X, XII, XIII, 1, 51*, 109, 123, 124, 142, 299

Measures of performance, 113, 318

Measuring instruments, X, 13, 50

Mechanical energy conversion, 17, 68*, 75, 375

Mechanical engineers, X, 52

Meter runs, 62

Minimum flow requirement, XI

Miscellaneous fittings and manifolds, 78

Miscellaneous pumping devices, 108, 125

Mist flow, XI

Mixed flow pumps, 106, 120

Mixed phase flow, 151

Mixed units, XII, 8*, 10, 80, 81, 117, 210, 251, 252, 255

Mixing rules, 157*, 167, 171

Mixture, VI, XI, 49, 95, 126, 131, 156*, 157*, 167, 171, 183, 186, 187, 199, 320, 324, 347, 350, 351, 352, 353, 354, 355, 357, 360, 370, 379, 385, 386

Index page numbers identified with an asterix are key citations, and as such, the reader may wish to consult these pages first.

INDEX

Moody, 35*, 36, 39, 50, 59, 181, 185, 229, 236, 258, 308, 309, 310, 344

Motivation, XII

Multiple choke points with ideal gases, adiabatic models, 337

N

Negative K factors, 265, 271

Newtonian fluid, 11, 23*, 25*, 26*

Newton-Raphson iteration, 199, 345*, 354, 371, 376, 387

Non-newtonian fluids, 23*, 24*, 26

NPSH, 110, 112, 113, 134, 136, 139, 294*, 295*, 297, 314, 318

O

Orifice plate, 262

Orifice plates, 15, 22, 51, 76*, 77, 98, 177, 202

Original Redlich-Kwong equation, 157

Ostwald and de Waele, 27

Overpressure, 122, 123*, 125

P

Peter Paige, 151, 183*, 184*, 185*, 187, 188, 193, 199, 319, 323, 345, 350, 356, 360, 362, 371, 387

Piezometric head, 43

Pipe taps, 72, 76*, 77

Pipe, valves and fittings, 57

Piping system, 3, 20, 33, 34, 116, 127, 155, 195*, 206, 207, 262, 263, 294, 298

Piston and cylinder pumps, 107*, 122, 123, 140

Polytropic model, 155

Positive displacement pumps, 102*, 103, 104, 105, 107, 121, 122, 125, 131, 134, 140, 141, 148, 149

Power failure, 198, 368

Power to the shaft, 288*, 289*, 293, 304

Prausnitz, 167*, 168, 171, 172, 320, 351, 352, 353, 357, 378, 384, 385

Pressure, IX, X, XI, 1, 2, 3, 7, 13, 15, 16, 17, 19, 20, 21, 22, 25, 26, 28, 30, 32, 34, 37, 38, 43, 44, 45, 46, 51, 52, 54, 55, 56, 58, 64, 65, 66, 69, 70, 71, 72, 74, 75, 76, 77, 78, 80, 82, 83, 84, 85, 86, 87, 88, 89, 90, 91, 92, 93, 94, 95, 100, 101, 102, 103, 108, 109, 110, 112, 113, 114, 115, 116, 118, 120, 121, 122, 123, 124, 125, 126, 127, 128, 130, 131, 132, 133, 134, 137, 139, 140, 141, 142, 143, 144, 145, 146, 147, 148, 151, 152, 153, 154, 156, 157, 162, 164, 165, 166, 167, 168, 170, 171, 172, 173, 175, 177, 178, 179, 180, 182, 183, 184, 185, 186, 187, 188, 190, 191, 195, 196, 198, 199, 200, 201, 202, 203, 204, 205, 206, 209, 210, 212, 213, 214, 215, 216, 218, 219*, 220*, 221, 222, 224, 227, 228, 229, 230, 231, 233, 234, 237, 240, 243, 247, 248, 250, 251, 252, 253, 254, 255, 256, 257, 259, 260, 263, 264, 265, 266, 267, 268, 269, 271, 274, 275, 282, 283, 287, 288, 290, 292, 293, 294, 295, 296, 297, 298, 299, 300, 301, 302, 304, 306, 308, 310, 311, 313, 314, 315, 316, 317, 320, 322, 323, 324, 325, 326, 327, 328, 329, 331, 332, 335, 336, 337, 338, 339, 340, 341, 343, 344, 348, 352, 353, 356, 357, 358, 363, 364, 368, 369, 370, 371, 373, 374, 379, 380, 382, 384, 387

Index page numbers identified with an asterix are key citations, and as such, the reader may wish to consult these pages first.

Pressure and temperature inequalities, 17

Pressure drop, IV, X, XI, 51, 179, 183*, 185, 188, 229, 230, 231, 233, 248, 250, 251, 315, 324, 325, 326, 328, 376

Pressure losses, IX, 69

Pressure relief devices, XI, 151

Prime, 110*, 122, 123, 133, 134

Prime movers, IX, X, 52

Priming, 110*, 121, 134, 295, 312

Process engineers, X, XII, 52*, 196, 364

Prony brake, 293

Pseudoplastics, 23*, 24, 27

Pulsation, 102, 103, 121*, 124*, 125

Pump efficiency, 112, 119*, 133, 274, 288, 302, 304, 305, 309, 317

Pumps, IX, XI, 34, 51, 52, 53, 98, 99*, 100, 101, 102, 103, 104, 105, 106, 107, 108, 109, 110, 111, 112, 113, 114, 115, 119, 120, 121, 122, 123, 124, 125, 126, 127, 128, 129, 130, 131, 132, 133, 134, 135, 138, 139, 140, 141, 142, 144, 148, 149, 158, 224, 273, 274, 275, 284, 287, 289, 290, 294, 295, 298, 299, 304, 305, 307, 310, 318

Pure fluids, XI, 370

Purpose of book, IX

R

Radial pumps, 106, 108, 119*

Real (as-built) data, 53

Reciprocating pump, 121*, 122, 299

Recovery, 51, 56, 66, 69*, 70, 72, 73, 76, 77, 85, 87, 89, 93

Redlich-Kwong model, 154

Relief header, XI, 188, 338

Repulsions, 4, 166, 168

Restriction orifice, 78, 94*, 95, 97

Reynolds, XII, 6*, 7, 8, 9, 10, 11, 22, 33, 36, 39, 387

Rheopectic, 24, 29

Rotary analog of Newton's law, 279

S

Safety relief vent header, 195, 363

Self-regulation, 135

Shock losses, 120*, 286, 287, 289

SI, XII, 16, 19, 32, 35, 37*, 40, 43, 45, 46, 48, 49, 55, 56, 64, 70, 81, 113, 117, 118, 159, 163, 209, 210, 211, 212, 214, 224, 229, 235, 240, 250, 254, 256, 263, 264, 278, 280, 291, 292, 294, 295, 302, 308, 312, 314, 328, 333, 351, 354, 355, 370

Skin friction, 17*, 35, 62, 161, 371

Slugging, XI, 89

Smooth pipes, 61, 231

Specific weight, 213*, 214, 220, 252, 298

Square edged and rounded entrances, 79

Stagnation condition, adiabatic processes, 172

Static discharge head, 116

Static suction head, 116

Straight pipe of uniform diameter, 57, 80

Suction lift, 105, 123*, 132, 133, 134, 149, 294, 295, 296, 299, 304, 307, 312

Sudden contraction, 63, 73*, 74, 75, 82, 87, 91

Sudden expansion, 63, 73*, 74, 75, 84, 87, 89, 93, 175, 326, 373

Swage, 2, 78*, 338

*Index page numbers identified with an asterix are key citations, and as such, the reader may wish to consult these pages first.

T

Terminology, IX, 39, 69*, 98, 200, 211, 225, 352, 377

Thermal energy is not an exact differential, 160

Thixotropic, 24

Time dependent behavior, 28

Time dependent fluids, 24

Time-independent behavior, 27

Total (system) mechanical energy losses, 79

Total dynamic head, 110, 115*, 116, 132, 135, 137, 138, 282, 283, 284, 285, 290, 291, 293

Total energy, mechanical energy and hydraulic energy grade lines, 80

Total static head, 116

Total suction head, 115*, 146, 147, 148, 292

Turbine, X, 17, 20, 21, 37, 38, 53, 56, 57, 99, 104, 106, 111, 119, 120, 121, 132, 133, 141*, 142*, 159, 180, 209, 240, 242, 259, 266, 270, 279, 318

Turbulent, 1, 4, 6*, 7, 10, 12, 13, 21, 22, 23, 33, 35, 36, 38, 42, 49, 50, 53, 55, 58, 62, 66, 68, 73, 85, 92, 161, 221, 223, 229, 230, 231, 251, 258, 281, 308, 313, 316, 380

Turndown, X, 7, 111, 113, 121, 125, 139, 142, 148*

Two-phase, XI, 186, 188

U

Units, XII, 159, 163*, 210, 253, 359

Unsatisfactory models, 154

Unstable, 1, 7*, 49, 258, 290

Utility, 52*, 99, 122, 127, 151, 175, 192, 193, 236, 254, 264, 297, 377

V

Vacuum devices, IX

Valve coefficient, C_v 253, 254

Valves as fittings, 75

Van der Waals model, 166

Velocity, XI, 1, 2, 8*, 9, 11, 12, 13, 14, 18, 19, 22, 25, 26, 33, 34, 35, 36, 37, 38, 39, 43, 44, 48, 49, 50, 55, 58, 59, 63, 64, 65, 66, 69, 70, 71, 72, 73, 74, 77, 81, 82, 83, 87, 88, 89, 90, 91, 93, 94, 101, 102, 106, 113, 114, 115, 117, 118, 130, 139, 144, 145, 146, 147, 148, 152, 153, 158, 159, 161, 162, 163, 165, 166, 172, 176, 177, 179, 180, 181, 182, 183, 189, 190, 192, 200, 201, 203, 204, 205, 210, 224, 226, 227, 228, 229, 230, 231, 233, 234, 237, 245, 246, 247, 248, 249, 250, 251, 254, 255, 256, 257, 259, 260, 261, 263, 264, 268, 269, 271, 275, 276, 277, 278, 279, 280, 281, 282, 283, 285, 287, 292, 293, 294, 295, 296, 299, 301, 307, 308, 310, 316, 321, 322, 325, 326, 327, 328, 329, 331, 332, 333, 334, 335, 336, 342, 351, 356, 359, 361, 374, 376, 379, 380, 381, 382

Venturi, XII, 74, 177, 202

Virial equation, 168*, 171, 172, 192

Viscous drag forces, 16, 239

W

Walk-down, 198

Weirs and flumes, XI

Index page numbers identified with an asterix are key citations, and as such, the reader may wish to consult these pages first.